Scale-Space Theory in Computer Vision

THE KLUWER INTERNATIONAL SERIES IN ENGINEERING AND COMPUTER SCIENCE

ROBOTICS: VISION, MANIPULATION AND SENSORS

Consulting Editor

Takeo Kanade

Other books in the series:

NEURAL NETWORK PERCEPTION FOR MOBILE ROBOT GUIDANCE, Dean A. Pomerleau
ISBN: 0-7923-9373-2

DIRECTED SONAR SENSING FOR MOBILE ROBOT NAVIGATION, John J. Leonard, Hugh F. Durrant-Whyte
ISBN: 0-7923-9242-6

A GENERAL MODEL OF LEGGED LOCOMOTION ON NATURAL TERRAINE, David J. Manko
ISBN: 0-7923-9247-7

INTELLIGENT ROBOTIC SYSTEMS: THEORY, DESIGN AND APPLICATIONS, K. Valavanis, G. Saridis
ISBN: 0-7923-9250-7

QUALITATIVE MOTION UNDERSTANDING, W. Burger, B. Bhanu
ISBN: 0-7923-9251-5

NONHOLONOMIC MOTION PLANNING, Zexiang Li, J.F. Canny
ISBN: 0-7923-9275-2

SPACE ROBOTICS: DYNAMICS AND CONTROL, Yangsheng Xu, Takeo Kanade
ISBN: 0-7923-9266-3

NEURAL NETWORKS IN ROBOTICS, George Bekey, Ken Goldberg
ISBN: 0-7923-9268-X

EFFICIENT DYNAMIC SIMULATION OF ROBOTIC MECHANISMS, Kathryn W. Lilly
ISBN: 0-7923-9286-8

MEASUREMENT OF IMAGE VELOCITY, David J. Fleet
ISBN: 0-7923-9198-5

INTELLIGENT ROBOTIC SYSTEMS FOR SPACE EXPLORATION, Alan A. Desrochers
ISBN: 0-7923-9197-7

COMPUTER AIDED MECHANICAL ASSEMBLY PLANNING, L. Homen de Mello, S. Lee
ISBN: 0-7923-9205-1

PERTURBATION TECHNIQUES FOR FLEXIBLE MANIPULATORS, A. Fraser, R. W. Daniel
ISBN: 0-7923-9162-4

DYNAMIC ANALYSIS OF ROBOT MANUPULATORS: A Cartesian Tensor Approach, C. A. Balafoutis, R. V. Patel
ISBN: 0-7923-9145-4

ROBOT MOTION PLANNING, J. Latombe
ISBN: 0-7923-9129-2

Scale-Space Theory in Computer Vision

Tony Lindeberg

Royal Institute of Technology
Stockholm, Sweden

KLUWER ACADEMIC PUBLISHERS
Boston/London/Dordrecht

A C.I.P. Catalogue record for this book is available from the Library of Congress.

ISBN 978-1-4419-5139-7

Published by Kluwer Academic Publishers,
P.O. Box 17, 3300 AA Dordrecht, The Netherlands.

Kluwer Academic Publishers incorporates
the publishing programmes of
D. Reidel, Martinus Nijhoff, Dr W. Junk and MTP Press.

Sold and distributed in the U.S.A. and Canada
by Kluwer Academic Publishers,
101 Philip Drive, Norwell, MA 02061, U.S.A.

In all other countries, sold and distributed
by Kluwer Academic Publishers Group,
P.O. Box 322, 3300 AH Dordrecht, The Netherlands.

Printed on acid-free paper

Foreword

The problem of *scale* pervades both the natural sciences and the visual arts. The earliest scientific discussions concentrate on visual perception (much like today!) and occur in Euclid's (*c.* 300 B.C.) *Optics* and Lucretius' (*c.* 100–55 B.C.) *On the Nature of the Universe.* A very clear account in the spirit of modern "scale-space theory" is presented by Boscovitz (in 1758), with wide ranging applications to mathematics, physics and geography. Early applications occur in the cartographic problem of "generalization", the central idea being that a *map* in order to be useful has to be a "generalized" (coarse grained) representation of the actual terrain (Miller and Voskuil 1964). Broadening the scope asks for progressive summarizing. Very much the same problem occurs in the (realistic) artistic rendering of scenes. Artistic generalization has been analyzed in surprising detail by John Ruskin (in his *Modern Painters*), who even describes some of the more intricate generic "scale-space singularities" in detail: Where the ancients considered only the merging of blobs under blurring, Ruskin discusses the case where a blob splits off another one when the resolution is decreased, a case that has given rise to confusion even in the modern literature.

It is indeed clear that *any* physical observation of some extended quantity such as mass density or surface irradiance presupposes a scale-space setting due to the inherent graininess of nature on the small scale and its capricious articulation on the large scale. What is the "right scale" does indeed depend on the problem, i.e., whether one needs to see the forest, the trees or the leaves. (Of course this list could be extended indefinitely towards the microscopic as well as the the mesoscopic domains, as has been done in the popular film *Powers of Ten* (Morrison and Morrison, 1984)). The physicist almost invariably manages to pick the right scale for the problem at hand *intuitively.* However, in many modern applications the "right scale" need not be obvious at all, and one really needs a principled mathematical analysis of the scale problem.

In applications such as *vision* the front end system has to process the radiance function blindly (since no meaning resides in the photons as such) and the problem of finding the right scale becomes especially acute. This is true for biological and artificial vision systems alike. Here a principled theory is mandatory and can *a priori* be expected to yield important insights and lead to mechanistic models. The modern scale-space theory has indeed led to an increased understanding of the low level operations and novel handles on ways to design algorithms for problems in machine vision.

In this book the author presents a commendably lucid outline of the theory of scale-space, the structure of low level operations in a scale-space setting and algorithmic schemes to use these structures such as to solve important problems in computer vision. The subjects range from a mathematical underpinning, over issues in implementation (discrete scale-space structures) to more open ended algorithmic methods for computer vision problems. The latter methods seem to me to point a way to a range of potentially very important applications. This approach will certainly turn out to be part of the foundations of the theory and practice of machine vision.

It was about time for somebody to write a monograph on the subject of scale-space structure and scale-space based methods, and the author has no doubt performed an excellent service to many in the field of both artificial and biological vision.

Utrecht, October 4th, 1993 *Jan Koenderink*

Preface

We perceive objects in the world as having structures both at coarse and fine scales. A tree, for instance, may appear as having a roughly round or cylindrical shape when seen from a distance, even though it is built up from a large number of branches. At a closer look, individual leaves become visible, and we can observe that the leaves in turn have texture at an even finer scale.

The fact that objects in the world appear in different ways depending upon the scale of observation has important implications when analysing measured data, such as images, with automatic methods. A straightforward way of exemplifying this is to note that every operation on image data must be carried out on a window, whose size can range from a single point to the whole image. The type of information we can get from such an operation is largely determined by the relation between structures in the image and the size of the window. Hence, without prior knowledge about what we are looking for, there is no reason to favour any particular scale. We should therefore try them all and operate at all window sizes.

These insights are not completely new in computer vision. Multi-scale representations of images in terms of pyramids were developed already around 1970. A main motivation then was to achieve computational efficiency by coarse-to-fine strategies. This approach was also supported by findings in neurophysiology about the primate visual system. However, it was soon discovered that relating structures from different levels in the multi-scale representation was far from trivial. Structures at coarse levels could sometimes not be assigned any direct interpretation, since they were hard to trace to finer scales. Despite considerable efforts to develop techniques for matching between scales, a theoretical foundation was missing.

In 1983, Witkin proposed that scale could be considered as a continuous parameter, thereby generalizing the existing notion of Gaussian pyramids. He noted the relation to the diffusion equation and hence found a well-founded way of relating image structures between different scales. Koenderink soon furthered the approach, which has been developed into what we now know as scale-space theory.

Since that work, we have seen the theory develop in many ways, and also realized that it provides a framework for early visual computations of a more general nature. The aim of this book is to provide a coherent overview of this recently developed theory, and to make material, which has earlier existed only in terms of research papers, available to a larger audience. The presentation provides an introduction into the general foundations of the theory and shows how it applies to essential problems in computer vision such as computation of image features and cues to surface shape. The subjects range from the mathematical foundation to practical computational techniques. The power of the methodology is illustrated by a rich set of examples.

I hope that this work can serve as a useful introduction, reference, and inspiration for fellow researchers in computer vision and related fields such as image processing, signal processing in general, photogrammetry, and medical image analysis. Whereas the book is mainly written in the form of a research monograph, the level of presentation has been adapted so that it can be used as a basis for advanced courses in these fields.

The presentation is organized in a logical bottom-up way, following the ordering of the processing modules in an imagined vision system. It is, however, not necessary to read the book in such a sequential manner. Several of the chapters are relatively self-contained, and it should be possible to read them independently. A guide to the reader describing the mutual dependencies is given in section 1.7 (page 22). I wish the reader a pleasant tour into this highly stimulating and challenging subject.

Stockholm, September 1993, *Tony Lindeberg*

Abstract

The presentation starts with a philosophical discussion about computer vision in general. The aim is to put the scope of the book into its wider context, and to emphasize why the notion of *scale* is crucial when dealing with measured signals, such as image data. An overview of different approaches to multi-scale representation is presented, and a number of special properties of scale-space are pointed out.

Then, it is shown how a mathematical theory can be formulated for describing image structures at different scales. By starting from a set of axioms imposed on the first stages of processing, it is possible to derive a set of canonical operators, which turn out to be derivatives of Gaussian kernels at different scales.

The problem of applying this theory computationally is extensively treated. A *scale-space theory* is formulated for *discrete signals,* and it demonstrated how this representation can be used as a *basis* for expressing a large number of *visual operations.* Examples are smoothed derivatives in general, as well as different types of detectors for image features, such as edges, blobs, and junctions. In fact, the resulting scheme for feature detection induced by the presented theory is very simple, both conceptually and in terms of practical implementations.

Typically, an object contains structures at many different scales, but locally it is not unusual that some of these "stand out" and seem to be more significant than others. A problem that we give special attention to concerns how to find such locally stable scales, or rather how to generate hypotheses about interesting structures for further processing. It is shown how the scale-space theory, based on a representation called the *scale-space primal sketch,* allows us to extract *regions of interest* from an image without prior information about what the image can be expected to contain. Such regions, combined with knowledge about the scales at which they occur constitute *qualitative information,* which can be used for *guiding and simplifying* other low-level processes.

Experiments on different types of real and synthetic images demonstrate how the suggested approach can be used for different visual tasks, such as image segmentation, edge detection, junction detection, and focus-of-attention. This work is complemented by a mathematical treatment showing how the behaviour of different types of image structures in scale-space can be analysed theoretically.

It is also demonstrated how the suggested scale-space framework can be used for computing direct cues to *three-dimensional surface structure*, using in principle only the same types of *visual front-end* operations that underlie the computation of image features.

Although the treatment is concerned with the analysis of visual data, the notion of scale-space representation is of much wider generality and arises in several contexts where measured data are to be analyzed and interpreted automatically.

Acknowledgments

This book is based on the author's thesis *Discrete Scale-Space Theory and the Scale-Space Primal Sketch*, presented at KTH (Royal Institute of Technology) in Stockholm in May 1991. The material has been updated and extended with respect to research conducted since then.

It is a pleasure to take this opportunity to express my deep gratitude to the following people for their important contributions in various ways:

- *Jan-Olof Eklundh* for introducing me into this inspiring field, for excellent supervision during my period as a PhD student, and for providing the stimulating and pleasant research environment to work in known as the CVAP group at KTH.
- *Jan Koenderink* for his outstanding contributions to the field and for serving as the opponent at my defense.
- *Kjell Brunnström* for the enjoyable collaboration with the work on junction classification.
- *Jonas Gårding* for the fruitful and inspiring collaboration on shape-from-texture and our many discussions.
- *Harald Winroth* with whom I shared office during a number of years. Our many talks and discussions have contributed substantially to the presentation of this work.
- *Fredrik Bergholm* for many discussions about scale-space as well as valuable comments on an endless number of manuscript drafts.
- *Demetre Betsis* for always being a reliable source of good advice.
- *Stefan Carlsson* for useful discussions, which served as a large source of inspiration to the discrete scale-space theory.
- *Ambjörn Naeve* and *Lars Svensson* for always sharing their intuition and insights into many aspects of mathematics.
- *Thure Valdsoo* for providing the interesting aerosol problem.
- *Matti Rendahl* for maintaining an excellent systems environment at our laboratory, and for always providing wizardly help when needed.

I would also like to thank my other colleagues at the Computational Vision and Active Perception Laboratory, CVAP, for their friendship and for the help they have given me in so many ways: *Magnus Andersson*, *Ann Bengtson*, *Kiyo Chinzei*, *Antonio Francisco*, *Akihiro Horii*, *Meng-Xiang Li*, *Anders Lundquist*, *Oliver Ludwig*, *Atsuto Maki*, *Peter Nordlund*, *Niklas Nordström*, *Göran Olofsson*, *Lars Olsson*, *Kourosh Pahlavan*, *Eva Setterlind*, *Tomas Uhlin*, and *Wei Zhang*.

I would like to express my gratitude to the other colleagues at KTH with whom I've had valuable interactions. In particular, I would like to mention *Michael Benedicts*, *Anders Björner*, *Germund Dahlquist*, *Lars Holst*, *Torbjörn Kolsrud*, *Heinz-Otto Kreiss*, *Anders Liljeborg*, *Bengt Lindberg*, *Jesper Oppelstrup*, and *Johan Philip*.

This work has also benefitted from discussions with other colleagues in the field, in particular the partners in our national, European, and transatlantic collaborations. Especially, I would like to thank *Luc Florack*, *Peter Johansen*, *Bart ter Haar Romeny*, *Stephen Pizer*, and *Richard Weiss* with all of whom I have had several highly interesting discussions.

In preparing this manuscript I would like to thank *Jan Michael Rynning* for valuable advice on LaTeX and typesetting, *Birgitta Krasse* for redrawing many of the figures, and *Mike Casey* at Kluwer Academic Publishers for his enthusiasm and patience with this manuscript.

Moreover, I would like to thank my parents *Paul* and *Inga-Lill* for always being there when I needed help in any way.

The research presented in this book has been made possible by a graduate position, "excellenstjänst," and a postgraduate position, "forskarassistenttjänst," provided by KTH together with financial support from the Swedish National Board for Industrial and Technical Development, NUTEK. Parts of the work have been performed under the ESPRIT-BRA projects InSight and VAP. This support is gratefully acknowledged.

All implementations have been made within the Candela and CanApp programming environment for image analysis developed at the Computational Vision and Active Perception Laboratory.

Contents

Appendix

1

Introduction and overview

In our daily life we use vision as one of our main sources of information about the outside world. Compared to a sense like hearing, the visual sense gives a richer description of the world. Compared to a sense like touch, it allows us to gather information about objects at greater distance and without affecting the objects themselves physically. Considering the apparent ease with which we obtain information about the world from the light that enters our eyes, an intellectual effort is required to appreciate that this is a non-trivial task.

Computer vision addresses this problem computationally; it deals with the problem of deriving meaningful and useful information from visual data. What should be meant by "meaningful and useful information" is, of course, dependent on the goal of the analysis, that is, the underlying *purpose* why we want to make use of visual information and process it with automatic methods. One reason may be that of machine vision—the desire to provide machines and robots with visual abilities. Typical tasks to be solved are object recognition, object manipulation, and visually guided navigation. The type of information that needs to be computed to address a problem depends strongly on the task. For example, the problem of recognizing objects from complex scenes is generally regarded as one of the more complicated problems in the field, while under certain conditions descriptors like time-to-collision can be computed with comparably simpler low-level operations. Other common applications of techniques from computer vision can be found in image processing, where one can mention image enhancement, visualization and analysis of medical data, as well as industrial inspection, remote sensing, automated cartography, data compression, and the design of visual aids, etc.

A more theoretical reason why computer vision is studied is the desire of understanding mathematical and physical principles underlying the inference of scene characteristics from brightness data. If insights into such basic principles can be gained, then they may help us with the tremendously inspiring challenge of understanding the workings of the biological visual systems, which accomplish their tasks in a way that is essential for the survival of most living creatures.

The problem of understanding vision has interested and puzzled researchers through the centuries. Still, some of the most basic questions that remain to be answered concern what type of image information is relevant for accomplishing different tasks, how this information should be extracted from the sensory data, and how such features can be related to properties of environment? An indication of the complexity of the vision problem can be obtained from the fact that the term "vision" has been very hard to define. Then, what definitions have been stated? To the question "What does it mean to see?" Marr (1982) answered

> ... vision is the *process* of discovering from images what is present in the world and where it is.

emphasizing that vision is an information-processing task. He also stressed that the issue of *internal representation* of information is of utmost importance. Only by representation can information be captured and made available to decision processes. The purpose of a representation is to make certain aspects of the information content *explicit,* that is, immediately accessible without any need for additional processing.

While Marr's definition captures several important aspects, the active and goal-oriented nature of vision is only implicit in this formulation. Clearly, the vision problem is undefined unless related to a task. The existence proofs of vision provided by nature, the biological vision systems, are usually not passively registering images of the world. Instead, biological vision is strongly tied to action, since the visual agent has to attend to and respond to dynamic changes in the outside world. It is also well-known in perception psychology that perception of pictures differs from perception of the three-dimensional world.

These are some of the main arguments behind the *active vision* methodology (Bajscy 1988; Aloimonos *et al.* 1988; Ballard 1991; Pahlavan *et al.* 1993), which has received increasing attention during recent years. In this paradigm, the ability of the vision system to selectively control the image acquisition process is emphasized. Moreover, the desired behaviour of the visual agent is put into focus. If the visual system is allowed to acquire more information in difficult situations, then several problems occurring in the analysis of given pre-recorded images can be avoided.

A simple example is the problem of too low a resolution. It can be circumvented by foveating interesting structures, or if necessary, moving closer to the interesting object. The active approach makes it possible to acquire additional information about three-dimensional structure from cues like accommodation distance, vergence angles, etc. An active moving observer also has the potential of avoiding unfortunate situations like accidentally aligned structures. It is sometimes argued that accidentally coinciding structures are very singular cases that never occur in practice, but in reality such situations turn out to show up rather frequently, when taking overview images of moderately complex scenes using cameras with normal resolution and opening angles.

There have been, and still are, different opinions in the computer vision community about how a visual system should be constructed. A long debate concerned the choice between bottom-up and top-down based reasoning. It has been argued by many authors that a visual system should be constructed in a modular way with different levels of processing. At the simplest level of abstraction three layers can be distinguished, usually denoted low-level, intermediate level, and high-level.

Although a natural implication of the active vision paradigm is that it may not be as easy to clearly separate out different processing levels as would be needed for a dogmatic interpretation of such a simple three-layer description, and although extreme stand points have been taken, such as "direct pick-up" (Gibson 1979), "labyrinthic design" (Aloimonos 1990), or "intelligence without representation" (Brooks 1991), one should be careful of not interpreting the active vision approach as excluding the need for competence theories, like concerning the computation of early retinotopic representations such as intrinsic images (Barrow and Tenenbaum 1978). The need for some kind of early low-level processing and representation for providing a sparse but rich set of primitives for other processing modules still remains highly motivated.

This book deals with a basic aspect of early image representation—the notion of *scale*. More specifically, the work deals with a certain type of approach, the use of *scale-space representation,* for analysing image data at the very lowest levels in the chain of information processing of a visual system. The aim is to operate directly on the raw pixel values without any type of pre-processing. The suggested methodology is intended as a first confrontation between the reasoning process and the raw image data. This part of the visual system is usually termed the *visual front-end.* No specific assumptions will be made about how higher-level processes are to operate on the output. Therefore, the approach is applicable to a variety of reasoning strategies.

Computer vision is a cross-disciplinary field with research methodologies from several scientific disciplines such as computer science, mathematics, neurophysiology, physics, and psychology. The approach taken here will be computational.[1] A theory and a framework will be proposed for how certain aspects of image information can be represented and analysed at the earliest processing stages of a machine vision system.

[1] Although there are neurophysiological and psychophysical evidence for the existence of processing at multiple scales in biological vision (Campbell and Robson 1977; (Wilson 1983; Young 1985, 1987; Jones and Palmer 1987), no claims will be made that the methodology proposed here describes how processing is done in human perception. The treatment is concerned with what visual information can be extracted by a computer. When biological vision is discussed, it is mainly as a source of inspiration.

1.1. Theory of a visual front-end

If we are to construct a machine vision system, the problem can be addressed in several ways. If the visual task is sufficiently domain specific, then it may be sufficient to come up with any set of algorithms that performs the given task up to some prescribed tolerance. On the other hand, if the aim is to construct a flexible system able to solve a large number of problems using visual information, then it may be advantageous to aim at a certain degree of generality in the design, so that similar low-level modules can be shared between several algorithms or processes for solving different visual tasks. If such modules also are to be constructed without built-in limitations that would restrict their applicability, then a natural requirement is that the first stages of processing should make as few irreversible decisions and be as uncommitted as possible.

This presentation follows the latter strategy. If the vision problem is approached without strong presumptions about what specific tasks are to be solved, then a fundamental question concerns what information should be extracted at the earliest stages, and what kinds of operations are natural to perform on the data that reach the visual sensor. Is *any* type of operation feasible? An axiomatic approach that has been adopted in order to restrict the space of possibilities is to assume that the very first stages of processing should be able to function without any direct knowledge about what can be expected to be in the scene. For an uncommitted vision system, the scale-space theory states that under certain conditions, there is a natural choice of first stage operations to perform in a visual front-end (this notion will be made more precise later). The output from these operations can then, in turn, be used as input to a large number of other visual modules. An attractive property of this type of approach is that it gives a uniform structure on the first stages of computation.

1.2. Goal

The main subject of this book is to give a mathematical description of such early visual operations. The goal the work aims at is a methodology, in which significant structures can be extracted from an image in a solely bottom-up way, and scale levels can be selected for handling those structures without any prior information. A short summary in terms of key words can be expressed as follows:

A *ranking of events* in order of significance will be suggested based on *volumes* of certain four-dimensional objects in a *scale-space* representation of the signal. In this scale-space, the *scale* dimension is treated as *equally important* as the spatial and grey-level coordinates. The associated *extraction* method is based on a *systematic parameter variation principle,* where *locally stable states* are detected and abstractions are determined from those.

It will be exemplified how *qualitative scale and region information* extracted in this way can be used for *guiding the focus-of-attention* and *tuning* other early visual processes so as to *simplify* their tasks. The general principle is to adapt the low-level processing to the local image structure. The main theme of the book is to construct a theoretical framework in which these operations can be formalized.

1.3. The nature of the problem

When given an image as obtained from a standard camera device, say a digitized video signal or a scanned photograph, all information is encoded in the pixel values represented as a matrix of numerical data. If this information is presented to a human observer with the pixel values coded as grey-level intensities, then the human will usually have no problems in perceiving and interpreting what the image represents.

However, if the same pattern of grey-level values is coded as decimal digits, or as a three-dimensional diagram with the grey-level values drawn as a function of the image coordinates, then the problem is no longer as easy for biological vision. A person not familiar with the field often underestimates the difficulties in designing algorithms for interpreting data on this numerical form. The problem with the matrix representation of the image is that the information is only *implicit* in the data.

1.3.1. Ill-posedness

A major subtask of a visual processing system is to *extract* meaningful information about the outside world from such a set of pixel values, which is the result of light measurements from a physical scene. The image data may either be given beforehand, like in image processing, or have been acquired by an active system, which has directed its attention towards some interesting structure. What is meant by meaningful is in turn given by the task the vision system has to solve.

In principle, this problem of deriving three-dimensional shape information about the scene is impossible to solve if stated as a pure mathematical problem. Assume first that a set of grey-level data is given. Then, there will always be an infinite number of scenes that could have given rise to the same result. To realize that this is the case, consider for example a photograph on a paper, or a slide projected onto a screen. We easily interpret such brightness distributions on flat surfaces as corresponding to three-dimensional objects with perceived depth variations.

In an active vision system additional cues may be available, like accommodation depth, vergence, etc. Nevertheless, it is always possible to present two cameras with (possibly time varying) brightness patterns that would give the system a completely false impression of the world. There are two basic reasons to this. The first is that we are not measuring direct properties of the world, but light emitted from it. The second is a dimensionality problem; we are trying to analyse a three-dimensional world using two-dimensional image data.

From this viewpoint the vision problem is *ill-posed*[2] in the sense of Hadamard, since it does not have any unique solution. A rigorous person without plenty of unspoiled optimism would probably take this as a very good motivation to study some other field of science, where the prerequisites could be more clearly stated and better suited for formal analysis. Nevertheless, despite this indeterminacy, the human visual system as well as other biological vision systems are capable of coping with the ill-posedness. Moreover, since vision is generally regarded as the highest developed of our senses, one can speculate that there must be some inherent properties in the image data reaching the retina that make the visual perception[3] possible.

1.3.2. Grouping

A main purpose of the low-level processing modules is to provide a reasonable set of primitives that can be used for further processing or reasoning modules. A fundamental problem in this context concerns what points in the image can be regarded as related to each other and correspond to objects in the scene, i.e., which pixels in the image can be assumed to belong together and form meaningful entities. This is the problem of primitive grouping or perceptual organization. Before any such grouping operations have been performed, the matrix of grey-level values is, from the viewpoint of interpretation, in principle only a set of numerical values laid out on a given discrete grid.

The grouping problem has been extensively studied in psychology, especially by the *Gestaltists* (Koffka 1935), as well as in computer vision (Lowe 1985; Ahuja and Tuceryan 1989), and it seems to be generally agreed upon that the existence of active grouping processes in human perception can be regarded as established. Witkin and Tenenbaum (1983) discuss this property:

> People are able to perceive structures in images, apart from the perception of three-dimensionality, and apart from the recognition of familiar objects. We impose organization on

[2]For a mathematical problem to be regarded as *well-posed,* Hadamard stated three criteria: (i) a solution should exist, (ii) the solution should be unique, and (iii) the solution should depend continuously on the input data. A well-posed problem is not necessarily *well-conditioned.*

[3]Of course, experiences and expectations are generally believed to play an important role in the perception process. However, also that information must be related to the incoming image data in some way. Moreover, the experiences must have been acquired (learned) in some way, at least partially based on visual data

> data ... even when we have no idea what it is we are organizing. What is remarkable is the degree to which such naively perceived structure survives more or less intact once a semantic context is established: the naive observer often sees essentially the same thing as an expert does. ... It is almost as if the visual system has some basis for guessing *what* is important without knowing *why*.

Although the gestalt school of psychology formulated rules as those of proximity, similarity, closure, continuation, symmetry, and familiarity, we still have no satisfactory understanding of how these mechanisms operate from a quantitative point of view.

1.3.3. Operator size

To be able to compute any type of representation from image data, it is necessary to extract information from it, and hence interact with the data using some operators. Some of the most fundamental problems in low-level vision and image analysis concern *what* operators to use, *where* to apply them, and *how large* they should be. If these problems are not appropriately addressed, then the task of interpreting the output results can be very difficult.

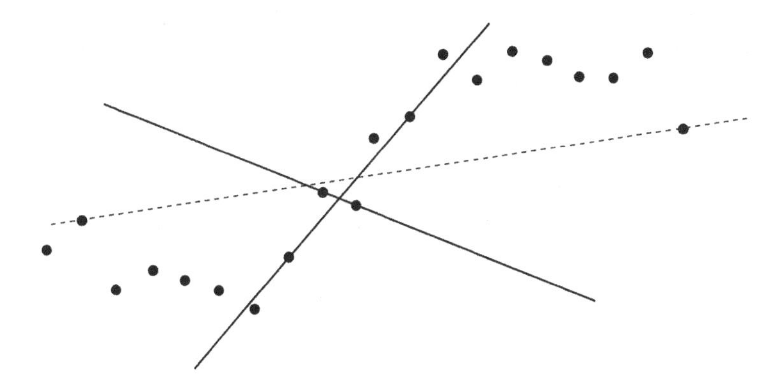

Figure 1.1. Illustration of the basic scale problem when computing gradients as a basis for edge detection. Assume that the dots represent (noisy) grey-level values along an imagined cross-section of an object boundary, and that the task is to find the boundary of the object. The lines show the effect of computing derivative approximations using a central difference operator with varying step size. Clearly, only a certain interval of step sizes is appropriate for extracting the major slope of the signal corresponding to the object boundary. Of course, this slope may also be interpreted as due to noise (or some other phenomena that should be neglected) if it is a part superimposed onto some coarser-scale structure (not visible here).

To illustrate this problem, consider the task of detecting edges. It is generally argued that this type of image feature represents important information, since under reasonably general assumptions, edges in an image can be assumed to correspond to discontinuities in depth, surface orientation, reflectance properties, or illumination phenomena in the scene. A standard way of extracting edges from an image is by gradient computation followed by some type of post-processing step, where "high values" should be separated from "low values," e.g., by detection of local maxima or by thresholding on gradient magnitude.

Consider, for simplicity, the one-dimensional case, and assume that the gradient is approximated by a central difference operator. More sophisticated approaches exist, but they will face similar problems. It is well-known that the selection of step size leads to a trade-off problem: A small step size leads to a small truncation error in the discrete approximation, but the sensitivity to fine-scale perturbations (e.g., noise) might be severe. Conversely, a large step size will, in general, reduce this sensitivity, but at the cost of an increased truncation error. In the worst case, a slope of interest can be missed and meaningless results be obtained if the difference quotient approximating the gradient is formed over a larger distance than the object considered in the image. See figure 1.1 for an illustration.

Although we shall here mainly be concerned with static images, the same kind of problem arises when dealing with image sequences. Similarly, models based on spatial derivatives ultimately rely on the computation of derivative approximations from measured data.

1.3.4. Scale

The problem falls back on a basic scale problem, namely that objects in the world and details in images, only exist as meaningful entities over limited ranges of scale,[4] in contrast to certain ideal mathematical entities like "point," "line," "step edge," or "linear slope," which appear in the same way at all scales of observation.

A simple example is the concept of a branch of a tree, which makes sense only at a scale from, say, a few centimeters to at most a few meters. It is meaningless to discuss the tree concept at the nanometer or

[4]An interesting philosophical question in this context concerns whether or not the scale property should be attributed to the actual physical objects themselves or just to our subjective way of perceiving and categorizing them. For example, a table made out of wood certainly has a fine-scale texture with underlying fibral and molecular structures that we usually suppress when dealing with it for every-day purposes. Obviously, such finer-scale properties will always be there, but we almost always automatically disregard them. One may speculate that such a organization at multiple scales may be one way of simplifying the representation of our extremely complicated environment into a hierarchical structure to cope with it efficiently.

the kilometer level. At those scales it is more relevant to talk about the molecules that form the leaves of the tree, or the forest in which the tree grows. Similarly, it is only meaningful to talk about a cloud over a certain range of coarse scales. At finer scales it is more appropriate to talk about the individual droplets, which in turn consist of water molecules, which consist of atoms, which consist of protons and electrons etc.

This fact is well-known in the experimental sciences. In physics, the world is described at several levels of scales, from particle physics and quantum mechanics at fine scales, through thermodynamics and solid mechanics dealing with every-day phenomena, to astronomy and relativity theory at scales much larger than those we are usually dealing with. The physical description depends strongly on the scale at which the world is modelled. In biology, the study of animals can only be performed over a certain range of coarse scales. An organism looks completely different seen through a microscope when individual cells become visible.

These examples demonstrate that the scale concept is of crucial importance if one aims at describing the structure of the world, or more specifically the structure of projections of the world to two-dimensional data sets. As Koenderink (1984) has emphasized, the problem of scale must be faced in any image situation. The extent of any real-world object is determined by two scales, the *inner scale* and the *outer scale.* The outer scale of an object or a feature may be said to correspond to the (minimum) size of a window that completely contains the object or the feature, while the inner scale may loosely be said to correspond to the scale at which substructures of the object or the feature begin to appear.

In a given image, only structures over a certain range of scales can be observed. This interval is delimited by two scales; the outer scale corresponding to the finite size of the image, and the inner scale given by the resolution. For a digital image the inner scale is determined by the pixel size, and for a photographic image by the grain size in the emulsion.

1.3.5. Multi-scale representation

While these qualitative aspects of scale have been well-known for a long time, the concept of scale has been very hard to formalize into a mathematical theory. It is only during the last few decades that tools have been developed for handling the scale concept in a formal manner. A driving force in this development has come from the need for developing robust algorithms in image processing, computer vision, and other fields related to automatic signal processing.

A methodology that has been proposed for handling the notion of scale in measured data is by representing measured signals at multiple scales. Since, in general, no particular levels of scale can be pre-supposed without strong *a priori* knowledge, the only reasonable solution is that the visual system must be able to handle image structures at *all* scales. The main idea of creating a *multi-scale representation* of a signal is by generating a one-parameter family of derived signals, where fine-scale information is successively suppressed. Then, a mechanism is required that systematically simplifies the data and removes finer-scale details, or high-frequency information. This operation, which will be termed *scale-space smoothing,* must be available at any level of scale.

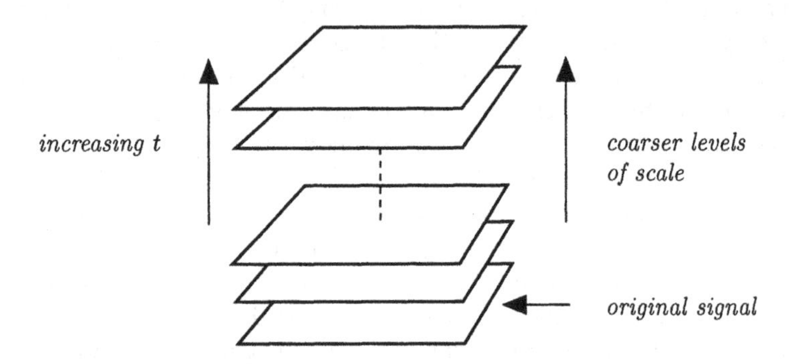

Figure 1.2. A multi-scale representation of a signal is an ordered set of derived signals intended to represent the original signal at different levels of scale.

Why should one represent a signal at multiple scales when all information is anyway in the original data? A major reason for this is to explicitly represent the multi-scale aspect of real-world data. Another aim is to suppress and remove unnecessary and disturbing details, such that later stage processing tasks can be simplified. More technically, the latter motivation reflects the common need for smoothing as a pre-processing step to many numerical algorithms as a means of noise suppression.

1.4. Scale-space representation

A methodology proposed by Witkin (1983) and Koenderink (1984) to obtain such a multi-scale representation of a measured signal is by embedding the signal into a one-parameter family of derived signals, the *scale-space,* where the parameter, denoted *scale parameter* $t \in \mathbb{R}_+$,[5] is intended to describe the current level of scale.

[5] $\mathbb{R}_+$ denotes the set of real non-negative numbers, and $\mathbb{R}_+ \backslash \{0\}$ the corresponding set excluding the zero point.

1.4.1. Scale-space for one-dimensional signals: Gaussian smoothing

Let us briefly review this procedure as it is formulated for one-dimensional continuous signals: Given a signal $f\colon \mathbb{R} \to \mathbb{R}$, the scale-space representation $L\colon \mathbb{R} \times \mathbb{R}_+ \to \mathbb{R}$ is defined such that the representation at "zero scale" is equal[6] to the original signal

$$L(\cdot;\ 0) = f(\cdot), \tag{1.1}$$

and the representations at coarser scales are given by convolution of the given signal with Gaussian kernels of successively increasing width

$$L(\cdot;\ t) = g(\cdot;\ t) * f. \tag{1.2}$$

In terms of explicit integrals, the result of the convolution operation '*' is written

$$L(x;\ t) = \int_{\xi=-\infty}^{\infty} g(\xi;\ t)\, f(x-\xi)\, d\xi, \tag{1.3}$$

where $g\colon \mathbb{R} \times \mathbb{R}_+ \backslash \{0\} \to \mathbb{R}$ is the (one-dimensional) Gaussian kernel

$$g(x;\ t) = \frac{1}{\sqrt{2\pi t}}\, e^{-x^2/2t}. \tag{1.4}$$

Figure 1.3 shows the result of smoothing a one-dimensional signal to different scales in this way. Notice how this successive smoothing captures the intuitive notion of fine-scale information being suppressed, and the signals becoming gradually smoother.

1.4.2. Diffusion formulation of scale-space

In terms of differential equations, the evolution over scales of the scale-space family L can be described by the (one-dimensional) *diffusion equation*

$$\partial_t L = \tfrac{1}{2}\, \nabla^2 L = \tfrac{1}{2}\, \partial_{xx} L. \tag{1.5}$$

In fact, the scale-space representation can equivalently be defined as the solution to (1.5) with initial condition $L(\cdot;\ 0) = f(\cdot)$.

This analogy also gives a direct physical interpretation of the smoothing transformation. The scale-space representation L of a signal f can be understood as the result of letting an initial heat distribution f evolve over time t in a homogeneous medium. Hence, it can be expected that fine-scale details will disappear, and images become more diffuse when the scale parameter increases.

[6] The notation $L(\cdot;\ 0) = f$ stands for $L(x;\ 0) = f(x)\ \forall x \in \mathbb{R}^N$.

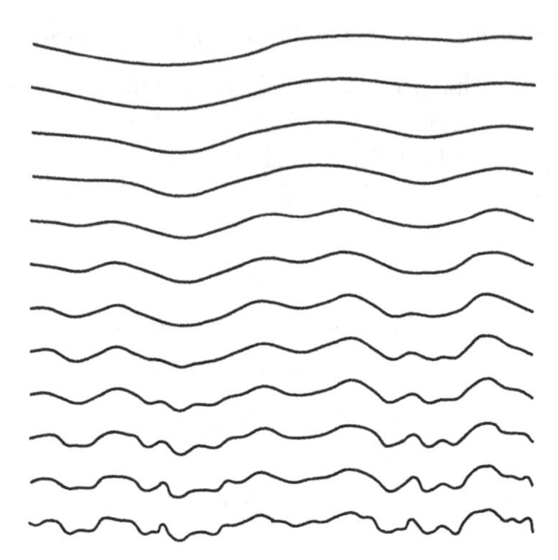

Figure 1.3. The main idea with a scale-space representation of a signal is to generate a one-parameter family of derived signals in which the fine-scale information is successively suppressed. This figure shows a signal that has been successively smoothed by convolution Gaussian kernels of increasing width. (Adapted from Witkin 1983).

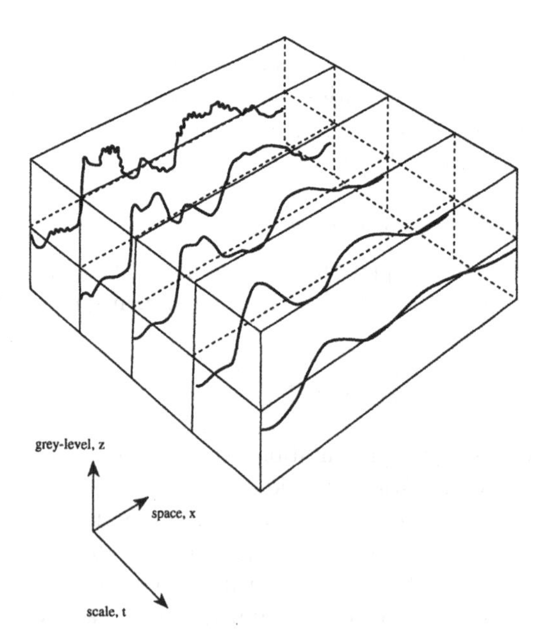

Figure 1.4. Schematic three-dimensional illustration of the scale-space representation of a one-dimensional signal.

1.4.3. Definition of scale-space: Non-creation of new structure

For a reader not familiar with the scale-space literature, the task of designing a multi-scale signal representation may at first glance be regarded as somewhat arbitrary. Would it suffice to carry out just *any* type of "smoothing operation"? This is, however, not the case. Of crucial importance when constructing a scale-space representation is that the transformation from a fine scale to a coarse scale really can be regarded as a simplification, so that fine-scale features disappear *monotonically* with increasing scale. If new artificial structures could be created at coarser scales, not corresponding to important regions in the finer-scale representations of the signal, then it would be impossible to determine whether a feature at a coarse scale corresponded to a simplification of some coarse-scale structure from the original image, or if it were just an accidental phenomenon, say an amplification of the noise, *created by the smoothing method—not the data.* Therefore, it is of utmost importance that artifacts are not introduced by the smoothing transformation when going from a finer to a coarser scale.

How should this property be formalized? When Witkin (1983) introduced the notion of scale-space, he was concerned with one-dimensional signals. He observed that the number of zero-crossings in the second derivative decreased monotonically with scale, and took that as a basic characteristic of the representation. In fact, this property holds for derivatives of arbitrary order, and also implies that the number of local extrema in any derivative of the signal cannot increase with scale. From this viewpoint, convolution with a Gaussian kernel possesses a strong smoothing property.

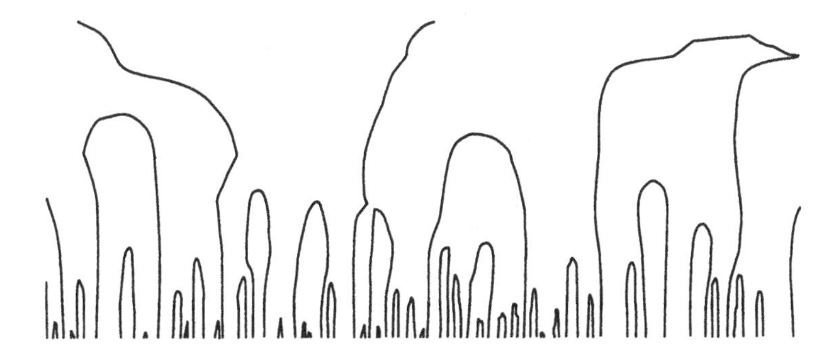

Figure 1.5. Since new zero-crossings cannot be created by the diffusion equation in the one-dimensional case, the trajectories of zero-crossings in scale-space (here, zero-crossings of the second derivative) form paths across scales that are never closed from below. (Adapted from Witkin 1983).

1.4.4. Uniqueness of the Gaussian

Later, when Koenderink (1984) extended the scale-space concept to two-dimensional signals, he introduced the notion of *causality,* which means that new level surfaces must not be created when the scale parameter is increased. Equivalently, it should always be possible to trace a grey-level value existing at a certain level of scale to a similar grey-level at any finer level of scale. By combining causality with the notions of *homogeneity* and *isotropy,* which essentially mean that all spatial points and all scale levels must be treated in a similar manner, he showed that the scale-space representation of a two-dimensional signal by necessity must satisfy the *diffusion equation*

$$\partial_t L = \tfrac{1}{2} \nabla^2 L = \tfrac{1}{2} (\partial_{xx} + \partial_{yy}) L. \tag{1.6}$$

Since convolution with the Gaussian kernel $g\colon \mathbb{R}^2 \times \mathbb{R}_+ \setminus \{0\} \to \mathbb{R}$

$$g(x, y;\ t) = \frac{1}{2\pi t} e^{-(x^2+y^2)/2t} \tag{1.7}$$

describes the solution of the diffusion equation at an infinite domain, it follows that the Gaussian kernel is the unique kernel for generating a scale-space. This formulation extends to arbitrary dimensions.

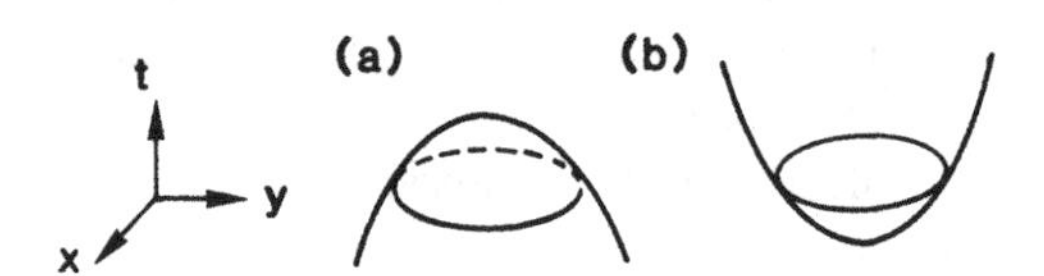

Figure 1.6. The causality requirement means that level surfaces in scale-space must point with their concave side towards finer scales; (a) the reverse situation (b) must never occur.

A similar result based on slightly different assumptions, was given by Yuille and Poggio (1986) concerning the zero-crossings of the Laplacian of the Gaussian. Related formulations have been expressed by Babaud *et al.* (1986), and by Hummel (1987).

Another formulation was stated by Lindeberg (1990), who showed that the property of not introducing new local extrema with increasing scale by necessity lead to the Gaussian kernel if combined with a *semi-group structure* on the family of convolution kernels.

Florack *et al.* (1992) have elegantly shown that the uniqueness of the Gaussian kernel for scale-space representation can be derived under weaker conditions, by combining the semi-group structure of a convolution operation with a *uniform scaling property* over scales.

A notable similarity between these (and other) results is that several different ways of choosing scale-space axioms give rise to the same conclusion. The transformation given by convolution with the Gaussian kernel possesses a number of special properties, which make it unique. From the similarities between the different scale-space formulations, it can be regarded as well-established that within the class of linear transformations, the scale-space formulation in terms of the diffusion equation describes the canonical way to construct a multi-scale image representation.

An extensive review of basic properties about scale-space and related multi-scale representations is given in chapter 2. Before proceeding to the next subject, let us consider two every-day analogies concerning the need for multi-scale representation.

1.4.5. The scale parameter delimits the inner scale of observation

With respect to the notions of inner and outer scale, increasing the scale parameter in scale-space has the effect of increasing the inner scale of an observation. To appreciate the usefulness for this type of operation, consider the well-known printing method called dithering. It is used for producing impressions of grey-level information when printing images using only one colour of the ink (typically black). One of the most common techniques is to produce a pattern of very small black discs of different size. While the original image usually does not contain such structures, by averaging this intensity pattern over a local spatial neighbourhood, the effect will be the impression of a grey-level corresponding to grey-tone information. In this respect, this printing method makes explicit use of the multi-scale processing capabilities of our vision system.

1.4.6. Symbolic multi-scale representation

Referring to the analogies with other fields of science, the need for multi-scale representation is well understood in cartography. Maps are produced at different degrees of abstraction. A map of the world contains the largest countries and islands, and possibly, some of the major cities, while towns and smaller islands appear at first in a map of a country. In a city guide, the level of abstraction is changed considerably to include streets and buildings, etc. In other words, maps constitute symbolic multi-scale representations of the world around us, although constructed manually and with very specific purposes in mind.

It is worth noting that an atlas usually contains a set of maps covering some region of interest. Within each map the outer scale typically scales in proportion with the inner scale. A single map is, however, usually not sufficient for us to find our way around the world. We need the ability to zoom in to structures at different scales; i.e., decrease or increase the inner scale of the observation according to the type of situation at hand.

1.5. Philosophies and ideas behind the approach

1.5.1. *Making information explicit*

The scale-space theory constitutes a well-founded framework for handling structures at different scales. However, the information in the scale-space embedding is only *implicit* in the grey-level values. The smoothed images in the raw scale-space representation contain no *explicit* information about the features in them or about the relations between features at different levels of scale.

One of the main goals of this book is to present such an explicit representation called the *scale-space primal sketch,* and to demonstrate that it enables extraction of significant image structures in such a way that the output can be used for guiding later stage processes and simplifying their tasks. The treatment will be concerned with intensity images, the grey-level landscape, and the chosen objects will be blobs, that is, bright regions on dark backgrounds or vice versa. However, the methodology applies to any bounded function and is therefore useful in many tasks occurring in computer vision, such as the study of level curves and spatial derivatives in general, depth maps, and histograms, point clustering and grouping, in one or in several dimensions. Moreover, the underlying principles behind its construction are general, and extend to other aspects of image structure.

1.5.2. *Scale and segmentation*

Many methods in computer vision and image analysis implicitly assume that the problems of scale detection and initial segmentation have already been solved. Models based on spatial derivatives ultimately rely upon the computation of derivative approximations, which means that they will face similar scale problems as were described in the discussion about edge detection from gradient data in section 1.3.3. Although we shall here be mainly concerned with static imagery, the same type of problems arise also when dealing with image data over time. In other words, when computing derivatives from measured data, we in general always fall back to the basic scale problem of selecting a filter mask size[7] for the approximation.

A commonly used technique for improving the results obtained in computer vision and other fields related to numerical analysis is by pre-processing the input data with some amount of smoothing or careful tuning of the operator size or some other parameters. In some situations the output may depend strongly on these processing steps. In certain algorithms these *tuning parameters* can be estimated; in other cases they are set manually. A robust image analysis method intended to work in an

[7] Observe that it is not the actual size of the filter mask that is important, but rather the *characteristic length* over which the difference approximation is computed.

autonomous robot situation must, however, be able to make such decisions automatically. How should this be done? I contend that these problems are in many situations nothing but disguised scale problems.

Also, to apply a refined mathematical model like a differential equation or some kind of deformable template, it is necessary to have some kind of qualitative initial information, e.g., a domain where the differential equation is (assumed to be) valid, or an initial region for applying the raw deformable template. Examples can be obtained from many "shape-from-X" methods, which in general assume that the underlying assumptions are valid in the image domain the method is applied to. A commonly used assumption is that of smoothness implying that the region in the image, to which the model is applied to, must correspond to, say, one physical object, or one facet of a surface. How should such regions be selected *automatically*? Many methods cannot be used unless this non-trivial part of the problem is solved.

How can we detect appropriate scales and regions of interest when there is no *a priori* information available? In other words, how can we detect the scale of an object and where to search for it before knowing what kind of object we are studying and before knowing where it is located. Clearly, this problem is intractable if stated as a pure mathematical problem. Nevertheless, it arises implicitly in many kinds of processes (e.g., dealing with texture, contours etc.), and seems to boil down to an intractable chicken-or-the-egg problem. The solution of the pre-attentive recognition problem seems to require the solution of the scale and region problems and vice versa.

The goal of this presentation is to demonstrate that such pre-attentive groupings can be performed in a bottom-up manner, and that it is possible to generate initial hypotheses about regions of interest as well as to give coarse indications about the scales at which the regions manifest themselves. The basic tools for the analysis will be scale-space theory, and a heuristic principle stating that blob-like structures which are stable in scale-space are likely candidates to correspond to significant structures in the image. Concerning scale selection, scale levels will be selected that correspond to local maxima over scales of a measure of blob response strength. (Precise definitions of these notions will be given later.) It will be argued that once such scale information is available, and once regions of interest have been extracted, later stage processing tasks can be simplified. This claim is supported by experiments on edge detection and classification based on local features.

1.5.3. Detection of image structure

The main features that arise in the (zero-order) scale-space representation of an image are smooth regions which are brighter or darker than the background and stand out from their surroundings. These will be termed blobs. The purpose of the suggested representation is to make these blobs explicit as well as their relations across scales. The idea is also that the representation should reflect the intrinsic shape of the grey-level landscape—it should not be an effect of some externally chosen criteria or tuning parameters. The theory should in a bottom-up fashion allow for a data-driven detection of significant structures, their relations, and the scales at which they occur. It will, indeed, be experimentally shown that the proposed representation gives perceptually reasonable results, in which salient structures are (coarsely) segmented out. Hence, this representation can serve as a guide to subsequent, more finely tuned processing, which requires knowledge about the scales at which structures occur. In this respect it can serve as a mechanism for focus-of-attention.

Since the representation tries to capture important image structures with a small set of primitives, it bears some similarity to the *primal sketch* proposed by Marr (1976, 1982), although fewer primitives are used. The central issue here, however, is to represent explicitly the scales at which different events occur. In this respect the work addresses problems similar to those studied by Bischof and Caelli (1988). They tried to parse scale-space by defining a measure of stability. Their work, however, was focused on zero-crossings of the Laplacian. Moreover, they overlooked the fact that the scale parameter must be properly treated when measuring significance or stability. Here, the behaviour of structures over scale will be analysed in order to give the basis of such measurements.

Of course, several other representations of the grey-level landscape have been proposed without relying on scale-space theory. Let us also note that Lifshitz and Pizer (1990) have studied the behaviour of local extrema in scale-space. However, we shall defer discussing relations to other work until the suggested methodology has been described.

1.5.4. Consistency over scales

The idea of scale-space representation of images, suggested by Witkin (1983) has, in particular, been developed by Koenderink and van Doorn (1984, 1986, 1992), Babaud *et al.* (1986), Yuille and Poggio (1986), Hummel (1987), Lindeberg (1990, 1993), and Florack *et al.* (1992). This work is intended to serve as a complement addressing computational aspects, and adding means of *making significant structures and scales explicit.*

The main idea of the approach is to *link* similar structures (here blobs) at different levels of scales in scale-space into higher-order objects (here four-dimensional objects called scale-space blobs), and to extract significant image features based on the appearance and lifetime of the higher-order objects in scale-space. A basic principle that will be used is that *significant image features must be stable with respect to variations in scale.*

Another important point within the work is that *the scale parameter is treated as being as equally important as the spatial and grey-level coordinates.* This is directly reflected in the fact that the primitives in the representation are objects having extent not only in space and grey-level, but also in scale.

1.6. Relations to traditional applied mathematics

In principle, we are to derive information from image data by operating on it with certain operators. An obvious question to ask is then why this problem could not be seen as an ordinary standard problem in numerical analysis and be solved with standard numerical techniques? Let us point out several reasons as to why the problem is hard.

1.6.1. Modelling, simulation, and inverse problem

Traditional numerical analysis is often concerned with the *simulation* of mathematical or physical models (for example, formulated as discrete approximations to continuous differential equations, which are rather good descriptions of the underlying reality). The problems are usually well-defined, the models can often be treated as exact, and the errors involved in these types of computations are mainly due to discretization and round-off errors.

In computer vision the situation is different. Given a signal, the task is to analyse and extract information from it. We are trying to solve an *inverse problem,* where the noise level is generally substantially higher[8] and the modelling[9] aspect is still open. With a precise model of the illumination situation as well as the reflectance properties of the surfaces in the environment, one could conceive solving for the surface geometry based on the physical light characteristics. However, it is well-known that this problem of *reconstructing the world* is extremely hard, to a large extent because it is very difficult to formulate an accurate and physically useful model for the image formation process, but also because such a model would require much additional *a priori* knowledge in order to be computationally tractable. Although further attempts to explore the situation in more detail are being made (Forsyth and Zissermann 1989; Nayar *et*

[8] A rule of thumb sometimes used in this context is that when derivatives of order higher than two are computed from raw image data, the amplitude of the amplified noise will often be of the same order of magnitude as the derivative of the signal or be even higher.

[9] The *geometry* of image formation is quite simple and well understood, but our knowledge about the complicated *physical phenomena* (comprising reflections, etc.), and how to model them from a computational viewpoint, is still rather vague. In addition, we have the problem of *representing* the enormous variety of *different situations* that can occur in the real world, as well as the question of how *cognitive* aspects should be incorporated into the process.

al. 1990), most shape-from-shading and similar algorithms still rely on very restrictive simplifying assumptions.

1.6.2. Scale and resolution

Other aspects are those of scale and resolution. In numerical analysis the accuracy can often be increased by a refinement in the grid sampling. The selection of a larger grid size is mainly motivated by efficiency reasons, since exact equations are usually simulated. In computer vision algorithms the number of grid points used for *resolving structures* in a given image is sometimes very low, which makes a difficult problem even more difficult. This restriction can be relaxed, however, in an active vision situation, as will be developed in section 11.3.

A more serious problem is that of scale. In most standard numerical problems the inner scale is zero, which means that the smaller the grid size that is being used, the higher will be the accuracy in computations (compare again with the example in section 1.3.3). In easy problems, the solutions asked for contain variations taking place on essentially a single scale. Problems having solutions with variations on different scales are more complicated and require more advanced algorithms for their solution. Examples can be obtained from *computer fluid dynamics,* where turbulence and very thin boundary layers are known to lead to very hard numerical problems. These fine-scale phenomena cannot always be fully resolved by discrete approximations, and in fact some type of (sometimes artificial) smoothing (dissipative terms) is often required. When the fine-scale phenomena are not properly dealt with, they can interfere with and disturb the coarse-scale phenomena that usually are the ones of interest in, for example, design applications. Moreover, the occurrences of *discontinuities* in the solutions, which are also very frequent in image data, are known to complicate the situation further.

The idea with scale-space representation is to separate out information at different scales. Note, that this may be a difficult problem, since in general, very little or no *a priori* knowledge can be expected about what types of structures the visual system is studying, or at what scales they occur.

1.6.3. Interpreting the results

If an operator is applied all over an image, then it will at best give reasonable answers in those regions in which the underlying assumptions for the method are valid (provided that the operator size has been appropriately tuned). However, the operator also gives *false alarms* in regions where the assumptions are not satisfied. One could say that such a uniform application of an operator enforces an answer in every point even though any well-defined answer does not exist. In general, it is hard to distinguish from the output of such an operation which responses can be trusted as correct and which ones should be rejected. Plain thresholding on the magnitude of the response is usually not sufficient. Therefore, a conservative strategy is to aim at deriving a sparse set of safe and reliable cues at the risk of "missing" a few that could be included rather than to try to compute "every" feature at the risk of including a large number of false responses. This is the motivation for trying to determine in advance where to apply[10] refined operations.

1.6.4. Approximation and regularization

It is sometimes argued that the main aims of approximation theory have already been accomplished. Nevertheless, one is confronted with serious problems when applying this theory to irregular and noisy measurement data like those obtained from images. Some of the most basic problems concern how to determine a region in space appropriate for fitting a model to the data, and how one should tune the associated parameters (such as the filter weights). An approach that has been extensively used in computer vision during the last decade is regularization. This technique has been applied to a variety of reconstruction problems (see Terzopoulos 1986; Terzopoulos *et al.* 1987, 1988; Kass *et al.* 1987; Witkin *et al.* 1987; Blake and Zisserman 1987; Pentland 1990; Aloimonos and Schulmann 1990). The basic methodology is to define a functional, which is a weighted combination of different error criteria, and then try to compute the function within some restricted space that minimizes it. These methods often contain a large number of parameters but the theory usually gives little or no information about how they should be set without manual intervention, although attempts have been made to learn them from examples. In addition there is a verification problem, since the algorithm is forced to always find a solution within the given space. How does one determine whether that function resembles the answer we actually want (the answer to the original problem)? The solution to a regularized problem is, in general, not equal to the solution to the original problem, not even if the input data are exact. To summarize, both these types of methods require a careful setting of their associated parameters, as well as the regions in space to which they should be applied.

1.6.5. Principles behind the work

A basic intention behind the approach taken here is to pre-process the data and to derive context information from it in such a way that the

[10]This is a problem arising mainly in an initialization phase of a reasoning process. When a time aspect is present, this problem is simplified, since context knowledge can be used for predictions about the future. It is generally argued that problems become easier once the boot-strapping step has been performed.

output from these types of operations can be well-defined. Although no claims are made that these problems have been solved, and even though further complications may appear on the way to the solution, I believe that the framework to be developed here represents a significant step toward posing the questions in a context where standard numerical techniques could be readily applied and give useful answers.

1.7. Organization of this book

The book deals with the fundamental problems that are associated with the use of scale-space analysis in early processing of visual information. More specifically some of the main questions it addresses are the following:

- How should the scale-space model be implemented computationally? The scale-space theory has been formulated for continuous signals, while realistic signals are discrete.

- Can the scale-space representation be used for extracting information? How should this be done?

- The scale-space representation in itself contains no information about preferred scales. In fact, without any *a priori* scale information all levels of scale must be treated similarly. Is it possible to determine a sparse set of appropriate scales for further processing?

- How can the scale-space concept interact with and cooperate with other processing modules?

- What can happen in scale-space? What is the behaviour of structure in scale-space? How do features evolve under scale-space smoothing? What types of bifurcation events can take place?

- Can cues to three-dimensional surface shape be computed directly from visual front-end operations?

The presentation is divided into four parts. We start by considering the basic theory of scale-space representation. A number of fundamental results on scale-space and related multi-scale representations are reviewed. The problem of how to formulate a *scale-space theory for discrete signals* is treated, as is the problem of how to compute image features within the Gaussian derivative framework.

Then, a representation called *the scale-space primal sketch* is presented, which is a formal representation of structures at multiple scales in scale-space aimed at the making information in the scale-space representation explicit. The theory behind its construction is analysed, and an algorithm is presented for computing the representation.

It is demonstrated how this representation can be integrated with other visual modules. Qualitative scale and region information extracted from the scale-space primal sketch can be used for *guiding other low-level processes and simplifying their tasks.*

Finally, it is shown how the suggested method for *scale selection* can be extended to other aspects of image structure, and how *three-dimensional shape cues* can be computed within the Gaussian derivative framework. Such information can then be used for adapting the shape of the smoothing kernel, to reduce the shape distorting effects of the scale-space smoothing, and thus increase the accuracy in the computed surface orientation estimates.

1.7.1. Guide to the reader

As a guide to the reader it should be remarked that it is not necessary to read this book in a sequential manner. While the ordering of the chapters follows the bottom-up chain of processing levels in an imagined vision system, the chapters are written so that it should be possible to read them independently and still get the major ideas without having to digest the preceding chapters. The following table describes the mutual dependencies.

Chapter	Contents	Background
2	Review of multi-scale analysis	
3, 4	Discrete scale-space theory	
5	Computing derivatives in scale-space	(3, 4)
6	Feature detection in scale-space	(5)
7	The scale-space primal sketch	
8	Theoretical analysis of scale-space	(7)
9	Algorithm for blob linking	7, 8
10	Extracting salient image structures	7
11	Guiding processes with scale-space	7, 10
12	Summary and discussion of chapters 7–11	7–11
13	Scale selection	
14	Shape computation	13
15	Non-uniform smoothing	(14)

The level of presentation varies depending on the subjects. Some chapters are highly mathematical, while others are more descriptive. For a reader who wants to avoid the mathematics at first, I recommend chapters 7, 10, and 11 for getting the basic ideas of the approach. Then, it may be natural to proceed with chapters 6, 13, and 14, where straightforward descriptions can be found of how to use the scale-space methodology for different types of early visual computations. The basic scale-space theory underlying these chapters is described in chapters 3–5, which give a detailed mathematical analysis of scale-space theory for discrete signals, and chapter 8, which shows how the behaviour of image structures in scale-space can be analysed.

Now, in the form of a long abstract, a brief overview will be given of some of the main results presented in each of the different parts.

1.7.2. Part I: Basic scale-space theory

Chapter 2: Review of multi-scale analysis. A summary is given of basic properties of scale-space and related multi-scale representations, notably, pyramids, wavelets, and regularization. A number of special properties of the scale-space representation are listed, and the different multi-scale approaches are compared.

Chapter 3: One-dimensional discrete scale-space theory. Which convolution kernels share the property of *never introducing new local extrema* in a signal? Qualitative properties of such kernels are pointed out, and a complete classification is given.

These results are then used for showing that there is only one reasonable way to define a scale-space for one-dimensional discrete signals, namely by discrete convolution with a family of kernels called the *discrete analogue of the Gaussian kernel.* This scale-space can equivalently be described as the solution to a *semi-discretized* version of the *diffusion equation.* The conditions that single out this scale-space are essentially non-creation of local extrema combined with a *semi-group* assumption and the existence of a *continuous* scale parameter. Similar arguments applied in the continuous case uniquely lead to the Gaussian kernel.

The commonly adapted technique with a sampled Gaussian may lead to undesirable effects (scale-space violations). This result exemplifies the fact that properties derived in the continuous case might be violated after discretization.

Chapter 4: Discrete scale-space theory in higher dimensions. The one-dimensional scale-space theory is generalized to discrete signals of arbitrary dimension. The treatment is based upon the assumptions that (i) the scale-space representation should be defined by convolving the original signal with a one-parameter family of symmetric smoothing kernels possessing a *semi-group property,* and (ii) *local extrema must not be enhanced* when the scale parameter is increased continuously.

Given these requirements, the scale-space representation must satisfy a *semi-discretized* version of the *diffusion equation.* In a special case the representation is given by convolution with the one-dimensional discrete analogue of the Gaussian kernel along each dimension.

Chapter 5: Computing derivatives in scale-space. It is shown how discrete derivative approximations can be defined so that scale-space properties hold exactly also in the discrete domain. A family of kernels is derived which constitute *discrete analogues* to the continuous Gaussian derivatives, and possesses an algebraic structure similar to that possessed by the derivatives of the traditional scale-space representation in the continuous domain.

The representation has theoretical advantages compared to other discretizations of scale-space theory in the sense that operators which *commute* before discretization commute after discretization. Some computational implications of this are that derivative approximations can be computed *directly* from smoothed data (without any need for repeating the smoothing operation), and this will give *exactly* the same result as convolution with the corresponding derivative approximation kernel. Moreover, a number of *normalization* conditions are automatically satisfied.

Chapter 6: Feature detection in scale-space. The proposed methodology leads to a conceptually simple scheme of computations for multi-scale low-level feature extraction, consisting of four basic steps; (i) *large support* convolution smoothing, (ii) *small support* difference computations, (iii) *point operations* for computing differential geometric entities, and (iv) *nearest neighbour operations* for feature detection.

Applications are given demonstrating how the proposed scheme can be used for edge detection and junction detection based on derivatives up to order three.

1.7.3. Part II: Theory of the scale-space primal sketch

Chapter 7: The scale-space primal sketch. A representation is presented for making explicit image structures in scale-space as well as the relations between image structures at different scales. The representation is based on blobs that are either brighter or darker than the background. At any scale in scale-space *grey-level blobs* are defined at that scale. Then, these grey-level blobs are *linked across scales* into objects called *scale-space blobs.* The relations between these blobs at different scales define a hierarchical data structure called the scale-space primal sketch, and it is proposed that the volume of a scale-space blob in scale-space constitutes a natural measure of blob significance.

To enable comparisons of significance between structures at different scales, it is necessary to measure significance in such a way that structures at different scales are treated in a uniform manner. It is shown how a definition of a transformed scale parameter, *effective scale*, can be expressed such that it gives intuitive results for both continuous and discrete signals. The volumes of the grey-level blobs must be transformed in a similar manner. That normalization is based on simulation results accumulated for a set of reference signals.

Chapter 8: Theoretical analysis of scale-space. It is demonstrated how the behaviour of image structures over scales can be analysed using elementary techniques from real analysis, singularity theory, and statistics.

The implicit function theorem describes how critical points form *trajectories* across scales when the scale parameter changes, and gives direct estimates of their drift velocity. Momentarily, the drift velocity may tend to infinity. Generically, this occurs in *bifurcation situations* only.

The qualitative behaviour of critical points at bifurcations is analysed, and the generic blob events are classified. A set of illustrative examples is presented, demonstrating how blobs behave in characteristic situations.

Chapter 9: Algorithm for blob linking. An algorithm is described for computing the scale-space primal sketch. It is based on detection of grey-level blobs at different levels of scale. On that output data an *adaptive* scale sampling algorithm operates and performs the actual linking of the grey-level blobs into *scale-space blobs* as well as the registration of the bifurcations and the blob events.

1.7.4. Part III: Applications of the scale-space primal sketch

Chapter 10: Extracting salient image structures. It is experimentally demonstrated how the scale-space primal sketch can be used for *extracting* significant blob-like structures from image data as well as associated scale levels for treating those. Such descriptors constitute coarse segmentation cues, and can serve as regions of interest to other processes.

The treatment is based on two basic assumptions; (i) in the absence of other evidence, structures, which are significant in scale-space, are likely to correspond to salient structures in the image, and (ii) in the absence of other evidence, scale levels can be selected where the blob response assumes its maximum over scales.

Chapter 11: Guiding processes with scale-space. It is demonstrated how the qualitative scale and region descriptors extracted by the scale-space primal sketch can be used for *guiding* other processes in early vision and for *simplifying* their tasks.

An integration experiment with *edge detection* is presented, where edges are detected at coarse scales given by scale-space blobs, and then tracked to finer scales in order to improve the localization. In *histogram analysis,* the scale-space primal sketch is used for automatic peak detection. More generally, such descriptors can be used for guiding the *focus-of-attention* of active vision systems. With respect to a test problem of *detecting and classifying junctions,* it is demonstrated how the blobs can be used for generating regions of interest, and for providing coarse context information (window sizes) for analysing those.

Finally, it is briefly outlined how the scale-space primal sketch can be applied to other visual tasks such as texture analysis, perceptual grouping and matching problems. Experiments on real imagery demonstrate that the proposed theory gives intuitively reasonable results.

Chapter 12: Summary and discussion of the scale-space primal sketch approach (chapters 7–11). Basic properties of scale-space representation and the scale-space primal sketch are pointed out, and relations to previous work are described. A summary is given of the basic ideas, and a few alternative approaches are discussed.

1.7.5. Part IV: Scale selection and shape computation

Chapter 13: Scale selection. A heuristic principle for scale selection is proposed stating that *local extrema over scales* of different combinations of *normalized scale invariant derivatives* are likely candidates to correspond to interesting structures. The resulting methodology lends itself naturally to two-stage algorithms; feature detection at coarse scales followed by feature localization at finer scales. Support is given by theoretical considerations and experiments on blob detection, junction detection, and edge detection.

Chapter 14: Shape computation by scale-space operations. The problem of *scale* in shape-from-texture is addressed. The need for (at least) two scale parameters is emphasized; a *local scale* describing the amount of smoothing used for suppressing noise and irrelevant details when computing primitive texture descriptors from image data, and an *integration scale* describing the size of the region in space over which the statistics of the local descriptors is accumulated.

The mechanism for scale selection outlined in chapter 13 is used for *adaptive* determination of the two scale parameters in a multi-scale texture descriptor, the *windowed second moment matrix*, which is defined in terms of Gaussian smoothing, first-order derivatives, and non-linear pointwise combinations of these. This texture description can then be combined with various assumptions about surface texture in order to estimate local surface orientation. Two specific assumptions, "weak isotropy" and "constant area," are explored in more detail. Experiments on real and synthetic reference data with known geometry demonstrate the viability of the approach.

Chapter 15: Non-uniform smoothing. Various generalizations of linear and rotationally symmetric Gaussian smoothing are briefly described.

A special approach of performing linear shape adaption in shape-from-texture is treated in more detail. It is demonstrated how an *affine scale-space representation* can be used for defining an image texture descriptor that possesses useful invariance properties with respect to linear transformations of the image coordinates.

Part I

Basic scale-space theory

2

Linear scale-space and related multi-scale representations

As pointed out in the introductory chapter, an inherent property of objects in the world and details in images is that they only exist as meaningful entities over certain ranges of scale. If one aims at describing the structure of unknown real-world signals, then a multi-scale representation of data is crucial.

This chapter gives a tutorial review of different multi-scale representations that have been developed by the computer vision community in order to handle image structures at different scales in a consistent manner. The basic idea is to embed the original signal into a one-parameter family of gradually smoothed signals, in which the fine-scale details are successively suppressed.

Under rather general conditions on the type of computations that are to be performed at the first stages of visual processing, in what can be termed *the visual front-end*, it can be shown that the Gaussian kernel and its derivatives are singled out as the only possible smoothing kernels. The conditions that specify the Gaussian kernel are, basically, linearity and shift-invariance combined with different ways of formalizing the notion that structures at coarse scales should correspond to simplifications of corresponding structures at fine scales—they should not be accidental phenomena created by the smoothing method. Notably, several different ways of choosing axioms for generating a scale-space give rise to the same conclusion.

During the last few decades a number of other approaches to multi-scale representations have been developed, which are more or less related to scale-space theory, in particular the theories of *pyramids*, *wavelets*, and *multi-grid methods*. Despite their qualitative differences, the increasing popularity of each of these approaches indicates that the crucial notion of *scale* is increasingly appreciated by the computer vision community and by researchers in other related fields.

Here, some basic properties of these representations will be described and compared. An interesting similarity with biological vision is that the scale-space operators closely resemble receptive field profiles registered in neurophysiological studies of the mammalian retina and visual cortex.

2.1. Early multi-scale representations

The general idea of representing a signal at multiple scales is not entirely new. Early work in this direction was performed by Rosenfeld and Thurston (1971), who observed the advantage of using operators of different sizes in edge detection, and Klinger (1971), Uhr (1972), and Hanson and Riseman (1974) concerning image representations using different levels of spatial resolution,[1] i.e., different amounts of subsampling. These ideas have been developed further, mainly by Burt and Crowley, to one of the types of multi-scale representations most widely used today, the *pyramid*.

In this chapter, an overview[2] will be given of some main concepts relating to this representation and other multi-scale approaches, as the theories have been developed to their current states. As background material, let us first briefly consider one of the predecessors to multi-scale representations, a hierarchical data structure called quad-tree.

2.2. Quad-tree

One of the earliest types of multi-scale representations of image data is the *quad-tree* considered by Klinger (1971). It is a tree-like representation of image data, where the image is recursively divided into smaller regions.

The basic idea is as follows: Consider, a discrete image f, for simplicity defined in some square region D of size $2^K \times 2^K$ for some $K \in \mathbb{Z}$, and define a measure Σ of the grey-level variation in any region $D' \subset D$ in this image. As measure one may, for example, take the variance of the grey-level values,

$$\Sigma(D') = E_{D'}(f^2) - (E_{D'}(f)), \tag{2.1}$$

where $E_{D'}(h)$ denotes the average of a function h over the region D',

$$E_{D'}(h) = \frac{\sum_{x \in D'} h(x)}{\sum_{x \in D'} 1}. \tag{2.2}$$

Let $D^{(K)} = D$. If $\Sigma(D^{(K)})$ is greater than some pre-specified threshold α, then split $D^{(K)}$ into p sub-images $D_j^{(K-1)}$ ($j = 1..p$) according to some

[1]The terms "scale" and "resolution" are sometimes used interchangeably in the vision literature, and their precise meanings are not always clear. The convention adopted in this presentation is to use resolution for the spatial density of grid points (the sampling density). Scale, on the other hand, stands for the characteristic length over which variations in the image take place or the characteristic length of the operators used for processing the image data.

[2]Only few indications of proofs will be given; the reader is referred to the original sources concerning details and further explanation. Inevitably, a short summary like this one will be biased toward certain aspects of the problem. Therefore, the author would like to apologize for any reference that has been left out.

rule. A simple choice is to let $p = 4$ and divide $D^{(K)}$ into four regions of same size. Then, apply the procedure recursively to all sub-images until convergence is obtained. A tree of degree p is generated, in which each leaf $D_j^{(k)}$ is a homogeneous block with $\Sigma(D_j^{(k)}) < \alpha$.

If there are strong intensity variations in the image, and if the threshold α is selected small, then in the worst case, each pixel may correspond to an individual leaf. On the other hand, if the image contains a small number of regions with relatively uniform grey-level, then a substantial data reduction can be obtained by representing the image by such a hierarchical data structure, in which every region is represented by its average grey-level.

Concerning grey-level data, this representation has been used in simple segmentation algorithms for image processing. In the "split-and-merge" algorithm, a splitting step is first performed according to the above scheme. Then, adjacent regions are merged if the variation measure of the union of the two regions is below the threshold. Another application (typically when $\alpha = 0$, and all pixels within a leaf region hence are required to have the same grey-level) concerns objects defined by uniform grey-levels (e.g. binary objects); see, for example, the book by Tanimoto and Klinger (1980) for more references on this type representation.

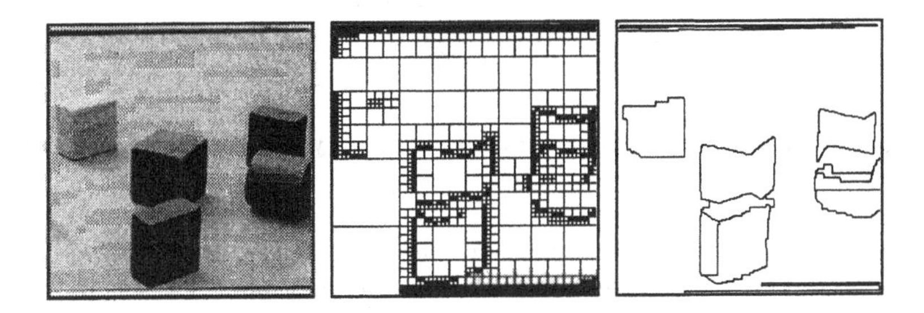

Figure 2.1. Illustration of quad-tree and the split-and-merge segmentation algorithm; (left) grey-level image, (middle) the leaves of the quad-tree, i.e., the regions after the split step that have a standard deviation below the given threshold, (right) regions after the merge step.

2.3. Pyramid representations

A pyramid representation of a signal is a set of successively smoothed *and* sub-sampled representations of the original signal organized in such a way that the number of pixels decreases with a constant factor (usually 2^N for an N-dimensional signal) from one layer to the next.

To illustrate the idea, assume again, for simplicity, that the size of the input image f is $2^K \times 2^K$ for some integer K, and let $f^{(K)} = f$.

Presumably, the simplest way to obtain a pyramid-like representation of a two-dimensional image is by letting each pixel value at a coarser scale $f^{(k-1)}$ be the average over a 2×2 neighbourhood of the pixel values in the finer scale representation $f^{(k)}$. If this operation is applied recursively, and if the corresponding representations are stacked on top of each other, then the result will be a pyramid-like data structure representing f at different resolutions.

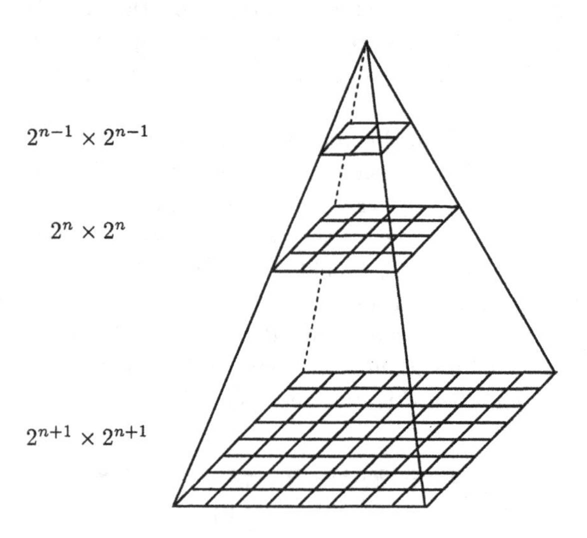

Figure 2.2. A pyramid representation is obtained by successively reducing the image size by combined smoothing and subsampling.

While this transformation rapidly decreases the image size, repeated application of such plain averaging and subsampling over 2×2 neighbourhoods is not the best choice of transformation for generating such a representation. The equivalent transformation describing the transformation from the original image $f^{(K)} = f$ to a coarser scale representation $f^{(K-i)}$ corresponds to box averaging over a $2^i \times 2^i$ neighbourhood followed by subsampling with a factor of 2^i along each direction. It is well-known that this operation leads to severe aliasing problems. It also leads to strong ringing effects in the Fourier domain.

2.3.1. *Low-pass pyramid generation: Smoothing and subsampling*

A more general approach to pyramid generation is to treat the transformation from a fine level to the next coarser level as a linear smoothing operation followed by subsampling. For simplicity, assume that the smoothing filter is separable, and that it thus can be described by some one-dimensional filter kernel $c\colon \mathbb{Z} \to \mathbb{R}$. Then, it is sufficient to study the following one-dimensional transformation

$$\begin{aligned} f^{(k-1)} &= \text{Reduce}(f^{(k)}) \\ f^{(k-1)}(x) &= \textstyle\sum_{n=-\infty}^{\infty} c(n)\, f^{(k)}(2x-n), \end{aligned} \qquad (2.3)$$

Here, the reduction operation from a fine scale level $f^{(k)}$ to the next coarser scale level $f^{(k-1)}$ has been expressed on a general form. Usually, c is selected as a low-pass kernel with finite support

$$c(n) = 0 \quad \text{if } |n| > N \text{ for some integer } N. \qquad (2.4)$$

This type of *low-pass pyramid* was proposed almost simultaneously by Burt (1981) and Crowley (1981). A main advantage with this representation is that the image size decreases exponentially with the scale level, and hence also the amount of computations required to process the data. The idea behind this construction is that if the filter coefficients $c(n)$ can be properly chosen, the representations at coarser scales (smaller k) should correspond to coarser scale structures in the image data.

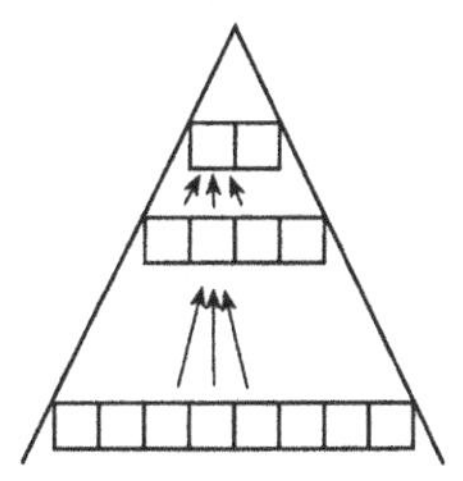

Figure 2.3. Illustration of the Reduce operation.

2.3.2. Choice of smoothing kernel: Conditions in the spatial domain

Some of the most obvious design criteria that have been proposed are

- positivity: $c(n) \geq 0$,
- unimodality: $c(|n|) \geq c(|n+1|)$,
- symmetry: $c(-n) = c(n)$, and
- normalization: $\sum_{n=-\infty}^{\infty} c(n) = 1$.

Another natural condition is that all pixels should contribute equally to all levels. In other words, any point that has an odd coordinate index should contribute equally much to the next coarser level as any point having an even coordinate value. Formally, this can be expressed as:

- equal contribution: $\sum_{n=-\infty}^{\infty} c(2n) = \sum_{n=-\infty}^{\infty} c(2n+1)$.

Equivalently, this condition means that the kernel (1/2, 1/2) of width two should occur as at least one factor[3] in the smoothing kernel.

2.3.3. Choice of smoothing kernel: Conditions in the frequency domain

While most authors agree on the previously stated design criteria, mutually exclusive conditions have been stated in the frequency domain. Motivated by the sampling theorem, Meer *et al.* (1987) suggested that the smoothing kernel c should approximate an ideal low-pass filter as closely as possible. Since there is no non-trivial finite support kernels with ideal low-pass properties, some approximation must be constructed. Meer *et al.* proposed to select the filters (of some fixed size) that minimize the difference between the Fourier transform of the smoothing kernel and the Fourier transform of an ideal bandpass kernel, and measured the error by the maximum norm in the frequency domain. This approach is termed "equi-ripple design."

An alternative is to require the kernel to be positive and *unimodal* also in the frequency domain. Then, any high-frequency signal is guaranteed to be suppressed more than any lower-frequency signal.

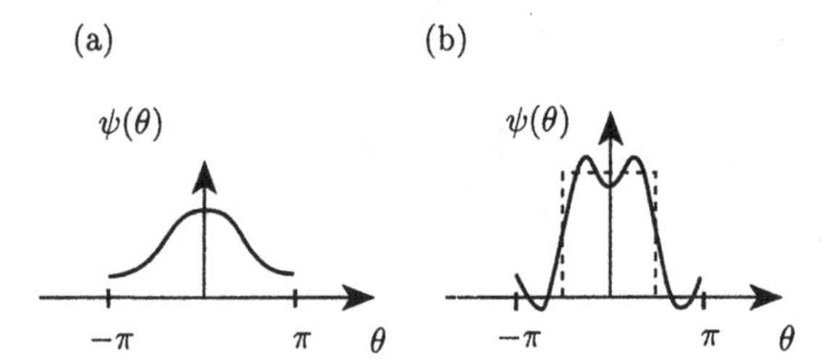

Figure 2.4. Requirements on the filter coefficients in the Fourier domain; (a) unimodality, and (b) equi-ripple approximation of an ideal low-pass kernel.

2.3.4. Choice of filter size

The choice of N gives rise to a trade-off problem. A larger value of N increases the number of degrees of freedom in the design, at the cost of increased computational work. A natural choice when $N = 1$ is the binomial filter

$$(\tfrac{1}{4},\ \tfrac{1}{2},\ \tfrac{1}{4}). \tag{2.5}$$

---

[3]This condition can be formally expressed in terms of the generating function $\varphi_c(z) = \sum_{n=-\infty}^{\infty} c(n)\, z^n$ associated with the filter coefficient sequence c. The equal contribution condition corresponds to $\sum_{n=-\infty}^{\infty}(-1)^n c(n) = 0$, which means that $\varphi_c(-1) = 0$, and the polynomial $(1+z)$ must occur as at least one factor in $\varphi_c(z)$. The function $(1+z)$ is the generating function of the binomial kernel $(1,1)$.

This is the unique filter of width 3 that satisfies the equal contribution condition. It is also the unique filter of width 3 for which the Fourier transform

$$\psi(\theta) = \sum_{n=-N}^{N} c(n) \exp(-in\theta) \tag{2.6}$$

is zero at $\theta = \pm\pi$. A property that is usually regarded as a negative of this kernel though is that when applied recursively, the equivalent convolution kernel (corresponding to the combined effect of repeated smoothing and subsampling) tends to function with a triangular shape.

Of course, there is a large class of other possibilities. Concerning kernels of width 5, the previously stated conditions in the spatial domain imply that the kernel has to be of the form

$$(\tfrac{1}{4} - \tfrac{1}{2}a, \ \tfrac{1}{4}, \ a, \ \tfrac{1}{4}, \ \tfrac{1}{4} - \tfrac{1}{2}a). \tag{2.7}$$

Burt and Adelson (1983) argued that a should be selected such that the equivalent smoothing function should be as similar to a Gaussian as possible. Empirically, they selected the value $a = 0.4$.

2.3.5. *Bandpass pyramids*

By considering a representation defined as the difference between two adjacent levels in a low-pass pyramid, one obtains a *bandpass pyramid*. This representation has been termed a "Laplacian pyramid" by Burt, and "DOLP" (Difference Of Low Pass Pyramid) by Crowley. It is defined by

$$\begin{aligned} L^{(k)} &= f^{(k)} - \text{Expand}(f^{(k-1)}) \\ L^{(0)} &= f^{(0)}, \end{aligned} \tag{2.8}$$

where Expand is an interpolation operator that in a sense constitutes the reverse of operation Reduce. Formally, the expansion operator can be expressed as

$$\begin{aligned} \tilde{f}^{(k)} &= \text{Expand}(f^{(k-1)}) \\ \tilde{f}^{(k)}(x) &= 2\sum_{n=-\infty}^{\infty} c(n)\, f^{(k-1)}(\tfrac{1}{2}(x-n)), \end{aligned} \tag{2.9}$$

where only the terms for which $x - n$ is even are to be included in the sum. This interpolation procedure means that the same weights are used for propagating grey-levels from a coarser to a finer sampling as are used when subsampling the signal.

2.3.6. *Applications: Feature detection and data compression*

The bandpass pyramid representation has been used for feature detection and data compression. Among features that can be detected are blobs (maxima), and peaks and ridges, etc. (Crowley *et al.* 1984, 1987).

The idea behind using such pyramids for data compression is that a bandpass filtered signal will in general be decorrelated, and have its grey-level histogram centered around the origin. If a coarse quantization of the grey-levels is sufficient for the purpose in mind (typically display), then a data reduction can be obtained by coding the quantized grey-levels by some standard data compression technique, such as variable length encoding. From the set of coded bandpass images $\{\tilde{L}^{(k)}\}$, an approximation of the original image $\tilde{f}^{(K)}$ can then be reconstructed by essentially reversing the construction in (2.8),

$$\begin{aligned}\tilde{f}^{(0)} &= \tilde{L}^{(0)} \\ \tilde{f}^{(k)} &= \tilde{L}^{(k)} + \text{Expand}(\tilde{f}^{(k-1)}).\end{aligned} \tag{2.10}$$

2.3.7. Further reading

There is a large literature on different aspects of pyramid representation. Besides the early works by Burt (1981), Burt and Adelson (1983), Crowley and his co-workers (1981, 1984, 1987), and Meer *et al.* (1987), there are also the books edited by Rosenfeld (1984), and Cantoni and Levialdi (1986). For a selection of other developments, one can especially mention the articles by Chehikian and Crowley (1991), Knudsen and Christensen (1991), Wilson and Bhalerao (1992), and Burt and Kolczynski (1993).

An interesting approach is the introduction of "oversampled pyramids," in which not every smoothing step is followed by a subsampling operation. Instead, several representations are generated at by successive smoothing of images at the same resolution. In this way, a denser sampling along the scale direction can be obtained.

2.3.8. Basic properties of pyramids

To summarize, the main advantages of the pyramid representations are that they lead to a *rapidly decreasing image size*, which reduces the computational work both in the actual computation of the representation, and in the subsequent processing. The memory requirements are small, and there exist commercially available implementations of pyramids in hardware. The main disadvantage with pyramids is that they are defined from an algorithmic process. This makes theoretical analysis complicated. Moreover, they correspond to quite a coarse quantization along the scale direction, which makes it algorithmically hard to relate (match) image structures across scales. Pyramids are not translationally invariant, which implies that the representation changes when the image is shifted.

It is worth noting that pyramid representations show a high degree of similarity with a type of numerical methods called *multi-grid methods*; see the book by Hackbusch (1985) for an extensive treatment of the subject.

2.4. Scale-space representation and scale-space properties

From the viewpoint of these multi-scale representations, the scale-space representation defined in section 1.4 can be seen as a very special type of multi-scale representation, which (i) comprises a *continuous scale parameter*, and (ii) preserves the *same spatial sampling* at all scales.

Moreover, while there is a large degree of freedom in the choice of smoothing kernels for pyramid generation, in scale-space representation the Gaussian kernel is singled out as the unique smoothing kernel for describing the transformation from a representation at a fine scale to a representation at a coarser scale.

2.4.1. Scale-space definition for continuous signals

For continuous signals of arbitrary dimension N, the scale-space concept is formally constructed as follows. Given a signal $f\colon \mathbb{R}^N \to \mathbb{R}$, the scale-space representation $L\colon \mathbb{R}^N \times \mathbb{R}_+ \to \mathbb{R}$ is defined such that the representation at zero scale is equal[4] to the original signal

$$L(\cdot;\ 0) = f, \tag{2.11}$$

and the representations at coarser scales are given by convolution with Gaussian kernels of increasing width,

$$L(\cdot;\ t) = g(\cdot;\ t) * f. \tag{2.12}$$

In terms of explicit integrals, the result of the convolution operation '*' is written

$$L(x;\ t) = \int_{\xi \in \mathbb{R}^N} g(\xi;\ t)\, f(x-\xi)\, d\xi, \tag{2.13}$$

where $x = (x_1, \ldots, x_N)^T \in \mathbb{R}^N$, and $g\colon \mathbb{R}^N \times \mathbb{R}_+ \backslash \{0\} \to \mathbb{R}$ is the N-dimensional Gaussian kernel

$$g(x;\ t) = \frac{1}{(2\pi t)^{N/2}} e^{-x^T x/2t}. \tag{2.14}$$

The result of smoothing a one-dimensional signal in this way is illustrated in figure 1.3 (page 12). Experimental results with two-dimensional images can be found in figure 6.4 (page 156), figure 7.9 (page 175), and figure 7.10 (page 176). Notice how this successive smoothing captures the notion of the signals becoming successively smoother, and the fine scale information being suppressed.

[4] The notation $L(\cdot;\ 0) = f$ stands for $L(x;\ 0) = f(x)\ \forall x \in \mathbb{R}^N$.

2.4.2. *Special properties of the Gaussian kernel*

Of course, there exists a variety of possible ways to construct a continuous one-parameter family of signals from a given signal. As we shall return to in section 2.5, it can, however, be shown that this particular way of defining a multi-scale representation constitutes a very special choice. It will be shown that under rather general assumptions about the structure of the computations performed at the first processing stages, the Gaussian kernel and its derivatives are singled out uniquely. The conditions that specify this uniqueness are basically, linearity, and shift-invariance combined with different ways of formalizing the notion that a representation at a coarse scale should really correspond to a simplification of any representation at a finer scale. New artificial structures must not be created by the smoothing transformation.

As a background to the treatment of uniqueness properties, let us now give an intuitive explanation of some consequences of the stated definition, and describe some basic properties that the scale-space representation satisfies. Some properties are special, due to the special properties of the Gaussian function. Then, we shall address the issue of uniqueness in section 2.5.

2.4.3. *Weighted averaging and finite aperture*

The reason the signals become smoother can be understood in several ways. A simple way is to look at the graphs of Gaussian kernels corresponding to different parameter values. Consider first the one-dimensional case. Then, the square root of the scale parameter $\sigma = \sqrt{t}$ is the standard deviation of the Gaussian kernel, and can be interpreted as a characteristic length descriptor. From this viewpoint, the value of the scale-space representation at a certain point in the scale-space representation of a signal can be seen as the result of averaging the original signal with symmetric weight functions of increasing width; see figure 2.5.

In higher dimensions, a natural descriptor of characteristic length of the smoothing kernel is its covariance matrix,

$$C(g(\cdot;\ t)) = M_2(g(\cdot;\ t)) - M_1(g(\cdot;\ t))\, M_1^T(g(\cdot;\ t)) = t\, I, \tag{2.15}$$

where M_2 is the second moment matrix,

$$M_2(g(\cdot;\ t)) = \int_{x \in \mathbb{R}^N} x x^T g(x;\ t)\, dx, \tag{2.16}$$

M_1 is the first moment vector

$$M_1(g(\cdot;\ t)) = \int_{x \in \mathbb{R}^N} x\, g(x;\ t)\, dx = 0, \tag{2.17}$$

and I is the $N \times N$ unit matrix.

Clearly, the transformation (2.13) represents a true averaging operation since the Gaussian kernel is normalized,

$$\int_{x \in \mathbb{R}^N} g(x;\ t)\, dx = 1. \tag{2.18}$$

Philosophically, this weighted averaging operation can be interpreted as a way of defining an *aperture* of a physical observation. To measure any real-world entity, it is necessary to integrate the entity over some finite (non-infinitesimal) window size. Ideal *point* measurements can never be performed in reality, since some finite amount of energy is needed to obtain a non-infinitesimal response that can be registered by a physical detector. The scale-space representation of a signal f at a certain scale t corresponds to the result of measuring the signal using a rotationally symmetric aperture function having the same characteristic length $\sigma = \sqrt{t}$ along all coordinate directions. This has the effect that structures having a characteristic length less than σ will be suppressed.[5]

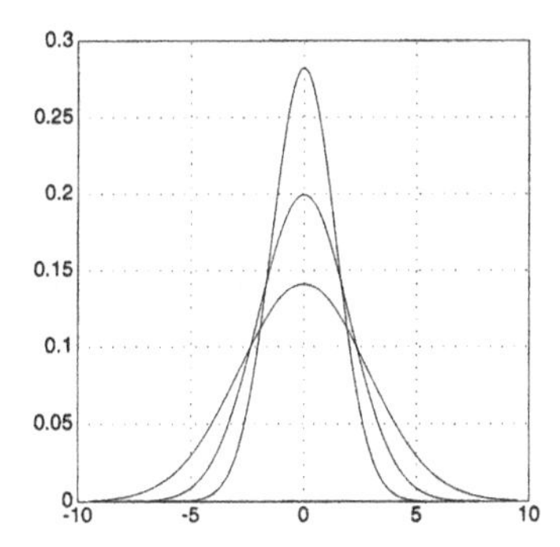

Figure 2.5. One-dimensional Gaussian kernels with different standard deviation $\sigma = \sqrt{t}$. The value of the scale-space representation of a signal at a certain point is the weighted average of the signal using the Gaussian weight function.

2.4.4. Semi-group property and cascade smoothing

Another important property of the scale-space representation stems from the *semi-group* property of the family of Gaussian kernels. The result of convolving a Gaussian kernel with a Gaussian kernel is another Gaussian kernel,

$$g(\cdot;\ t) * g(\cdot;\ s) = g(\cdot;\ t + s). \tag{2.19}$$

[5]Some care must be taken when expressing such a statement. As we shall see later, adjacent structures (e.g., extrema) can be arbitrary close after arbitrary large amounts of smoothing, although the likelihood for the distance between two adjacent structures to be less than some value ϵ decreases with increasing scale.

In terms of scale-space representation, this means that a representation at a coarse scale $L(\cdot;\ t_2)$ can be computed from a representation at a finer scale $L(\cdot;\ t_1)$ by convolution with a Gaussian kernel with parameter value $t_2 - t_1 > 0$

$$L(\cdot;\ t_2) = g(\cdot;\ t_2 - t_1) * L(\cdot;\ t_1). \tag{2.20}$$

This is the so-called *cascade smoothing* property of the scale-space representation. It means that signals at coarse scales really can be interpreted as smoother than corresponding signals at finer scales, since the transformation from any fine level to any coarse level is of the same type as the transformation from the original signal to a non-zero level in scale-space.

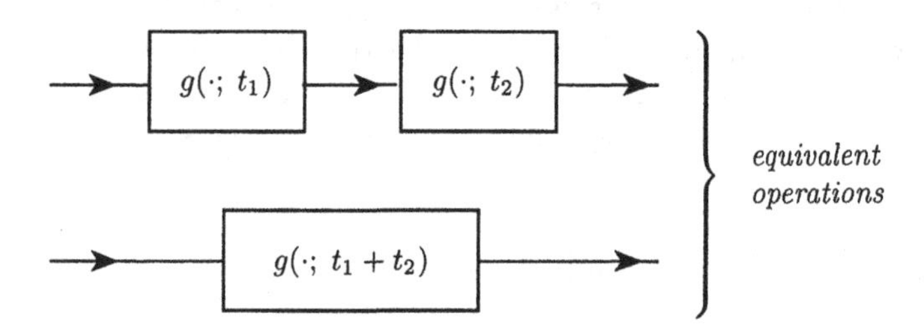

Figure 2.6. The semi-group property means that convolution with $g(\cdot;\ t_1 + t_2)$ is equivalent to convolution with $g(\cdot;\ t_1)$ followed by convolution with $g(\cdot;\ t_2)$.

These relations become even more obvious in the Fourier domain. Given any function $h\colon \mathbb{R}^N \to \mathbb{R}$, define the Fourier transform $\hat{h}\colon \mathbb{R}^N \to \mathbb{C}$ by

$$\hat{h}(w) = \int_{x\in\mathbb{R}} h(x)\, e^{-i\omega^T x}\, dx. \tag{2.21}$$

Then, the semi-group property assumes the form

$$\hat{g}(\omega;\ t)\, \hat{g}(\omega;\ s) = \hat{g}(\omega;\ t + s). \tag{2.22}$$

Using

$$\hat{L}(\omega;\ t) = \hat{g}(\omega;\ t) \hat{f}(\omega), \tag{2.23}$$

the cascade smoothing property can then be written

$$\begin{aligned} \hat{L}(\omega;\ t_2) &= \hat{g}(\omega;\ t_2)\, \hat{f}(\omega) \\ &= \hat{g}(\omega;\ t_2 - t_1)\, \hat{g}(\omega;\ t_1)\, \hat{f}(\omega) \\ &= \hat{g}(\omega;\ t_2 - t_1)\, \hat{L}(\omega;\ t_1). \end{aligned} \tag{2.24}$$

2.4.5. Separability

A technically useful property of the Gaussian kernel is its separability. The N-dimensional Gaussian kernel $g\colon \mathbb{R}^N \to \mathbb{R}$ can be written as the product of N one-dimensional kernels $g_1 : \mathbb{R} \to \mathbb{R}$. Let again $x = (x_1, \ldots, x_N) \in \mathbb{R}^N$. Then,

$$g(x;\ t) = \prod_{i=1}^{N} g_1(x_i;\ t). \tag{2.25}$$

In terms of the explicit expressions for the Gaussian kernels, it holds that

$$\frac{1}{(2\pi t)^{N/2}} e^{-x^T x/2t} = \prod_{i=1}^{N} \frac{1}{(2\pi t)^{1/2}} e^{-x_i^2/2t}. \tag{2.26}$$

The separability is important in terms of computational efficiency, at least when implementing the smoothing operation by convolutions in the spatial domain.

Assume that the smoothing operation is implemented by some discrete approximation, and that the discrete filter mask has width M along each dimension. Then, the separability means that the number of operations for every point decreases from M^N in the non-separable case to MN operations for separable discrete convolution.

The separability also simplifies theoretical analysis, although one should be very careful when considering the generalization of one-dimensional results to higher dimensions.

2.4.6. Diffusion equation and Markov process

In terms of differential equations, the scale-space family L is described by the *diffusion equation*

$$\partial_t L = \frac{1}{2} \nabla^2 L = \frac{1}{2} \sum_{i=1}^{N} \partial_{x_i x_i} L. \tag{2.27}$$

This is the well-known physical equation that describes how the heat distribution L evolves over time t in a homogeneous medium with uniform conductivity (Fourier 1955; Widder 1975; Strang 1986).

This analogy gives a direct physical interpretation. The scale-space representation of a signal f corresponds to the result of letting an initial heat distribution f evolve over time t in a infinite homogeneous medium.

This relation constitutes an equivalent way to *define* the scale-space representation L of a signal f. The scale-space is the solution to (2.27) on an infinite N-dimensional domain with initial condition $L(\cdot;\ 0) = f(\cdot)$.

2.4.7. Maximum principle: Non-enhancement of local extrema

A strong smoothing property of this representation is the *non-enhancement of local extrema*, which corresponds to the strong *maximum principle* for parabolic differential equations:

- If at a certain scale $t_0 \in \mathbb{R}_+$ a point $x_0 \in \mathbb{R}$ is a local maximum for the mapping $x \mapsto L(x;\ t_0)$, then the Laplacian $\nabla^2 L(x_0;\ t_0)$ at this point is negative, which means that $\partial_t L(x_0;\ t_0) < 0$.
- On the other hand, if at a certain scale $t_0 \in \mathbb{R}_+$ a point $x_0 \in \mathbb{R}$ is a local minimum for the mapping $x \mapsto L(x;\ t_0)$, then the Laplacian $\nabla^2 L(x_0;\ t_0)$ at this point is positive, which means that $\partial_t L(x_0;\ t_0) > 0$.

In other words, the operation (2.27) has the effect of suppressing small local variations. In terms of the physical analogy with heat distributions, this result means that a hot spot will not become warmer, and a cold spot will not become cooler.

2.4.8. Scaling property

Since the scale-space representation is to be used for analysing structures at different scales, it is of interest to analyse its behaviour under rescalings of the input signal. Given an original signal $f\colon \mathbb{R}^N \to \mathbb{R}$, define a rescaled signal $f'\colon \mathbb{R}^N \to \mathbb{R}$ by

$$f(x) = f'(sx), \tag{2.28}$$

where $x \in \mathbb{R}^N$, and introduce new variables,

$$\begin{aligned} x' &= sx, \\ t' &= s^2 t, \end{aligned} \tag{2.29}$$

where $s \in \mathbb{R}_+$ is a scaling factor. Then, the scale-space representations L of f and L' of f'

$$L'(\cdot;\ t') = g(\cdot;\ t') * f'. \tag{2.30}$$

are equal

$$L(x;\ t) = L'(x';\ t'), \tag{2.31}$$

which means that the scale-space representation has a nice behaviour under rescalings of the spatial domain. This result can be proved by a change of variables in (2.13), and the easily verified scaling property of the N-dimensional Gaussian kernel

$$g(\xi;\ t) = s^N g(s\xi;\ s^2 t). \tag{2.32}$$

2.4.9. Scale-space derivatives: Infinite differentiability

Introduce *multi-index notation* in the following way. Let $n = (n_1, \ldots, n_N)^T \in \mathbb{Z}_+^N$, where $n_i \in \mathbb{Z}_+$, and let $x = (x_1, \ldots, x_N)^T \in \mathbb{R}^N$. Then, define x^n by

$$x^n = x_1^{n_1} x_2^{n_2} \ldots x_N^{n_N}, \tag{2.33}$$

and the derivative operator ∂_{x^n} of order

$$|n| = n_1 + n_2 + \cdots + n_N \tag{2.34}$$

by

$$\partial_{x^n} = \partial_{x_1^{n_1}} \partial_{x_2^{n_2}} \ldots \partial_{x_N^{n_N}}. \tag{2.35}$$

From this scale-space representation, multi-scale spatial derivatives can be defined by

$$L_{x^n}(\cdot;\ t) = \partial_{x^n} L(\cdot;\ t) = g_{x^n}(\cdot;\ t) * f, \tag{2.36}$$

where g_{x^n} denotes a (possibly mixed) partial derivative of the Gaussian kernel of order $|n|$. In terms of explicit integrals, the convolution operation (2.36) is written

$$\begin{aligned} L_{x^n}(x;\ t) &= \int_{x' \in \mathbb{R}^N} g_{x^n}(x - x';\ t)\, f(x')\, dx' \\ &= \int_{x' \in \mathbb{R}^N} g_{x^n}(x';\ t)\, f(x - x')\, dx'. \end{aligned} \tag{2.37}$$

The output from these operations are called the *scale-space derivatives* of f at scale t. This representation has a strong regularizing property. If f is bounded from above by some polynomial, i.e. if there exist some constants $C_1, C_2 \in \mathbb{R}_+$ such that

$$|f(x)| \leq C_1 \, (1 + x^T x)^{C_2}, \tag{2.38}$$

then because of the exponential decrease of the Gaussian function, the integrals in (2.37) are guaranteed to converge for any $t > 0$. Hence, (2.37) provides a well-defined way to construct multi-scale derivatives of any function f bounded by some polynomial, although the function itself may not be differentiable of any order.

In this sense, the scale-space representation of a signal shows a high degree of similarity with generalized functions and Schwartz distribution theory (1951), although it is neither needed nor desired to explicitly compute the limit case when the (scale) parameter t tends to zero. Thus, for any further considerations, the scale-space representation of a signal f given by (2.12) can for every $t > 0$ be treated as infinitely differentiable.

2.4.10. *Scale-space properties transfer to scale-space derivatives*

Most of the above stated properties of the zero-order[6] scale-space representation transfer to the scale-space derivatives. Since the convolution operation commutes with the differentiation operator, the scale-space derivatives satisfy the diffusion equation, and hence all the scale-space properties that the scale-space representation of an arbitrary signal satisfies. For example, also derivatives can be expected to become successively smoother. The derivatives also satisfy the non-enhancement property of local extrema, and the *cascade smoothing property*

$$g(\cdot;\ t_1) * g_{x^n}(\cdot;\ t_2) = g_{x^n}(\cdot;\ t_2 + t_1). \tag{2.39}$$

The latter result is a special case of the more general statement

$$g_{x^m}(\cdot;\ t_1) * g_{x^n}(\cdot;\ t_2) = g_{x^{m+n}}(\cdot;\ t_2 + t_1), \tag{2.40}$$

whose validity follows directly from the commutative property of convolution and differentiation.

Concerning rescalings of the input signal, the situation is only slightly different. While under rescalings of the spatial coordinates, the Gaussian kernel rescales according to (2.32), the rescaling rule for Gaussian derivatives can be verified to be

$$g_{x^n}(x;\ t) = s^{N+|n|}\, g_{x^n}(sx;\ s^2 t), \tag{2.41}$$

where $|n|$ denotes the order of differentiation. Differentiation of (2.31) obviously gives

$$L_{x^n}(x;\ t) = \partial_{x^n} L'(x';\ t') = s^{|n|} L'_{x'^n}(x';\ t'). \tag{2.42}$$

On the other hand, if dimensionless coordinates ξ are introduced by

$$\xi = \frac{x}{\sqrt{t}}, \quad \xi' = \frac{x'}{\sqrt{t'}}, \tag{2.43}$$

and corresponding normalized (dimensionless) derivative operators (see also chapter 13)

$$\partial_{\xi^n} = t^{n/2} \partial_{x^n}, \quad \partial_{\xi'^n} = t'^{n/2} \partial_{x'^n}, \tag{2.44}$$

then it clearly holds that

$$L_{\xi^n}(x;\ t) = L'_{\xi'^n}(x';\ t'). \tag{2.45}$$

This means that these *normalized scale-space derivatives* will have a nice behaviour under spatial rescalings of original signal, and that structures at different scales will be treated in a similar manner.

[6] Here, zero-order representation denotes the original signal before differentiation.

2.5. Uniqueness of scale-space representation

As pointed out in section 1.3.5, the main idea behind the construction of scale-space representation is that fine scale information should be successively suppressed when the scale parameter increases. It is also the aim that this process should reveal inherent properties of the signal—the "smoothing method" in itself must not generate spurious structures.

This condition has been formalized in different ways by different authors. In this section, we shall give a summary of a number of such formulations.

2.5.1. Sufficiency: Non-creation of local extrema/zero-crossings; 1D

When Witkin introduced the term scale-space, he was concerned with one-dimensional signals, and observed that new local extrema cannot be created in this family. Since differentiation commutes with convolution,

$$\partial_{x^n} L(\cdot;\ t) = \partial_{x^n} (g(\cdot;\ t) * f) = g(\cdot;\ t) * \partial_{x^n} f, \tag{2.46}$$

this non-creation property applies also to any nth-order spatial derivative computed from the scale-space representation.

Recall that an extremum in L is equivalent to a zero-crossing in the first derivative L_x. The non-creation of new local extrema means that the zero-crossings in any derivative of L form closed curves across scales, which will never be closed from below; see figure 1.5 (page 13). Hence, in the one-dimensional case, the zero-crossings form paths across scales, with a set of inclusion relations that form a tree-like data structure, termed "interval tree."

An interesting empirical observation made by Witkin was that he noted a marked correspondence between the length of the branches in the interval tree and perceptual saliency. In chapter 10 we shall extend this observation to a principle for actually detecting significant image structures from the scale-space representation.

2.5.2. Necessity: Causality, homogeneity, and isotropy

The observation that new local extrema cannot be created with increasing scale shows that Gaussian convolution satisfies certain sufficiency requirements for being a smoothing operation. The first proof of the *necessity* of Gaussian smoothing for scale-space representation was given by Koenderink (1984), who also gave a formal extension of the scale-space theory to higher dimensions.

He introduced the concept of *causality*, which means that new level surfaces $\{(x, y;\ t) \in \mathbb{R}^2 \times \mathbb{R}\colon L(x, y;\ t) = L_0\}$ must not be created in the scale-space representation when the scale parameter is increased. By combining causality with the notions of *isotropy* and *homogeneity*, which essentially mean that all spatial positions and all scale levels must be treated in a similar manner, he showed that the scale-space representation must satisfy the diffusion equation

$$\partial_t L = \tfrac{1}{2}\nabla^2 L. \tag{2.47}$$

Since the Gaussian kernel is the Green's function of the diffusion equation at an infinite domain, it follows that the Gaussian kernel is the unique kernel for generating the scale-space. A similar result holds in one dimension, and as we shall see later, also in higher dimensions.

The technique used for proving this necessity result was by studying the level surface through any point in scale-space for which the grey-level function assumes a maximum with respect to the spatial coordinates. If no new level surface is to be created with increasing scale, then the level surface must point with its concave side towards decreasing scales; see figure 1.6. This gives rise to a sign condition on the curvature of the level surface, which when expressed in terms of derivatives of the scale-space representation with respect to the spatial and scale coordinates assumes the form (2.47). Since the points at which the extrema are attained cannot be assumed to be known *a priori*, this condition must hold in any point, which proves the result.

In the *one-dimensional* case, this level surface condition becomes a level curve condition, and is equivalent to the previously stated non-creation of local extrema. Since any nth-order derivative of L also satisfies the diffusion equation

$$\partial_t L_{x^n} = \frac{1}{2}\nabla^2 L_{x^n}, \tag{2.48}$$

it follows that new zero-crossing curves in L_x cannot be created with increasing scale, and hence, no new maxima.

A similar result was given by Yuille and Poggio (1986) concerning the zero-crossings of the Laplacian of the Gaussian. Related formulations have been expressed by Babaud *et al.* (1986), and by Hummel (1987).

2.5.3. Necessity: Semi-group and non-creation of local extrema; 1D

Lindeberg (1990, 1991) considered the problem of characterizing one-dimensional kernels having the property that they do not increase the number of local extrema in any signal under convolution. If this condition is combined with the requirement that the family of convolution kernels should be a *semi-group*, and that the kernels should be normalized, then the Gaussian kernel is uniquely specified (see section 3.5). These arguments, which can be equivalently expressed in terms of zero-crossings, can also be used for deriving a corresponding scale-space theory for discrete signals (see chapter 3).

Although complete classification results can be obtained in this way, the approach based on non-creation of local extrema (or zero-crossings) is limited to one-dimensional signals. It can be shown that there are no non-trivial kernels in higher dimensions that never increase the number of local extrema or zero-crossings (see chapter 4).

2.5.4. Necessity: Scale invariance

A formulation by Florack *et al.* (1992) shows that the uniqueness of the Gaussian kernel for scale-space representation can be derived under weaker conditions, essentially by combining the earlier mentioned linearity, shift-invariance, and semi-group conditions with *scale invariance.* The basic argument is taken from physics; physical laws must be independent of the choice of fundamental parameters. In practice, this corresponds to what is known as dimensional analysis; a function that relates physical observables must be independent of the choice of dimensional units. Notably, this condition comprises no direct measure of "structure" in the signal; the non-creation of new structure is only implicit in the sense that physical observable entities that are subjected to scale changes should be treated in a self-similar manner.

Some more technical requirements must be used to prove the uniqueness. The solution must not tend to infinity when the scale parameter increases. Moreover, either *rotational symmetry* or *separability* in Cartesian coordinates needs to be imposed to guarantee uniform treatment of the different coordinate directions. Since the proof of this result is valid in arbitrary dimensions and not very technical, a simplified[7] version of it will be reproduced, which nevertheless contains the basic steps.

2.5.4.1. Necessity proof from scale invariance

Recall that any linear and shift-invariant operator can be expressed as a convolution operator. Hence, assume that the scale-space representation $L\colon \mathbb{R}^N \times \mathbb{R}_+ \to \mathbb{R}$ of any signal $f\colon \mathbb{R}^N \to \mathbb{R}$ is constructed by convolution with some one-parameter family of kernels $h\colon \mathbb{R}^N \times \mathbb{R}_+ \to \mathbb{R}$

$$L(\cdot;\ t) = h(\cdot;\ t) * f. \tag{2.49}$$

In the Fourier domain ($\omega \in \mathbb{R}^N$), this can be written

$$\hat{L}(\omega;\ t) = \hat{h}(\omega;\ t)\hat{f}(\omega). \tag{2.50}$$

[7]Here, it is assumed that the semi-group is of the form $g(\cdot;\ t_1)*g(\cdot;\ t_2) = g(\cdot;\ t_1+t_2)$, and that the scale values measured in terms of t should be added by regular summation. This is a so-called *canonical semi-group.* More generally, Florack *et al.* (1992) consider semi-groups of the form $g(\cdot;\ \sigma_1^p) * g(\cdot;\ \sigma_2^p) = g(\cdot;\ \sigma_1^p + \sigma_2^p)$ for some $p \geq 1$, where the scale parameter σ is assumed to have dimension length. By combining rotational symmetry with separability in Cartesian coordinates, they show that these conditions uniquely fixate the exponent p to be two. For one-dimensional signals though, this parameter will then be undetermined.

A result in physics, called the Pi-theorem, states that if a physical process is scale independent, then it should be possible to express the process in terms of dimensionless variables. Here, the following dimensions and variables occur[8]

$$\begin{array}{ll} [\text{luminance}]: & \hat{L},\ \hat{f} \\ [\text{length}]^{-1}: & \omega,\ 1/\sqrt{t}. \end{array}$$

Natural dimensionless variables to introduce are hence, $\hat{L}/\hat{f}$ and $\omega\sqrt{t}$. Using the Pi-theorem, a necessary requirement for scale invariance is that (2.50) can be expressed on the form

$$\frac{\hat{L}(\omega;\ t)}{\hat{f}(\omega;\ t)} = \hat{h}(\omega;\ t) = \hat{H}(\omega\sqrt{t}) \tag{2.51}$$

for some function $\hat{H}\colon \mathbb{R}^N \to \mathbb{R}$. A necessary requirement on $\hat{H}$ is that $\hat{H}(0) = 1$. Otherwise $\hat{L}(\omega;\ 0) = \hat{f}(\omega)$ would be violated.

If h is required to be a semi-group with respect to the scale parameter, then the following relation must hold in the Fourier domain

$$\hat{h}(\omega;\ t_1)\,\hat{h}(\omega;\ t_2) = \hat{h}(\omega;\ t_1 + t_2), \tag{2.52}$$

and consequently in terms of $\hat{H}$,

$$\hat{H}(\omega\sqrt{t_1})\,\hat{H}(\omega\sqrt{t_2}) = \hat{H}(\omega\sqrt{t_1 + t_2}). \tag{2.53}$$

Assume first that $\hat{H}$ is rotationally symmetric, and introduce new variables $v_i = u_i^T u_i = (\omega\sqrt{t_i})^T(\omega\sqrt{t_i}) = \omega^T\omega\, t_i$. Moreover, let $\tilde{H}\colon \mathbb{R} \to \mathbb{R}$ be defined by $\tilde{H}(u^T u) = \hat{H}(u)$. Then, (2.53) assumes the form

$$\tilde{H}(v_1)\,\tilde{H}(v_2) = \tilde{H}(v_1 + v_2). \tag{2.54}$$

This expression can be recognized as the definition of the exponential function, which means that

$$\tilde{H}(v) = \exp(\alpha v) \tag{2.55}$$

for some $\alpha \in \mathbb{R}$, and

$$\hat{h}(\omega;\ t) = \hat{H}(\omega\sqrt{t}) = \tilde{H}(\omega^T\omega\, t) = e^{\alpha\,\omega^T\omega\, t}. \tag{2.56}$$

Concerning the sign of α, it is natural to require $\lim_{t\to\infty} \hat{h}(\omega;\ t) = 0$ rather than $\lim_{t\to\infty} \hat{h}(\omega;\ t) = \infty$. This means that α must be negative,

[8] Since the dimension is preserved under a Fourier transform, it follows that $\hat{L}$ and $\hat{f}$ have the same dimension as L and f, i.e., [luminance]. The scale parameter $t = \sigma^2$ has dimension [length]^2, and the frequency ω has dimension [length]$^{-1}$.

and we can without loss of generality set $\alpha = -1/2$, in order to preserve consistency with the previous definitions of the scale parameter t. Hence, the Fourier transform of the smoothing kernel is uniquely determined as the Fourier transform of the Gaussian kernel

$$\hat{g}(\omega;\ t) = e^{-\omega^T \omega t/2}. \tag{2.57}$$

Alternatively, the assumption about rotational invariance can be replaced by separability. Assume that $\hat{H}$ in (2.53) can be expressed on the form

$$\hat{H}(u) = \hat{H}(u^{(1)}, u^{(2)}, \ldots, u^{(N)}) = \prod_{i=1}^{N} \bar{H}(u^{(i)}) \tag{2.58}$$

for some function $\bar{H}\colon \mathbb{R} \to \mathbb{R}$. Then, (2.53) assumes the form

$$\prod_{i=1}^{N} \bar{H}(v_1^{(i)}) \prod_{i=1}^{N} \bar{H}(v_2^{(i)}) = \prod_{i=1}^{N} \bar{H}(v_1^{(i)} + v_2^{(i)}), \tag{2.59}$$

where new coordinates $v_j^{(i)} = (u_j^{(i)})^2$ have been introduced. Similarly to above, it must hold for any $\omega \in \mathbb{R}^n$, and hence under independent variations of the individual coordinates, which gives

$$\bar{H}(v_1^{(i)})\, \bar{H}(v_2^{(i)}) = \bar{H}(v_1^{(i)} + v_2^{(i)}), \tag{2.60}$$

for any $v_1^{(i)}, v_2^{(i)} \in \mathbb{R}$. This means that $\bar{H}$ must be an exponential function, and that $\hat{h}$ must be the Fourier transform of the Gaussian kernel.

2.5.5. *Necessity: Operators derived from the scale-space representation*

The previous results show that the Gaussian kernel is the unique kernel for generating a (linear) scale-space. An interesting problem, concerns what operators are natural to apply to this representation.

In early work, Koenderink and van Doorn (1987) advocated the use of the multi-scale *N-jet signal representation,* that is the set of spatial derivatives of the scale-space representation up to some (given) order N. Then, in (Koenderink and van Doorn 1992) they considered the problem of deriving linear operators from the scale-space representation, which are to be invariant under scaling transformations. Inspired by the relation between the Gaussian kernel and its derivatives, here in one dimension,

$$\partial_{x^n} g(x;\ t) = (-1)^n \frac{1}{(2t)^{n/2}} H_n(\frac{x}{\sqrt{2t}})\, g(x;\ t), \tag{2.61}$$

which follows from the well-known relation between the derivatives of the Gaussian kernel and the Hermite polynomials H_n

$$\partial_{x^n}(e^{-x^2}) = (-1)^n H_n(x)\, e^{-x^2}, \tag{2.62}$$

they considered the problem of deriving operators with a similar scaling behaviour. Starting from the *Ansatz*

$$\psi^{(\alpha)}(x;\ t) = \frac{1}{(2t)^{\alpha/2}}\,\varphi^{(\alpha)}(\frac{x}{\sqrt{2t}})\,g(x;\ t), \tag{2.63}$$

where the superscript (α) describes the "order" of the function, they considered the problem of determining all functions $\varphi^{(\alpha)}\colon \mathbb{R}^N \to \mathbb{R}$ such that $\psi^{(\alpha)}\colon \mathbb{R}^N \to \mathbb{R}$ satisfies the diffusion equation. Interestingly, it can be shown that $\varphi^{(\alpha)}$ must then satisfy the time-independent Schrödinger equation

$$\nabla^2\varphi(\xi) + ((2\alpha + N) - \xi^T\xi)\,\varphi(\xi) = 0, \tag{2.64}$$

where $\xi = x/\sqrt{2t}$. This is the physical equation that governs the quantum mechanical free harmonic oscillator. It is well-known from mathematical physics that the solutions $\varphi^{(\alpha)}$ to this equation are the Hermite functions, that is Hermite polynomials multiplied by Gaussian functions. Since the derivative of a Gaussian kernel is a Hermite polynomial times a Gaussian kernel, it follows that the solutions $\psi^{(\alpha)}$ to the original problem are the derivatives of the Gaussian kernel.

This result provides a formal statement that Gaussian derivatives are natural operators to derive from scale-space.

2.5.6. Other special properties of the Gaussian kernel

Let us conclude this discussion concerning continuous signals by listing a number of other special properties of the Gaussian kernel.

Given any one-dimensional function h with Fourier transform $\hat{h}$ define the normalized second moments (variances) Δx and $\Delta\omega$ in the spatial and the Fourier domain respectively by

$$\Delta x = \frac{\int_{x\in\mathbf{R}} x^T x |h(x)|^2 dx}{\int_{x\in\mathbf{R}} |h(x)|^2 dx} \tag{2.65}$$

$$\Delta\omega = \frac{\int_{\omega\in\mathbf{R}} \omega^T \omega |\hat{h}(\omega)|^2 d\omega}{\int_{\omega\in\mathbf{R}} |\hat{h}(\omega)|^2 d\omega} \tag{2.66}$$

These entities describe the "spread" of the distributions of h and $\hat{h}$ respectively. Then, the uncertainty relation states that

$$\Delta x \Delta\omega \geq \tfrac{1}{2}. \tag{2.67}$$

A remarkable property of the Gaussian kernel is that it is the only real kernel that gives equality in this relation. Moreover, the Gaussian kernel is the only rotationally symmetric kernel that is separable in Cartesian coordinates.

The Gaussian kernel is also the frequency function of the normal distribution. The central limit theorem in statistics states that under rather general requirements on the distribution of a stochastic variable, the distribution of a sum of a large number of such stochastic variables asymptotically approaches a normal distribution, when the number of terms in the sum increases.

2.6. Summary and retrospective

As we have seen, the uniqueness of the Gaussian kernel for scale-space representation can be derived in a variety of different ways, non-creation of new level curves in scale-space, non-creation of new local extrema, non-enhancement of local extrema, and scale invariance. Similar formulations can be stated both in the spatial domain and in the frequency domain. The essence of these results is that the scale-space representation is given by a (possibly semi-discretized) parabolic differential equation corresponding to a second-order differential operator with respect to the spatial coordinates, and a first-order differential operator with respect to the scale parameter.

2.6.1. Previous use of Gaussian smoothing

Admittedly, Gaussian smoothing has been used before Witkin and Koenderink formally defined the scale-space concept. Marr (1976) and Marr and Hildreth (1980) made use of difference of Gaussians (DOG), which are approximations to the Laplacian of the Gaussian, at different scales. Their motivation for using the Gaussian kernel was that it is the function that minimizes the inequality (2.67) associated with the uncertainty relation, and that the Gaussian kernel in this respect gives the best trade-off between the conflicting goals of localizing image structures in the spatial and frequency domains simultaneously.

Moreover, binomial kernels have been used for pyramid generation, often motivated by the fact that when applied repeatedly, they rapidly approach the Gaussian kernel.

One of the most important contributions with Witkin's scale-space formulation, however, was the systematic way to *relate* and *interconnect* such representations and image structures at different scales. As Koenderink (1984) phrased it concerning the different levels in a multi-scale representation:

> The challenge is to understand the image really on all these levels *simultaneously*, and not as an unrelated set of derived images at different levels of blurring ...

Adding a *scale dimension* onto the original data set, as is done in the one-parameter embedding, provides a formal way to express this interrelation.

2.6.2. Is Gaussian smoothing the only reasonable choice?

A natural question then arises: Does this approach constitute the *only* reasonable way to perform the low-level processing in a vision system, and are the Gaussian kernels and their derivatives the only smoothing kernels that can be used? Of course, this question is impossible to answer without any further specification of the purpose of the representation and what tasks the visual system is to accomplish. In any sufficiently specific application it should be possible to design a smoothing filter that in some sense has a "better performance" than the proposed Gaussian derivative model. For example, it is well-known that scale-space smoothing leads to shape distortions at edges by smoothing across object boundaries, and also in surface orientation estimates computed by algorithms like shape-from-texture. Hence, it should be emphasized that the theory developed here is rather aimed at describing *basic principles* of the very first stages of low-level processing in an *uncommitted* visual system aimed at handling a large class of different situations, and in which no or very little *a priori* information is available.

Then, once initial hypotheses about the structure of the world have been generated within this framework, the intention is that it should be possible to invoke more refined processing, which can compensate for these distortion effects, and adapt to the current situation and the task at hand. From the viewpoint of such non-uniform scale-space approaches, the linear scale-space model based on the rotationally symmetric Gaussian kernel provides a natural starting point for such analysis.

2.6.3. Relations to biological vision

In fact, a certain degree of agreement[9] can be obtained with the result from this solely theoretical analysis and the experimental results of biological evolution. Neurophysiological studies by Young (1985, 1987) have shown that there are receptive fields in the mammalian retina and visual cortex, whose measured response profiles can be very well modelled by Gaussian derivatives. For example, Young models cells in the mammalian

[9] Another interesting similarity concerns the spatial layout of receptive fields over the visual field. If the scale-space axioms are combined with the assumption of a fixed readout capacity from the visual front-end, then it is straightforward to show that there is a natural distribution of receptive fields (of different scales and different spatial position) over the retina such that the minimum receptive field size grows linearly with eccentricity, that is the distance from the center of the visual field (Lindeberg and Florack 1992). There are several results in psychophysics, neuroanatomy and electrophysiology in agreement with such a linear increase (Koenderink and van Doorn 1978; van de Grind *et al.* 1986; Bijl 1991).

retina by kernels termed differences of offset Gaussians (DOOG). Each such kernel, basically, corresponds to the Laplacian of the Gaussian with an added Gaussian offset term. He also reports cells in the visual cortex, whose receptive field profiles agree with Gaussian derivatives up to order four.

Of course, far-reaching conclusions should not be drawn from such a qualitative similarity, since there are also other functions, like Gabor functions that satisfy the recorded data up to the tolerance of the measurements. Nevertheless, it is interesting to note that receptive field profiles similar to the Laplacian of the Gaussian have been reported to be dominant in the retina. A possible explanation concerning the construction of derivatives of other orders from the output of these operators can be obtained from the observation that the original scale-space representation can always be reconstructed from this data if Laplacian derivatives are available at all other scales. If the scale-space representation tends to zero at infinite scale, then it follows from the diffusion equation that

$$
\begin{aligned}
L(x;\, t) &= -(L(x;\, \infty) - L(x;\, t)) && (2.68)\\
&= -\int_{t'=t}^{\infty} \partial_t L(x;\, t')dt' && (2.69)\\
&= -\frac{1}{2}\int_{t'=t}^{\infty} \nabla^2 L(x;\, t')dt'. && (2.70)
\end{aligned}
$$

Observe the similarity with the method (2.10) for reconstructing the original signal from a bandpass pyramid.

What remains to be understood is if there are any particular theoretical advantages of computing the Laplacian of the Gaussian in the first step. Of course, such an operation suppresses any linear illumination gradients. Let us summarize by contending that spatial derivatives of the Gaussian can be approximated by differences of Gaussian kernels at different spatial position, and it is therefore, at least in principle, possible to construct any spatial derivative from this representation. Remaining questions concerning the plausibility with respect to biological vision are left to the reader's speculation.

We shall now briefly review some other types of representations with multi-scale interpretation, and then describe how the scale-space model relates to these.

2.7. Wavelets

A type of multi-scale representation that has attracted great interest in signal processing, numerical analysis, and mathematics during recent years is *wavelet representation* (Daubechies 1988), which dates back to Strömberg (1983) and Meyer (1988). A (two-parameter) family of translated and dilated (scaled) functions

$$h_{a,b}(x) = |a|^{-1/2}\, h((x-b)/a) \quad (a, b \in \mathbb{R}, a \neq 0) \tag{2.71}$$

defined from a single function $h\colon \mathbb{R} \to \mathbb{R}$ is called a *wavelet*. Provided that h satisfies certain admissibility conditions

$$\int_{\omega=-\infty}^{\infty} \frac{|\hat{h}(\omega)|^2}{|\omega|}\, d\omega < \infty, \tag{2.72}$$

then the representation $\mathcal{W}f\colon \mathbb{R}\backslash\{0\} \times \mathbb{R} \to \mathbb{R}$ given by

$$(\mathcal{W}f)(a,b) = < f, h_{a,b} > = |a|^{-1/2} \int_{x\in\mathbb{R}} f(x)\, h((x-b)/a)\, dx \tag{2.73}$$

is called the *continuous wavelet transform* of $f\colon \mathbb{R} \to \mathbb{R}$. From this background, scale-space representation can be considered to be a *special case of continuous wavelet representation*, where the scale-space axioms imply that the function h *must* be selected as a derivative of the Gaussian kernel. In other words, scale-space representation can be seen as giving a canonical choice of wavelet representation.

In traditional wavelet theory though, the zero-order derivative is not permitted. It does not satisfy the admissibility condition, which in practice implies that

$$\int_{x=-\infty}^{\infty} h(x)\, dx = 0. \tag{2.74}$$

There are several developments of this theory concerning different special cases. A particularly well-studied problem is the construction of *orthogonal wavelets* for discrete signals, which permit a compact non-redundant multi-scale representation of the image data. This representation was suggested for image analysis by Mallat (1989, 1992). We will not attempt to review any of that theory here. Instead, the reader is referred to the rapidly growing literature on the subject.

2.8. Regularization

According to Hadamard, a problem is said to be well-posed if: (i) a solution exists, (ii) the solution is unique, and (iii) the solution depends continuously on the input data. It is well-known that several problems in computer vision are ill-posed; one example is differentiation. A small disturbance in a signal,

$$f(x) \mapsto f(x) + \varepsilon \sin \omega x, \tag{2.75}$$

where ε is small and ω is large, can lead to an arbitrarily large disturbance in the derivative

$$f_x(x) \mapsto f_x(x) + \omega\varepsilon \cos \omega x, \tag{2.76}$$

provided that ω is sufficiently large relative to $1/\epsilon$.

Regularization is a technique that has been developed for transforming ill-posed problems into well-posed ones. See Tikhonov and Arsenin (1977) for an extensive treatment of the subject. Torre and Poggio (1986) describe this issue with application to one of the most intensely studied subproblems in computer vision, edge detection, and develop how regularization can be used in this context. One example of regularization concerning the problem "given an operator $\mathcal{A}$ and data y, find z such that $\mathcal{A}z = y$" is the transformed problem "find z that minimizes the following functional"

$$\min_z \; (1-\lambda)\,||\mathcal{A}z - y||^2 + \lambda\,||\mathcal{P}z||^2, \tag{2.77}$$

where $\mathcal{P}$ is a stabilizing operator penalizing variations in z, and $\lambda \in [0,1]$ is a regularization parameter controlling the compromise between the degree of regularization of the solution and closeness to the given data. Variation of the regularization parameter gives solutions with different degree of smoothness; a large value of λ may give rise to a smooth solution, while a small value increases the accuracy at the cost of larger variations in the estimate. Hence, this parameter has a certain interpretation in terms of spatial scale in the result. (It should be observed, however, that the solution to the regularized problem is in general not a solution to the original problem, not even in the case of ideal noise-free data.)

In the special case when $\mathcal{P} = \partial_{xx}$, and the measured data points are discrete, the solution of the problem of finding $S\colon \mathbb{R} \to \mathbb{R}$ that minimizes

$$\min_S \; (1-\lambda) \sum (f_i - S(x_i))^2 + \lambda \int |S_{xx}(x_i)|^2 dx \tag{2.78}$$

given a set of measurements f_i is given by approximating cubic splines; see de Boor (1978) for an extensive treatment of the subject. Interestingly, this result was first proved by Schoenberg (1946), who also proved the classification of Pólya frequency functions and sequences, which are the natural concepts in mathematics that underlie the scale-space kernels that will be considered in chapter 3. Torre and Poggio made the observation that the corresponding smoothing filters are very close to Gaussian kernels.

The strong regularization property of scale-space representation can be appreciated in the introductory example. Under a small high-frequency disturbance in the original signal $f(x) \mapsto f(x) + \varepsilon \sin \omega x$, the propagation of the disturbance to the first-order derivative of the scale-space representation is given by

$$L_x(x;\ t) \mapsto L_x(x;\ t) + \varepsilon\, \omega e^{-\omega^2 t/2} \cos \omega x. \tag{2.79}$$

Clearly, this disturbance can be made arbitrarily small provided that the derivative of the signal is computed at a sufficiently coarse scale t in scale-space.

2.9. Relations between different multi-scale representations

2.9.1. Multi-scale vs. multi-resolution

The main difference between the scale-space representation and the pyramid representation is that the scale-space representation is defined by smoothing and preserves the same spatial sampling at all scales, while in the pyramid representation the main objective is to reduce the number of grid points from one layer to the next. To reduce the aliasing problems some pre-filtering must be performed before the sub-sampling step is carried out. Different operators have been proposed for this task (Burt and Adelson 1983; Crowley *et al.* 1984; Meer *et al.* 1987).

Hence, a multi-resolution representation will be efficient in the sense that it leads to a rapidly decreasing image size, while a scale-space representation successively becomes more redundant as the scale parameter increases. The decreasing image size in pyramids reduces the computational work both in the actual computation of the representation and in the subsequent processing. The memory requirements are small, and there are commercially available implementations of pyramids in hardware. An orthogonal wavelet representation is in fact non-redundant while a scale-space representation can be said to be maximally redundant.

On the other hand, in a scale-space representation, the representations at all levels of scale are immediately accessible without any need for further computations. The task of operating on the data will be successively simplified, since a feature existing at a coarse scale will in general correspond to a larger number of grid points than a feature at a fine scale. In pyramid representations, however, this relation remains unchanged—there is a fixed relation between the scale parameter and the resolution. Moreover, in contrast to the pyramids and the wavelets, the scale-space representation is invariant to translations.

Another important property with the scale-space representation is that the behaviour of structure across scales can be analytically described with a simple formalism. By definition, it is given as the solution to the diffusion equation, which means that features at different scales can be related to each other in a precise manner (see chapter 8).

Moreover, the pyramid representations imply a fixed sampling step in scale or resolution that cannot be decreased, while the scale-space concept possesses a *continuous* scale parameter. Therefore, one can expect the task of following or tracking features across scales to be easier in a multi-scale than in a multi-resolution representation, since refinements of the scale sampling can be performed whenever required. Finally, it is sometimes argued that the pyramid representations undersample the signals along the scale direction.

2.9.2. Related multi-scale representations

Among other types of representations involving multiple scales one can mention the Gabor functions (1946), as well as the related work by Granlund (1978) and his co-workers (Knutsson and Granlund 1983; Bigün and Granlund 1987). The Mirage model developed by Watt (1988) is based on the absolute values of the Laplacian of the Gaussian at multiple scales.

The concept of non-linear anisotropic diffusion has been studied by, among others, Grossberg (1984), Perona and Malik (1990), Nordström (1990), Shah (1991), Whitaker and Pizer (1993), and Alvarez *et al.* (1992).

Of interest is also the previously mentioned wavelet theory (Strömberg 1983; Meyer 1988; Daubechies 1988), which has been applied to image analysis by Mallat (1989, 1992). These ideas are strongly related to *multi-grid methods* (Hackbusch 1985), which have received an increasing interest in numerical analysis together with techniques based on hierarchical basis functions for finite element spaces (Yserentant 1986; Szeliski 1990).

Another interesting early work was done by Ehrich and Lai (1978). They did not directly rely on multiple scales, but a different type of hierarchical signal representation based on the inclusion of extremal regions into each other. Related approaches have been considered in mathematical morphology (Serra 1982).

Multi-scale representations of *curves* have been studied by Bengtsson and Eklundh (1991), who define a sequence of polygons approximating the original data with varying accuracy, Asada and Brady (1986), who define a curvature primal sketch, Mokhtarian and Mackworth (1986, 1988), who smooth the coordinate functions of a parameterized curve, Lowe (1988), who suggests a way to compensate for the shrinking problems in that type of smoothing, and Kimia *et al.* (1990), who use a reaction-diffusion equation approach.

2.9.3. Information content in scale-space

Originating from the computational vision model by Marr and Hildreth (1980), in which zero-crossings of the Laplacian play a dominant role, substantial efforts have been spent on analysing the information content of those features in scale-space. Problems that have been treated concern whether the original signal can be reconstructed using the evolution properties over scales of these zero-crossing curves. The main result is that such reconstruction is possible (up to certain constants), although unstable unless regularized in some sense, see Hummel (1986), Yuille and Poggio (1988), and Hummel and Moniot (1989).

An interesting result is presented by Johansen *et al.* (1986), who show that if also negative scales are considered, then a band-limited one-dimensional signal is up to a multiplicative constant determined by its "top points," that is, the points in scale-space where bifurcations between critical points occur.

When considering the entire scale-space representation, it is, of course, obvious that the original signal can be "reconstructed" from its scale-space representation. The philosophy that underlies this presentation is to use the scale-space representation for *making explicit* certain aspects of the information content, rather than for deriving any "minimal and complete" representation, from which the signal can be reconstructed. In this respect the approach differs strongly from orthogonal wavelet representation.

2.9.4. Special properties of scale-space representation

There have been thorough investigations about the theoretical properties of this representation. As was described above, the fundamental property of not introducing "new" or "artificial" structure has been given different formulations by different authors. The behaviour of structures under this type of smoothing has been analysed by Koenderink and van Doorn (1986). Other studies concerning edges have been made by Bergholm (1987), Clark (1988), and Lindeberg (1992, 1993); (see chapter 8).

These properties together make the scale-space representation special. One should therefore be careful of not using the term "scale-space" for other possible types of representation with a multi-scale characters, such as those that can be obtained, by varying regularization parameters and error criteria in optimization methods, unless similar theoretical properties can be proved.

3

Scale-space for 1-D discrete signals

The scale-space theory has been developed and well-established for continuous signals and images. However, it does not tell anything about how the implementation should be performed computationally in real-life problems, i.e., for discrete signals and images. In principle, there are two possible approaches:

- Apply the results from the continuous scale-space theory by discretizing the occurring equations. For example, the convolution integral (1.3) can be approximated by a sum using standard numerical methods. Or, the diffusion equation (1.5) can be discretized in space with the ordinary five-point operator forming a set of coupled ordinary differential equations, which can be further discretized in scale. If the numerical methods are chosen with care, reasonable approximations to the continuous numerical values can certainly be expected. But it is not guaranteed that the original scale-space conditions, however formulated in a discrete situation, will be preserved.

- Formulate a genuinely discrete theory by postulating suitable axioms.

The goal of the first part of this book is to address the second item, and to develop a scale-space theory for discrete signals. We will start with a one-dimensional signal analysis. In this case it turns out to be possible to characterize exactly what kernels can be regarded as smoothing kernels, and a complete and exhaustive treatment will be given. One among many questions which will be answered is this: If one performs repeated averaging, does one then get scale-space behaviour? A family of kernels will be presented, which is the discrete analogue of the Gaussian family of kernels. The set of arguments, which in the discrete case uniquely leads to this family of kernels, does in the continuous case uniquely lead to the traditional Gaussian family of kernels.

The structure of the problem in higher dimensions is more complex, since it is harder to formulate what should be meant by non-creation of structure. However, by slight modification of the arguments used in the one-dimensional case, an answer will be given as to how the scale-space for two-dimensional discrete signals should be constructed. In the separable case, the scale-space reduces to representation given by separated convolution with the presented one-dimensional discrete analogue of the Gaussian kernel. The representation obtained in this way has computational advantages compared to the commonly adopted approach, where the scale-space is based on different versions of the sampled Gaussian kernel. One of many spin-off products that comes up is a well-conditioned and efficient method to compute Gaussian derivatives. It is well-known that, for example, the implementation of the Laplacian of the Gaussian has lead to computational problems (Grimson and Hildreth 1985).

The theory developed in this presentation does also have the attractive property that it is linked to the continuous theory through a discretized version of the diffusion equation. This means that continuous results may be transferred to the discrete implementation provided that the discretization is done correctly. However, the important point with the scale-space concept to be outlined here is that the properties we want from a scale-space hold not only in the ideal theory but also in the discretization,[1] since the discrete nature of the problem has been taken into account already in the theoretical formulation of the scale-space representation. Therefore, I believe that the suggested way to implement the scale-space theory really describes the proper way to do it.

The presented results should have implications for image analysis as well as other disciplines related to digital signal processing.

3.1. Scale-space axioms in one dimension

A multi-scale representation is a family of derived signals intended to represent the original signal at various levels of scale. Each member of the family should be associated with a value of a scale parameter intended to somehow describe the current level of scale. The scale parameter, here denoted by t, may be either discrete ($t \in \mathbb{Z}_+$) or continuous ($t \in \mathbb{R}_+$). This gives two different types of discrete scale-spaces—discrete signals with a discrete scale parameter and discrete signals with a continuous scale parameter. In both cases we start from the following basic assumptions:

[1] In a practical implementation, of course, also truncation and rounding errors due to finite precision have to be faced. The idea with this approach is to improve the algorithms by including at least the discretization effects in the theory. In ordinary numerical analysis for simulation of physical phenomena it is almost always possible to reduce these effects by increasing the density of grid points, if the current grid is not fine enough to give a prescribed accuracy in the result. However, in image analysis and computer vision one is usually locked to some fixed maximal resolution, beyond which additional image data are not available.

- Every representation should be generated by a linear and shift-invariant transformation of the original signal. Therefore, the smoothing operator can be expressed as a convolution operator.
- An increasing value of the scale parameter t should correspond to coarser levels of scale and signals with less structure. In particular, $t = 0$ should represent to the original signal.
- All signals should be real-valued functions $\mathbb{Z} \to \mathbb{R}$ defined on the same infinite grid; in other words no pyramid representations will be used.

The essential requirement is that a signal at a coarser level of scale should contain less structure than a signal at a finer level of scale. If one regards the number of local extrema as one measure of the amount of structure, it is thus necessary that the number of local extrema in space does not increase when going from a finer to a coarser level of scale. It can be shown that the family of functions generated by convolution with the Gaussian kernel possesses this property in the continuous case. Here, it is stated it as the basic axiom for the one-dimensional analysis:

DEFINITION 3.1. (DISCRETE SCALE-SPACE KERNEL; 1D)
A one-dimensional discrete kernel $K: \mathbb{Z} \to \mathbb{R}$ is said to be a scale-space kernel if for all signals $f_{in}: \mathbb{Z} \to \mathbb{R}$ the number of local extrema in the convolved signal $f_{out} = h * f_{in}$ does not exceed the number of local extrema in the original signal.

A minor complication is involved in this statement. If either f_{in} or f_{out} would happen to be non-generic and have a plateau, then the question must be raised about how many local extrema the plateau should be counted as. At this moment we will not go into the details of those peculiar cases. A plateau is counted as one local maximum (minimum) if there are strictly smaller (larger) values bounding it both at the left and at the right (see figure 3.1). An accurate treatment will be given in section 3.3.

Figure 3.1. Examples illustrating the definition of local extremum; (a) a local maximum (generic case), (b) a plateau counted as one local maximum, and (c) a plateau not counted as a local extremum.

An important observation to be made is that this definition equivalently can be expressed in terms of zero-crossings just by replacing the string "local extrema" with "zero-crossings." The result follows from the facts that a local extremum in a discrete function f is equivalent to a zero-crossing in its first difference Δf, defined by $(\Delta f)(x) = f(x+1) - f(x)$, and that the difference operator commutes with the convolution operator.

However, the stated definition has further consequences. It means that the number of local extrema (zero-crossings) in any nth-order difference of the convolved signal cannot be greater than the number of local extrema (zero-crossings) in the nth-order difference of the original signal. Actually, the result can be generalized to arbitrary linear operators.

PROPOSITION 3.2. (GENERAL SMOOTHING PROPERTY OF DISCRETE SCALE-SPACE KERNELS) *Let $K: \mathbb{Z} \to \mathbb{R}$ be a discrete scale-space kernel and let $\mathcal{L}$ be an operator (from the space of real-valued discrete functions to itself) that commutes with K. Then, for any $f: \mathbb{Z} \to \mathbb{R}$ (such that the involved quantities exist) the number of local extrema (zero-crossings) in $\mathcal{L}(h * f)$ cannot exceed the number of local extrema (zero-crossings) in $\mathcal{L}(f)$.*

Proof. Let $f' = \mathcal{L}(f)$. Since K is a discrete scale-space kernel, the number of local extrema (zero-crossings) in $K * f'$ cannot be larger than the number of local extrema (zero-crossings) in f'. Since K and $\mathcal{L}$ commute, it holds that $K * f' = h * \mathcal{L}(f) = \mathcal{L}(K * f)$, and the result follows. $\square$

This shows that not only the function, but also all its "derivatives" will become smoother. Accordingly, convolution with a discrete scale-space kernel can really be regarded as a smoothing operation.

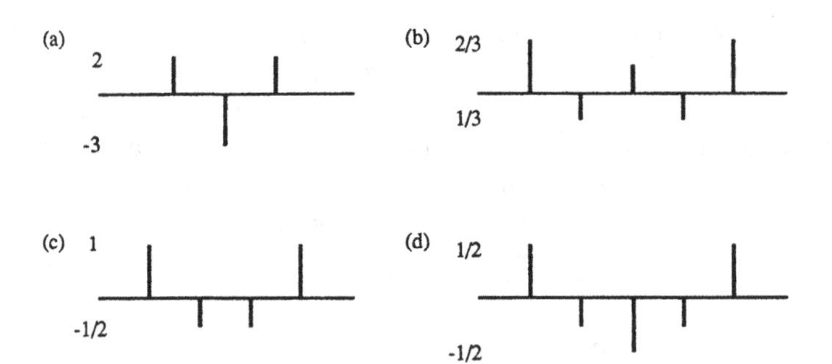

Figure 3.2. (a) Input signal. (b) Convolved with $(\frac{1}{3}, \frac{1}{3}, \frac{1}{3})$. (c) Convolved with $(\frac{1}{2}, \frac{1}{2})$. (d) Convolved with $(\frac{1}{4}, \frac{1}{2}, \frac{1}{4})$.

To realize that the number of local extrema or zero-crossings can increase even in a rather uncomplicated situation, consider the input signal

$$f_{in}(x) = \begin{cases} -3 & \text{if } n = 0, \\ 2 & \text{if } n = \pm 1, \\ 0 & \text{otherwise,} \end{cases} \tag{3.1}$$

and convolve it with the kernels $(\frac{1}{3}, \frac{1}{3}, \frac{1}{3})$, $(\frac{1}{2}, \frac{1}{2})$ and $(\frac{1}{4}, \frac{1}{2}, \frac{1}{4})$. The results are shown in figures 3.2 (b), (c), and (d) respectively.

It can be seen that both the number of local extrema and the number of zero-crossings have increased for the first kernel, but not for the two latter ones. Thus, an operator which naively can be apprehended as a smoothing operator, might actually give a less smooth result. Further, it can really matter if the averaging is performed over three instead of two points and on how the averaging is performed.

3.2. Properties of scale-space kernels

For the reader to become familiar with the consequences of the definition of scale-space kernel, it will now be illustrated what this scale-space property means. To start with, a few general qualitative requirements of scale-space kernels will be pointed out, which are necessarily induced by the given axiom. It will also be shown by elementary techniques that the two latter kernels in figure 3.2 indeed are discrete scale-space kernels.

3.2.1. Positivity and unimodality in the spatial domain

By considering the impulse response it is possible to draw some qualitative conclusions about the properties of a discrete scale-space kernel. Let the input function be the discrete delta function

$$f_{in}(x) = \delta(x) = \begin{cases} 1 & \text{if } x = 0, \\ 0 & \text{otherwise.} \end{cases} \tag{3.2}$$

Then, the output signal will be identical to the kernel

$$f_{out}(x) = (h * \delta)(x) = K(x). \tag{3.3}$$

The function $\delta(x)$ has exactly one local maximum and no zero-crossings. Therefore, in order to be a scale-space kernel K must not have more than one extremum and no zero-crossings. Thus,

PROPOSITION 3.3. (POSITIVITY)
All coefficients of a scale-space kernel must have the same sign.

PROPOSITION 3.4. (UNIMODALITY)
The coefficient sequence of a scale-space kernel $\{K(n)\}_{n=-\infty}^{\infty}$ must be unimodal.[2]

Without loss of generality, the rest of the treatment can therefore be restricted to positive sequences where $K(n) \geq 0$ for all $n \in \mathbb{Z}$.

[2] A real sequence is said to be unimodal if it is first ascending (descending) and then descending (ascending).

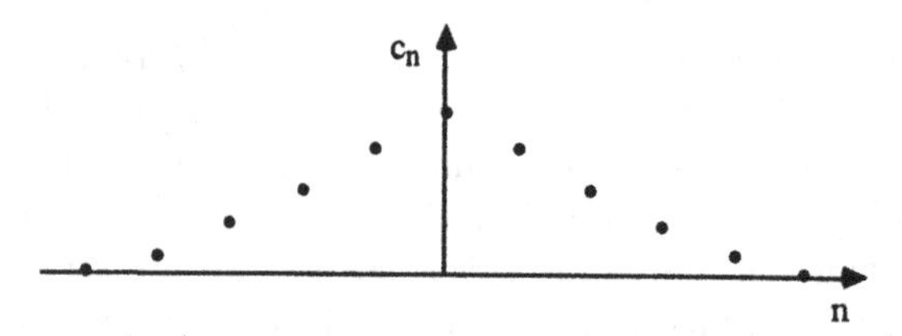

Figure 3.3. The filter coefficient sequence $K(n)_{n=-\infty}^{\infty}$ of a discrete scale-space kernel must be positive and unimodal.

It seems reasonable to require[3] that $K \in l_1$, i.e. that $\sum_{n=-\infty}^{\infty} |K(n)|$ is finite. If f_{in} is bounded and $K \in l_1$, then the convolution is well-defined and the Fourier transform of the filter coefficient sequence exists. This property also makes it possible normalize the coefficients such that $\sum_{n=-\infty}^{\infty} K(n) = 1$. In particular, the filter coefficients $K(n)$ must then tend to zero as n tends to infinity.

3.2.2. Generalized binomial kernels

Consider a two-kernel with only two non-zero filter coefficients:

$$K^{(2)}(n) = \begin{cases} p & \text{if } n = 0 \\ q & \text{if } n = -1 \\ 0 & \text{otherwise} \end{cases} \tag{3.4}$$

Assume that $p \geq 0$, $q \geq 0$ and $p + q = 1$.

It is easy to verify that the number of zero-crossings (local extrema) in $f_{out} = K^{(2)} * f_{in}$ cannot exceed the number of zero crossings (local extrema) in f_{in}. This result follows from the fact that convolution of f_{in} with $K^{(2)}$ is equivalent to the formation a weighted average of the sequence $\{f_{in}(x)\}_{x=-\infty}^{\infty}$, see figure 3.4. The values of the output signal can be constructed geometrically and will fall on straight lines connecting the values of the input signal. The offset along the x-axis is determined by the ratio $q/(p+q)$. It is obvious that no additional zero-crossings can be introduced by this transformation. Thus, any kernel on the form (3.4) is a discrete scale-space kernel. Directly from the definition of a scale-space kernels it follows that:

LEMMA 3.5. (REPEATED APPLICATION OF SCALE-SPACE KERNELS) *If two kernels K_a and K_b are scale-space kernels then also $K_a * K_b$ is a scale-space kernel*

[3]Some regularity requirement must be imposed on the input signal as well. Throughout the considerations the following general convention will be used. If nothing else is explicitly stated, it is assumed that f_{in} is sufficiently regular such that the involved quantities exist and are well-defined.

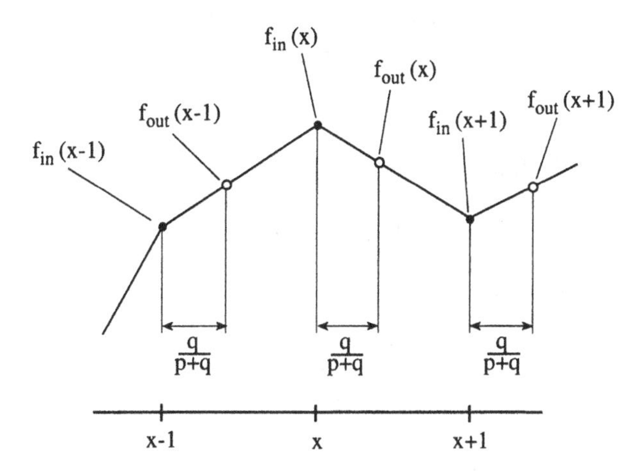

Figure 3.4. To convolve a signal f_{in} with a two-kernel $K^{(2)}(n)$ is equivalent to the formation of a weighted average of the sequence $\{f_{in}(x)\}_{x=-\infty}^{\infty}$. It is obvious that no new zero-crossings can be introduced by this transformation.

Repeated application of this result yields:

PROPOSITION 3.6. (REPEATED AVERAGING AND SCALE-SPACE KERNELS) *All kernels K on the form $*_{i=1}^{n} K_i^{(2)}$, with $K_i^{(2)}$ according to* (3.4), *are discrete scale-space kernels.*

The filter coefficients generated in this way can be regarded as a kind of generalized binomial coefficients. The ordinary binomial coefficients are obtained, except for a scaling-factor, as a special case if all p_i and q_i are equal. Another formulation of proposition 3.6 in terms of generating functions is also possible.

PROPOSITION 3.7. (GENERATING FUNCTION OF GENERALIZED BINOMIAL KERNELS) *All kernels with the generating function*

$$\varphi_K(z) = \sum_{n=-\infty}^{\infty} K(n)\, z^n \tag{3.5}$$

on the form

$$\varphi_K(z) = C\, z^k \prod_{i=1}^{N} (p_i + q_i z), \tag{3.6}$$

where $p_i > 0$, $q_i > 0$ and $k \in \mathbb{Z}$, are discrete scale-space kernels.

Proof. The generating function of a kernel on the form (3.4) is

$$\varphi_{K_i^{(2)}}(z) = p_i + q_i z. \tag{3.7}$$

Since convolution in the spatial domain corresponds to multiplication of generating functions, proposition 3.6 gives that

$$\varphi_h(z) = \varphi_{K_1^{(2)}}(z)\,\varphi_{K_2^{(2)}}(z)\,\ldots\,\varphi_{K_N^{(2)}}(z) \tag{3.8}$$

is the generating function of a scale-space kernel. A constant scaling-factor C or a translation $\varphi_{transl}(z) = z^k$ cannot affect the number of local extrema. Therefore, these factors can be multiplied onto $\varphi_h(z)$ without changing the scale-space properties. □

Another way to express this result is as follows:

PROPOSITION 3.8. (SUFFICIENT CONDITION FOR SCALE-SPACE KERNELS) *Let* $\{c_n\}_{n=-\infty}^{\infty}$ *be the coefficients of a discrete kernel with finite support* ($c_n = 0$ *if* $|n| > N_{supp}$)*. Then, a sufficient condition for the kernel to be a scale-space kernel is that all roots of the generating function*

$$\varphi(z) = \sum_{n=-\infty}^{\infty} c_n\, z^n \tag{3.9}$$

are real and non-positive.

Proof. Let $k = -m$, $N = n + m$ in (3.6). If all roots of $\varphi(z)$ are real and negative, then (3.6) must be the factorization of (3.9). □

3.2.3. Positivity and unimodality in the Fourier domain

The Fourier transform of a symmetric sequence on the form (3.6) has some interesting properties. In its most general form, the generating function of such a sequence can be written

$$\varphi_K(z) = c \prod_{\nu=1}^{N} (p_\nu + q_\nu z)(p_\nu + q_\nu z^{-1}). \tag{3.10}$$

Consider one factor $(p_\nu + q_\nu z)(p_\nu + q_\nu z^{-1})$. Its Fourier transform is

$$\psi_K(\theta) = \varphi_K(e^{i\theta}) = (p_\nu + q_\nu e^{i\theta})(p_\nu + q_\nu e^{-i\theta}) = p_\nu^2 + q_\nu^2 + 2p_\nu q_\nu \cos\theta \tag{3.11}$$

On the interval $[-\pi, \pi]$ this function is non-negative. It assumes its maximum value $(p_\nu + q_\nu)^2$ for $\theta = 0$ and its minimum value $(p_\nu - q_\nu)^2$ for $\theta = \pm\pi$. $\psi_K(\theta)$ is monotonically increasing on $[-\pi, 0]$ and monotonically decreasing on $[0, \pi]$, in other words unimodal; see figure 3.5. It is easy to show that any finite product of non-negative increasing (decreasing) functions is also increasing (decreasing). Consequently, the Fourier transform of a symmetric kernel on the form (3.6) is non-negative and unimodal on the interval $[-\pi, \pi]$.

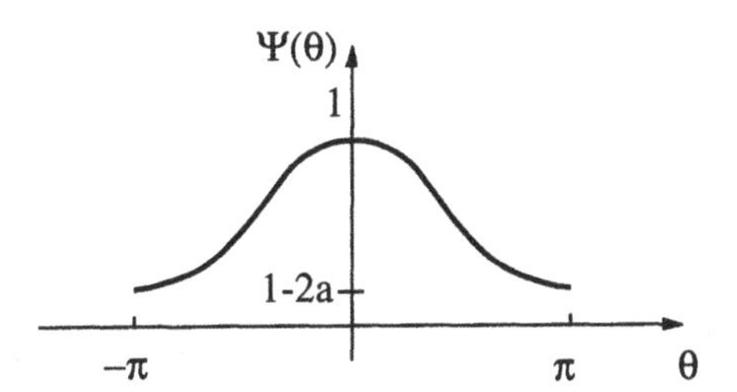

Figure 3.5. The Fourier transform of a (normalized) symmetric three-kernel with the coefficients $(a/2,\ 1-a,\ a/2)$ is $\psi(\theta) = 1 - a(1-\cos\theta)$. If $0 \leq a \leq 1/2$ then this function is non-negative and unimodal on the interval $[-\pi, \pi]$. In the special case when $a = 1/2$, the Fourier transform tends to zero at the end points of the interval.

In this section, results will be derived showing that the Fourier transform of any symmetric scale-space kernel must possess these properties. The proofs, which sometimes are of a rather technical nature, can be skipped by the hasty reader without loss of continuity.

3.2.3.1. No real negative eigenvalues of the convolution matrix

If the convolution transformation $f_{out} = h * f_{in}$ is represented on matrix form $\bar{f}_{out} = C\bar{f}_{in}$, then a matrix with constant values along the diagonals $C_{i,j} = K(i-j)$ appears. Such a matrix is called a Toeplitz matrix. If this matrix has a real and negative eigenvalue, then the corresponding kernel cannot be a scale-space kernel.

PROPOSITION 3.9. (NO REAL NEGATIVE EIGENVALUES OF THE CONVOLUTION MATRIX) *Let $K: \mathbb{Z} \to \mathbb{R}$ be a discrete kernel with finite support and filter coefficients $c_n = K(n)$. If for some dimension N, the $N \times N$ convolution matrix*

$$C^{(N)} = \begin{pmatrix} c_0 & c_{-1} & \cdots & c_{2-N} & c_{1-N} \\ c_1 & c_0 & c_{-1} & \cdots & c_{2-N} \\ \vdots & \ddots & \ddots & \ddots & \vdots \\ c_{N-2} & \cdots & c_1 & c_0 & c_{-1} \\ c_{N-1} & c_{N-2} & \cdots & c_1 & c_0 \end{pmatrix} \tag{3.12}$$

has a negative eigenvalue with a corresponding real eigenvector, then K cannot be a scale-space kernel. In particular, if the kernel is symmetric then all eigenvalues must be real and non-negative.

Proof. Because of proposition 3.3 it is sufficient to study kernels having only non-negative filter coefficients. Assume that $C^{(N)}$ has a real negative eigenvalue for some dimension N and a corresponding real eigenvector

$\bar{v}$. Let I_N be the index set $[1..N]$. Create an input signal f_{in}, which is equal to the components of $\bar{v}$ for $x \in I_N$ and zero otherwise. Convolve this signal with the kernel. Then, for $x \in I_N$ the values of $K * f_{in}$ will be equal to the corresponding components of $C^{(N)}\bar{v}$ (see figure 3.6). Since $\bar{v}$

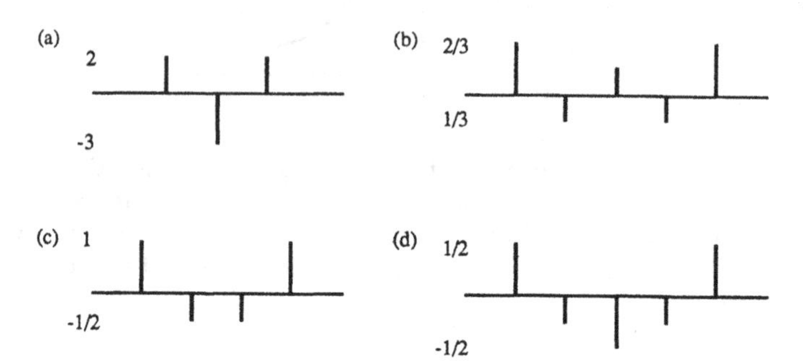

Figure 3.6. (a) The eigenvector $\bar{v}$. (b) The components of $C^{(N)}\ \bar{v}$ having indices $1 \ldots N$. (c) The components of $K * f_{in}$.

is an eigenvector with a negative eigenvalue, the components of $C^{(N)}\ \bar{v}$ and $\bar{v}$ have opposite signs. This means that $\bar{v}$, $C^{(N)}\ \bar{v}$ and $K * f_{in}$ all have the same number of internal zero-crossings provided that we only observe the components in I_N.

The reversal of these components and the positivity of the filter coefficients guarantee that at least one additional zero-crossing will occur in the output signal. Let α denote the index of the first non-zero component of f_{in}. If $f_{in}(\alpha)$ is positive (negative), then due to the negative eigenvalue $K * f_{in}(\alpha)$ will be negative (positive). Since the filter coefficients are non-negative the first non-zero component of $K * f_{in}$ (at position β) will have the same sign as $f_{in}(\alpha)$, i.e. positive (negative). Consequently, we have found at least one additional zero-crossing in $K * f_{in}$ between these two positions (α and β). The same argument can be carried out at other end point producing another scale-space violation. This shows that K cannot be a scale-space kernel. □

3.2.3.2. Positivity in the frequency domain

The eigenvalues of a Toeplitz matrix are closely related to the the Fourier transform of the corresponding sequence of coefficients (Grenander and Szegö 1959; Gray 1972). A theorem by Toeplitz (1911) relates the eigenvalues[4] of an infinite Toeplitz matrix C with elements $C_{i,j} = c_{i-j}$ to the the values of the generating function of the sequence of filter weights. Assume that $\varphi(z) = \sum_{n=-\infty}^{\infty} c_n z^n$ is convergent in the ring $r < |z| < R$, where

[4] λ is said to be an eigenvalue of an infinite matrix C if the matrix $C - \lambda I$ has no bounded inverse. I denotes the unit matrix.

$0 < r < 1 < R$. Then, the eigenvalues of C coincide with the set of complex values that $\varphi(z)$ assumes on the unit circle $|z| = 1$. This property makes it possible to derive an interesting corollary from proposition 3.9.

PROPOSITION 3.10. (NON-NEGATIVE FOURIER TRANSFORM)
The Fourier transform $\psi_K(\theta) = \sum_{n=-\infty}^{\infty} K(n)e^{-in\theta}$ of a symmetric discrete scale-space kernel K with finite support is non-negative.

Proof. Let $\lambda_1^{(N)}$ denote the smallest eigenvalue of the convolution matrix of dimension N, and let m denote the minimum value[5] the Fourier transform ψ_K assumes on $[-\pi, \pi]$. As a consequence of a theorem by Grenander and Szegö (1959) about the asymptotic distribution of eigenvalues of a finite Toeplitz matrix, it follows that

$$\lim_{N\to\infty} \lambda_1^{(N)} = m, \quad \lambda_1^{(N)} \geq m. \tag{3.13}$$

If m is strictly negative, then as $\lim_{N\to\infty} \lambda_1^{(N)} = m$ it follows that $\lambda_1^{(N)}$ will be negative for some sufficiently large N. According to proposition 3.9, the kernel cannot be a scale-space kernel. □

3.2.3.3. Unimodality in the frequency domain

If a linear transformation is to be regarded as a smoothing transformation, it turns out to be necessary that the low-frequency components are not suppressed more than the high-frequency components. This means that the Fourier transform must not increase when the absolute value of the frequency increases. The occurring unimodality property is easiest to establish for circular convolution. In that case, the convolution matrix becomes circulant,[6] which means that its eigenvalues and eigenvectors can be determined analytically.

PROPOSITION 3.11. (UNIMODAL FOURIER TRANSFORM; WRAP-AROUND)
Let $\{c_n\}_{n=-\infty}^{\infty}$ be the filter coefficients of a symmetric discrete kernel with $c_n = 0$ if $|n| > N$. For all integers $M \geq N$ it is required that the transformation given by multiplication with the $(2M+1) \times (2M+1)$ symmetric circulant matrix $C_C^{(M)}$ (3.14), defined by $(C_C^{(M)})_{i,j} = c_{i-j}$ $(i, j \in [0..M])$ and circulant extension, should be a scale-space transformation. Then, necessarily the Fourier transform $\psi(\theta) = \sum_{n=-\infty}^{\infty} c_n e^{-in\theta}$ must be unimodal on $[-\pi, \pi]$.

Proof. The core in the proof is to show that if a kernel has a non-unimodal Fourier transform, then there exists some low-frequency component that

[5]Due to the symmetry of the kernel, $\psi_K(\theta)$ assumes only real values. The minimum value exists, since $\psi(\theta)$ is a continuous function and the interval $[-\pi, \pi]$ is compact.

[6]In a circulant matrix each row is a circular shift of the previous row, except for the first row which is a circular shift of the last row.

disappears faster than some other high-frequency component. By considering a superposition of two such components, we will show that repeated application of the convolution operator will eventually lead to an increase in the number of local extrema when the low-frequency component has died out and the high-frequency component dominates, see figure 3.7.

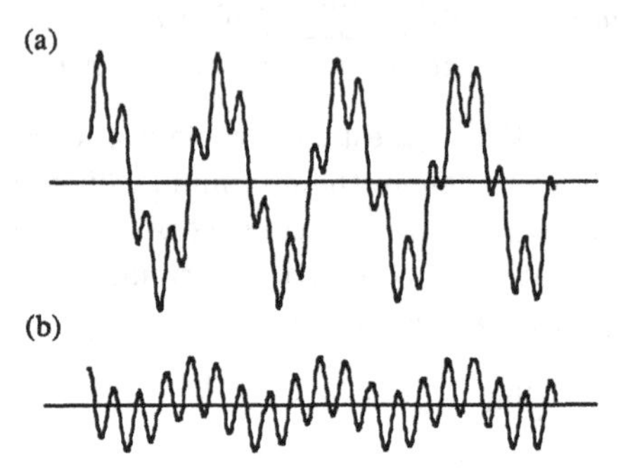

Figure 3.7. (a) Input signal consisting of a low-frequency component of high amplitude and a high-frequency component of low amplitude. (b) In the output signal the low-frequency component has been suppressed while the high-frequency component remains unchanged. As can be seen, additional zero-crossings have been introduced.

Let us introduce a temporary definition. If $\bar{x}$ is a vector of length L, let $V(\bar{x})$ denote the number of zero-crossings in the sequence of components $x_1, x_2, \dots x_L, x_1$. By verification, one shows that the eigenvalues λ_m and the eigenvectors $\bar{v}_m$ of the convolution matrix

$$C_C^{(M)} = \begin{pmatrix} c_0 & c_1 & \cdots & c_N & & & c_N & \cdots & c_1 \\ c_1 & c_0 & c_1 & & c_N & & & \ddots & \vdots \\ \vdots & & \ddots & & & \ddots & & & c_N \\ c_N & & & \cdot & & & \ddots & & \\ & c_N & & & \cdot & & & c_N & \\ & & \ddots & & & \cdot & & & c_N \\ c_N & & & \ddots & & & \ddots & & \vdots \\ \vdots & \ddots & & & c_N & & & c_0 & c_1 \\ c_1 & \cdots & c_N & & & c_N & \cdots & c_1 & c_0 \end{pmatrix} \tag{3.14}$$

are (unless otherwise stated $m, k \in [-M..M]$).

$$\lambda_m = \sum_{n=-M}^{M} c_n e^{-2\pi i m n/(2M+1)} = \sum_{n=-N}^{N} c_n e^{-2\pi i m n/(2M+1)}, \tag{3.15}$$

$$(\bar{v}_m)_k = \sin\left(\frac{2\pi mk}{2M+1}\right) \quad (m < 0), \tag{3.16}$$

$$(\bar{v}_m)_k = \cos\left(\frac{2\pi mk}{2M+1}\right) \quad (m \geq 0). \tag{3.17}$$

It can be noted that $V(\bar{v}_m)$ increases as $|m|$ increases. Further, the eigenvalues

$$\lambda_m = \psi(2\pi m/(2M+1)) \tag{3.18}$$

of $C_C^{(M)}$ are uniformly sampled values of the Fourier transform, and a larger value of $|m|$ corresponds to a larger absolute value of the argument to ψ. Now, assume that the Fourier transform is not unimodal. (Without loss of generality it can be presupposed that ψ is non-negative on $[-\pi, \pi]$, because otherwise, according to proposition 3.10, the kernel cannot be a scale-space kernel.) Then, as ψ is a continuous function of θ it is possible to find some sufficiently large integer M' such that there exist

$$\theta_\alpha = \frac{2\pi\alpha}{2M'+1}, \quad \theta_\beta = \frac{2\pi\beta}{2M'+1} \tag{3.19}$$

satisfying $\psi(\theta_\beta) > \psi(\theta_\alpha)$ for some integers $\beta > \alpha$ in $[0, M']$.

To summarize, $C_C^{(M')}$ has eigenvalues $\lambda_\beta > \lambda_\alpha$ and corresponding eigenvectors with $V(\bar{v}_\beta) > V(\bar{v}_\alpha)$. It will now be shown that this situation leads to a scale-space violation.

The scale-space properties are not affected by a scaling factor. Therefore, we can equivalently study $B = \frac{1}{\lambda_\beta} C_C^{(M')}$. For both eigenvectors, define the smallest and largest absolute values v_{absmin} and v_{absmax} by

$$v_{absmin} = \min_{k \in [1..N]} |v_k|, \quad v_{absmax} = \max_{k \in [1..N]} |v_k|. \tag{3.20}$$

Let $\bar{x} = c\bar{v}_\alpha + \bar{v}_\beta$, where c is chosen large enough such that $V(\bar{x}) = V(\bar{v}_\alpha)$. This can always be achieved if $|c|\, v_{\alpha,absmin} > v_{\beta,absmax}$, since then the components of $\bar{x}$ and $\bar{v}_\alpha$ will have pairwise same signs. ($v_{\alpha,absmin}$ will be strictly positive as all components of $\bar{v}_\alpha$ are non-zero.) Then, consider $B\bar{x} = \frac{1}{\lambda_\beta}(c\lambda_\alpha \bar{v}_\alpha + \lambda_\beta \bar{v}_\beta)$, and study

$$B^k \bar{x} = c\left(\frac{\lambda_\alpha}{\lambda_\beta}\right)^k \bar{v}_\alpha + \bar{v}_\beta. \tag{3.21}$$

For a fixed value of c, it is always possible to find a sufficiently large value of k such that $V(B^k \bar{x}) = V(\bar{v}_\beta)$. In a similar manner to above, it can be verified that the condition

$$|c| \left|\frac{\lambda_\alpha}{\lambda_\beta}\right|^k v_{\alpha,absmax} < v_{\beta,absmin} \tag{3.22}$$

suffices. Consequently, $V(B^k\bar{x}) > V(\bar{x})$, which shows that the transformation induced by B^k is not a scale-space transformation. Therefore, B cannot be a scale-space kernel since at least one scale-space violation must have occurred in the series of k successive transformations. □

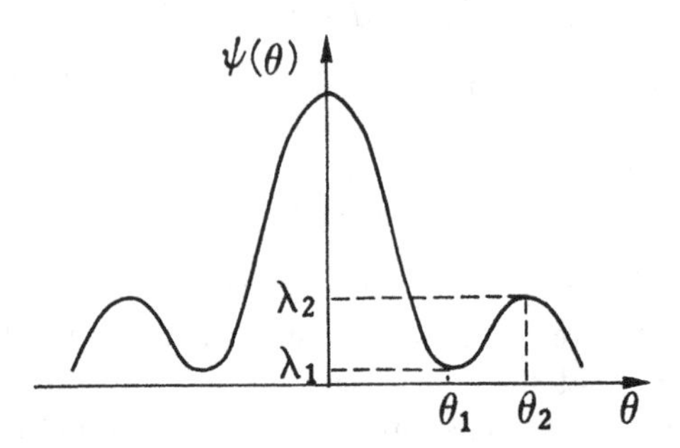

Figure 3.8. If the Fourier transform is not unimodal on $[-\pi, \pi]$, i.e if there exist $\theta_2 > \theta_1$ in $[0, \pi]$ such that $\psi(\theta_2) > \psi(\theta_1)$ then the corresponding transformation cannot be a scale-space transformation.

This result can be extended to comprise non-circular convolution as well. The basic idea is to construct an input signal consisting of several periods of the signal leading to a scale-space violation in the proof of proposition 3.11. Then, the convolution effect on the "interior" periods will be identical to effect on one period by circular convolution. If the signal consists of a sufficient number of periods the boundary effects will be negligible compared to the large number of scale-space violations occurring in the inner parts. The formal details are somewhat technical and can be found in (Lindeberg 1990, 1991).

PROPOSITION 3.12. (UNIMODAL FOURIER TRANSFORM; GENERAL CASE) *The Fourier transform $\psi_K(\theta) = \sum_{n=-\infty}^{\infty} K(n)e^{-in\theta}$ of a symmetric discrete scale-space kernel K with finite support is unimodal on the interval $[-\pi, \pi]$ (with the maximum value at $\theta = 0$).*

3.2.4. Kernels with three non-zero elements

For a three-kernel $K^{(3)}$ with exactly three non-zero consecutive elements $c_{-1} > 0$, $c_0 > 0$ and $c_1 > 0$, it is possible to determine the eigenvalues of the convolution matrix and the roots of the characteristic equation analytically. It is easy to verify that the eigenvalues λ_μ of the convolution matrix

$$C^{(N)}((c_{-1}, c_0, c_1)) = \begin{pmatrix} c_0 & c_{-1} & & & & & \\ c_1 & c_0 & c_{-1} & & & & \\ & c_1 & c_0 & c_{-1} & & & \\ & & \cdot & \cdot & \cdot & & \\ & & & \cdot & \cdot & \cdot & \\ & & & & c_1 & c_0 & c_{-1} \\ & & & & & c_1 & c_0 \end{pmatrix} \tag{3.23}$$

are all real and equal to

$$\lambda_\mu = c_0 - 2\sqrt{c_{-1}c_1}\cos(\frac{\mu\pi}{N+1}), \tag{3.24}$$

and that the roots of generating function $\varphi_{K^{(3)}}(z) = c_{-1}z^{-1} + c_0 + c_1 z$ are

$$z_{1,2} = \frac{-c_0 \pm \sqrt{c_0^2 - 4c_{-1}c_1}}{2c_1}. \tag{3.25}$$

From (3.24) it can be deduced that if $c_0 < 2\sqrt{c_{-1}c_1}$, then for some sufficiently large N at least one eigenvalue of $C^{(N)}$ will be negative. Thus, according to proposition 3.9 the kernel cannot be a scale-space kernel. However, if $c_0^2 \geq 4c_{-1}c_1$, then both the roots of $\varphi_{K^{(3)}}$ will be real and negative. This means that the generating function can be written on the form (3.6), and the kernel is a scale-space kernel. Consequently, a complete classification is obtained for all possible values of c_{-1}, c_0 and c_1. It can be concluded that:

PROPOSITION 3.13. (CLASSIFICATION OF GENERAL THREE-KERNELS) *A three-kernel with elements (c_{-1}, c_0, c_1) is a scale-space kernel if and only if $c_0^2 \geq 4c_{-1}c_1$, i.e., if and only if it can be written as the convolution of two two-kernels with positive elements.*

The corresponding result in the symmetric case $c_{-1} = c_1$ is:

COROLLARY 3.14. (CLASSIFICATION OF SYMMETRIC THREE-KERNELS) *A symmetric three-kernel with elements (c_1, c_0, c_1) is a scale-space kernel if and only if $c_0 \geq 2c_1 \geq 0$.*

The necessity of this property can also be shown directly from the positivity and unimodality properties in the spatial and Fourier domains. Observe that the usual binomial kernel with coefficients $(\frac{1}{4}, \frac{1}{2}, \frac{1}{4})$ is actually a boundary case.

At this moment one may ask if these results can be generalized to hold for kernels with arbitrary numbers of non-zero filter coefficients, i.e., if all discrete scale-space kernels with finite support have a generating function on the form (3.6). This question will be answered in the next section.

3.3. Kernel classification

So far an axiom has been postulated in terms of local extrema, or equivalently zero-crossings, and some of its consequences have been investigated for transformations expressed as linear convolutions with shift-invariant kernels. We have seen that the sequence of filter coefficients must be positive and unimodal and that its sum should be convergent. For symmetric kernels the Fourier transform must be positive and unimodal on $[-\pi, \pi]$.

In this section, a complete characterization of scale-space kernels will be performed. It turns out that several interesting results can be derived from the literature on a subject of mathematics called *total positivity.* Sometimes, the proofs of the important theorems are of a rather complicated nature for a reader with an engineering background. Therefore, the presentation will not be burdened with proofs, and only a brief background to the theory will be given with a few summarizing results.

The pioneer in the subject of variation-diminishing transforms was I. J. Schoenberg. He studied the subject in a series of papers from 1930 to 1953. Later, the theory of total positivity was covered in a monumental monograph by Karlin (1968). A more recent paper by Ando (1987) reviews the field using skew-symmetric vector products and Schur complements of matrices as major tools. The questions raised in this treatment constitute a new application of these not very well-known, but very powerful, results.

3.3.1. Background

Consider first a general linear transformation of discrete signal $f_{in}\colon \mathbb{Z} \to \mathbb{R}$, where the kernel does not need to be shift-invariant

$$f_{out}(x) = \sum_{y=-\infty}^{\infty} K(x, y)\, f_{in}(y). \tag{3.26}$$

Let $\bar{x} = (x_1, x_2, \ldots, x_n)$ be a vector of n real numbers. Two notions of sign changes in vectors will be used (Karlin 1968; Ando 1987). Denote by $V^-(\bar{x})$ the (minimum) number of sign changes obtained in the sequence $x_1, x_2, \ldots, x_n$ if all zero terms are deleted, and by $V^+(\bar{x})$ the maximum number of sign changes possible in the sequence $x_1, x_2, \ldots, x_n$ if each zero value is allowed to be replaced by either $+1$ or -1. A special convention is used saying that the number of sign changes in the null vector is -1.

The interesting sequences and kernels will defined in terms of minors of the transformation matrix. Given a kernel $K\colon X \times Y \to \mathbb{R}$, form minors of arbitrary order r by selections of $x_1 < x_2 < \cdots < x_r$ from $X \subset \mathbb{Z}$ and of $y_1 < y_2 < \cdots < y_r$ from $Y \subset \mathbb{Z}$. The determinant of the resulting matrix with components $\{K(x_i, y_j)\}_{i,j=1\ldots r}$ is called "a minor of order r" and is denoted by

$$K\begin{pmatrix} x_1, x_2, \ldots, x_r \\ y_1, y_2, \ldots, y_r \end{pmatrix} = \begin{vmatrix} K(x_1,y_1) & K(x_1,y_2) & \cdot & \cdot & K(x_1,y_r) \\ K(x_2,y_1) & K(x_2,y_2) & \cdot & \cdot & K(x_2,y_r) \\ \cdot & \cdot & & & \cdot \\ \cdot & \cdot & & & \cdot \\ K(x_r,y_1) & K(x_r,y_2) & \cdot & \cdot & K(x_r,y_r) \end{vmatrix}$$

A basic concept when dealing with variation-diminishing properties is sign-regularity:

SIGN-REGULARITY: *A discrete kernel $K: \mathbb{Z} \times \mathbb{Z} \to \mathbb{R}$ is said to be sign-regular (SR_∞) if all its r-order minors have same sign for every order r from 1 through ∞, i.e. if there exists a sequence of constants $\varepsilon_1, \varepsilon_2, \ldots$ each $+1$ or -1 such that*

$$\varepsilon_r K\begin{pmatrix} x_1, x_2, \ldots, x_r \\ y_1, y_2, \ldots, y_r \end{pmatrix} \geq 0 \qquad (3.27)$$

for all choices of $x_1 < x_2 < \cdots < x_r$ and $y_1 < y_2 < \cdots < y_r$ from $\mathbb{Z}$.

In other words, sign-regularity means that it is impossible to find two minors of same order having opposite signs. If strict inequality holds for all r, then K is said to be strictly sign-regular (SSR_∞). General linear transformations (not necessarily shift-invariant) possessing variation-diminishing properties in the sense that they never increase the number of sign changes in a vector, can be fully characterized in terms of sign-regularity.

CLASSIFICATION OF GENERAL VARIATION-DIMINISHING TRANSFORMATIONS I: *Let A be an $n \times m$ real matrix with $n \geq m$. Then, the linear map A from $\mathbb{R}^m$ to $\mathbb{R}^n$ diminishes variations in sign in the sense that*

$$V^+(A\bar{x}) \leq V^-(\bar{x}) \quad \textit{for all } \bar{x} \in \mathbb{R}^m,\ \bar{x} \neq \bar{0} \qquad (3.28)$$

if and only if A is strictly sign-regular.

The original proof of this powerful theorem, forming the foundation of the theory for variation-diminishing transforms, is given by Schoenberg (1953). Another formulation is possible (Ando 1987) if A is known to be of full rank.

CLASSIFICATION OF GENERAL VARIATION-DIMINISHING TRANSFORMATIONS II: *Let A be an $n \times m$ real matrix of rank m. Then*

$$V^-(A\bar{x}) \leq V^-(\bar{x}) \qquad (3.29)$$

holds for all $\bar{x} \in \mathbb{R}^m$ ($\bar{x} \neq \bar{0}$) if and only if A is sign-regular.

It can be noted that the condition (3.29) is equivalent to the formulation expressed in definition 3.1. Consequently, sign-regularity and full rank are the necessary and sufficient conditions for a kernel to be a potential scale-space kernel. A narrower class of transformations is obtained if all minors are required to be non-negative (Karlin 1968).

TOTAL POSITIVITY: *A discrete kernel* $K\colon \mathbb{Z} \times \mathbb{Z} \to \mathbb{R}$ *is said to be totally positive* (TP_∞) *if all its minors are nonnegative; i.e. if*

$$K\begin{pmatrix} x_1, x_2, \ldots, x_p \\ y_1, y_2, \ldots, y_p \end{pmatrix} \geq 0 \tag{3.30}$$

$$x_1 < x_2 < \cdots < x_p, \quad y_1 < y_2 < \cdots < y_p, \quad p = 1, 2, \ldots, \infty.$$

An important subclass of totally positive kernels appears if the discrete kernel is required to be shift-invariant i.e. if $K(x, y)$ can be written as $k(x - y) = c_{x-y}$.

PÓLYA FREQUENCY SEQUENCE: *A sequence* $\{c_n\}_{n=-\infty}^{\infty}$ *is said to be a Pólya frequency sequence if all minors of the infinite Toeplitz matrix*

$$C = \begin{pmatrix} \cdot & \cdot & \cdot & \cdot & \cdot & \cdot & \cdot \\ \cdot & \cdot & \cdot & \cdot & \cdot & \cdot & \cdot \\ \cdot & \cdot & c_0 & c_{-1} & c_{-2} & \cdot & \cdot \\ \cdot & \cdot & c_1 & c_0 & c_{-1} & \cdot & \cdot \\ \cdot & \cdot & c_2 & c_1 & c_0 & \cdot & \cdot \\ \cdot & \cdot & \cdot & \cdot & \cdot & \cdot & \cdot \\ \cdot & \cdot & \cdot & \cdot & \cdot & \cdot & \cdot \end{pmatrix} \tag{3.31}$$

are non-negative.

The importance of the Pólya frequency sequences becomes apparent when the generating function is required to be convergent, which for example holds if the sum of the filter coefficients is convergent.

NORMALIZED PÓLYA FREQUENCY SEQUENCE: *A Pólya frequency sequence* $\{c_n\}_{n=-\infty}^{\infty}$ *is said to be a normalized if its generating function* $\varphi(z) = \sum_{n=-\infty}^{\infty} c_n z^n \neq 0$ *converges in an annulus* $r < |z| < R$ $(0 < r < 1 < R)$.

According to a theorem by Schoenberg (1948), sign-regularity combined with the Toeplitz structure implies total positivity. Consequently,

CLASSIFICATION OF VARIATION-DIMINISHING CONVOLUTION TRANSFORMATIONS: *The convolution transformation*

$$f_{out}(x) = \sum_{n=-\infty}^{\infty} c_n f_{in}(x - n)$$

is variation-diminishing i.e.

$$V^-(f_{out}) \leq V^-(f_{in})$$

holds for all $\bar{f}_{in}$ if and only if the sequence of filter coefficients $\{c_n\}_{n=-\infty}^{\infty}$ is a normalized Pólya frequency sequence.

In other words, every shift-invariant discrete scale-space kernel corresponds to a normalized Pólya frequency sequence.

There exists a remarkably explicit characterization theorem for the generating function of a PF_∞-sequence. It has been proved in several steps by Edrei and Schoenberg (Schoenberg 1953; Karlin 1968).

CLASSIFICATION OF PÓLYA FREQUENCY SEQUENCES: *An infinite sequence $\{c_n\}_{n=-\infty}^{\infty}$ is a Pólya frequency sequence if and only if its generating function $\varphi(z) = \sum_{n=-\infty}^{\infty} c_n z^n$ is of the form*

$$\varphi(z) = c\, z^k\, e^{(q_{-1}z^{-1}+q_1 z)} \prod_{i=1}^{\infty} \frac{(1+\alpha_i z)(1+\delta_i z^{-1})}{(1-\beta_i z)(1-\gamma_i z^{-1})} \tag{3.32}$$

$$c > 0, \quad k \in \mathbb{Z}, \quad q_{-1}, q_1, \alpha_i, \beta_i, \gamma_i, \delta_i \geq 0, \quad \sum_{i=1}^{\infty}(\alpha_i+\beta_i+\gamma_i+\delta_i) < \infty.$$

The sequence $\{c_n\}_{n=-\infty}^{\infty}$ is normalized if and only if it in addition holds that $\beta_i < 1$ and $\gamma_i < 1$ (Karlin 1968).

3.3.2. Classification of discrete scale-space kernels

The results from the previous section allows for a complete classification of what kernels are scale-space kernels. To summarize, two criteria can be stated; one in terms of minors of the convolution matrix and one in terms of the generating function of the convolution kernel.

THEOREM 3.15. (CLASSIFICATION OF DISCRETE SCALE-SPACE KERNELS) *A discrete kernel $K\colon \mathbb{Z} \to \mathbb{R}$ is a scale-space kernel if and only if the corresponding sequence of filter coefficients $\{K(n)\}_{n=-\infty}^{\infty}$ is a normalized Pólya frequency sequence, i.e. if and only if all minors of the infinite matrix*

$$\begin{pmatrix} \cdot & \cdot & \cdot & \cdot & \cdot & \cdot & \cdot \\ \cdot & \cdot & \cdot & \cdot & \cdot & \cdot & \cdot \\ \cdot & \cdot & K(0) & K(-1) & K(-2) & \cdot & \cdot \\ \cdot & \cdot & K(1) & K(0) & K(-1) & \cdot & \cdot \\ \cdot & \cdot & K(2) & K(1) & K(0) & \cdot & \cdot \\ \cdot & \cdot & \cdot & \cdot & \cdot & \cdot & \cdot \\ \cdot & \cdot & \cdot & \cdot & \cdot & \cdot & \cdot \end{pmatrix} \tag{3.33}$$

are non-negative.

THEOREM 3.16. (CLASSIFICATION OF DISCRETE SCALE-SPACE KERNELS) *An discrete kernel $K\colon \mathbb{Z} \to \mathbb{R}$ is a discrete scale-space kernel if and only if its generating function $\varphi_K(z) = \sum_{n=-\infty}^{\infty} K(n) z^n$ is of the form*

$$\varphi_K(z) = c\, z^k\, e^{(q_{-1}z^{-1}+q_1 z)} \prod_{i=1}^{\infty} \frac{(1+\alpha_i z)(1+\delta_i z^{-1})}{(1-\beta_i z)(1-\gamma_i z^{-1})} \tag{3.34}$$

$$c > 0, \quad k \in \mathbb{Z}, \quad q_{-1}, q_1, \alpha_i, \beta_i, \gamma_i, \delta_i \geq 0,$$

$$\beta_i, \gamma_i < 1, \quad \sum_{i=1}^{\infty} (\alpha_i + \beta_i + \gamma_i + \delta_i) < \infty.$$

(Note that the Fourier transform of the kernel is obtained by replacing z by $e^{-i\theta}$.) The product structure of this expression corresponds to the previously mentioned property that if K_a and K_b are scale-space kernels, then $K_a * K_b$ is also a scale-space kernel. The meanings of the leading factors C and z^k are just rescaling and translation. In $(1 + \alpha_i z)$ and $(1 + \delta_i z^{-1})$ we recognize rewritten versions of the generating functions of two-kernels. The factors in the denominator are Taylor expansions of geometric series, which correspond to moving average processes of the forms $f_{out}(x) = f_{in}(x) + \beta_i f_{out}(x-1)$ and $f_{out}(x) = f_{in}(x) + \gamma_i f_{out}(x+1)$. The exponential factor describes infinitesimal smoothing. Its interpretation will become clearer in the next section, when the discrete scale-space with a continuous scale parameter is derived. To conclude:

COROLLARY 3.17. (PRIMITIVE SMOOTHING TRANSFORMATIONS) *For discrete signals $\mathbb{Z} \to \mathbb{R}$ there are exactly five primitive types of linear and shift-invariant smoothing transformations, of which the last two are trivial:*

- *two-point weighted average or generalized binomial smoothing*

$$\begin{aligned} f_{out}(x) &= f_{in}(x) + \alpha_i\, f_{in}(x-1) \quad (\alpha_i \geq 0), \\ f_{out}(x) &= f_{in}(x) + \delta_i\, f_{in}(x+1) \quad (\delta_i \geq 0), \end{aligned} \tag{3.35}$$

- *moving average or first-order recursive filtering*

$$\begin{aligned} f_{out}(x) &= f_{in}(x) + \beta_i\, f_{out}(x-1) \quad (0 \leq \beta_i < 1), \\ f_{out}(x) &= f_{in}(x) + \gamma_i\, f_{out}(x+1) \quad (0 \leq \gamma_i < 1), \end{aligned} \tag{3.36}$$

- *infinitesimal smoothing or diffusion smoothing, (see section 3.4.2 for further explanation),*
- *rescaling,*
- *translation.*

Moreover, it holds that

COROLLARY 3.18. (DECOMPOSITION PROPERTY OF DISCRETE SCALE-SPACE KERNELS) *A convolution transformation is a smoothing transformation with discrete scale-space properties if and only if it can be decomposed into primitive transformations that are all smoothing transformations possessing scale-space properties.*

This means that the inverse statement of lemma 3.5 is true, and that once a non-smoothing transformation has been performed, that step it is impossible to fully compensate for by further smoothing. Of course, one could in general expect that such further smoothing leads to a signal with a smaller number of local extrema. However, there will always exist some signals for which this is not possible.

For kernels with finite support q_{-1}, q_1, β_i and γ_i must be zero and the infinite product must be replaced with a finite one. Then, the generating function will be reduced to $\varphi_K(z) = c\, z^k \prod_{i=1}^{N}(1+\alpha_i z)(1+\delta_i z^{-1})$, for some finite N, which except for rescaling and translation is the generating function of the class of generalized binomial kernels in proposition 3.6 and 3.7. Hence,

THEOREM 3.19. (CLASSIFICATION OF DISCRETE SCALE-SPACE KERNELS WITH FINITE SUPPORT) *The kernels on the form $*_{i=1}^{n} K_i^{(2)}$, with $K_i^{(2)}$ according to* (3.4)*, are (except for rescaling and translation) the only discrete scale-space kernels with finite support.*

An immediate consequence of this result is that *convolution with a finite scale-space kernel can be decomposed into convolution with kernels having two strictly positive consecutive filter coefficients.* This gives further emphasis to the statement that the generalized binomial kernels are, except for a trivial translation, the only discrete scale-space kernels with finite support. In the symmetric case, the generating function can be further reduced to $\varphi_K(z) = c \prod_{i=1}^{N}(1+\alpha_i z)(1+\alpha_i z^{-1})$, which shows that

COROLLARY 3.20. (SYMMETRIC DISCRETE SCALE-SPACE KERNELS WITH FINITE SUPPORT) *Every symmetric discrete scale-space kernel can be decomposed into convolutions with symmetric three-kernels of type*

$$(a_i, b_i, a_i) \quad \text{where } b_i \geq 2a_i > 0. \tag{3.37}$$

In other words, every symmetric discrete scale-space kernel with finite support has a Fourier transform of the form

$$\psi_K(\theta) = \prod_{i=1}^{N}(b_i + 2a_i\cos(\theta)). \tag{3.38}$$

The representation (3.34), which gives a catalogue of all one-dimensional discrete smoothing kernels, can sometimes be very convenient for further analysis. For example, starting from (3.34) it is almost trivial to show that the Fourier transform of a symmetric discrete scale-space kernel is unimodal and non-negative on the interval $[-\pi, \pi]$. Due to the symmetry it holds that $q_{-1} = q_1$, $\alpha_\nu = \delta_\nu$ and $\beta_\nu = \gamma_\nu$. As a first step replace z with $e^{-i\theta}$ (which gives the Fourier transform). Then, it can be easily shown that each one of the factors $e^{(q_{-1}z^{-1}+q_1 z)}$, $(1+\alpha_\nu z)(1+\delta_\nu z^{-1})$ and $((1-\beta_\nu z)(1-\gamma_\nu z^{-1}))^{-1}$ is a non-negative and unimodal function of θ on $[-\pi, \pi]$. The remaining details are left to the reader.

3.4. Axiomatic construction of discrete scale-space

3.4.1. Discrete scale-space with discrete scale parameter

With the classification result from the previous section in mind, an apparent way to obtain a multi-scale representation of a discrete signal f is by defining a set of discrete functions L_i ($i \in [0..n]$), where $L_0 = f$ and each coarser level is calculated by convolution from the previous one

$$L_i = K_{i \leftarrow i-1} * L_{i-1} \quad (i = 1 \ldots n). \tag{3.39}$$

The kernels $K_{i \leftarrow i-1}$ should be appropriately selected scale-space kernels corresponding to suitable amounts of blurring. The scale-space condition for each kernel guarantees that signals at coarser levels of scale (larger value of i) do not contain more structure than signals at finer levels of scale. This leads to a sampled scale-space with a *discrete* scale parameter. Combined with a sub-sampling operator it provides a possible theoretical basis for pyramid representations.[7] However, there is still one problem that needs to be solved. How should the kernels or scale-levels be selected *a priori* in order to give a sufficiently dense sampling in scale?

3.4.2. Discrete scale-space with continuous scale parameter

The goal of this section is to tie together scale-space kernels corresponding to different degrees of smoothing in a systematic manner such that a *continuous* scale parameter can be introduced. The concept of a continuous scale parameter is of considerable importance, since the multi-scale representation will no longer be locked to a fixed set of pre-determined discrete scale levels. It makes it possible to smooth signals with arbitrary amounts of blurring, which will certainly make it easier to locate and trace events in scale-space. Of course, it is impractical to generate the representations at all levels of scale in a real implementation. However,

[7] When dealing with pyramids there are other problems arising due to the decreasing number of grid points, aliasing and the fixed scale sampling, which may also influence the design criteria. Those issues are not covered by this treatment.

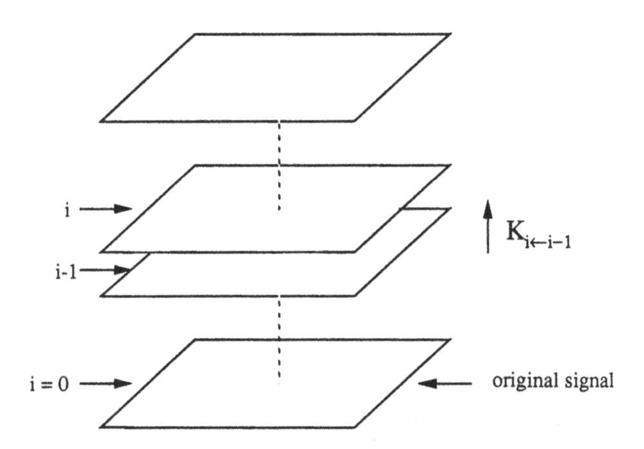

Figure 3.9. Given the classification of discrete scale-space kernels it is straightforward, at least in principle, to construct a scale-space representation associated with a discrete scale parameter: Start from the original signal and select a set of kernels $K_{i+1 \leftarrow i}$, each one describing the transformation from a scale level i to the next coarser level $i+1$, where every such kernel should be a discrete scale-space kernel. Then is is guaranteed that any coarser level of scale j does not contain more local extrema any a finer level of scale i provided that $j \geq i$.

the important idea is that, in contrast to the pyramid representations where the scale levels are fixed, *any* level of scale can be calculated if desired, which allows for *adaptive scale sampling.*

In this part of the presentation, we will not consider the question about how to choose a suitable set of scale levels in a practical case (such analysis will be given in section 8.1.1 and section 9.2). Imagine for instance that the task is to trace events, like local extrema, zero-crossings, edges, or convex and concave regions, as the blurring proceeds in scale-space. To analyze this scale-space behaviour, the continuum of multi-scale representations must be sampled at some levels of scale. It is certainly a non-trivial problem to make an appropriate selection of these levels, and it seems plausible that the sampling rate along the scale direction should depend upon the signal under study. If in some scale interval the representation varies relatively smoothly, then it should be possible to use a larger scale step than if it were strongly varying. Thus, one is led to methods that automatically regulate the scale step, based on the local structure of the signal as a function of the spatial and scale coordinates. The point with a scale-space with a continuous scale parameter is that it provides a theoretical framework for the development of such algorithms, in which the scale steps can be varied arbitrarily. There is no need for selecting any set of scale levels *in advance.* Instead, the decision can be left open to the actual situation.

3.4.2.1. Scale-space formulation

Given any signal $f\colon \mathbb{Z} \to \mathbb{R}$, start from the axioms given in section 3.1, and postulate that the scale-space $L\colon \mathbb{Z} \times \mathbb{R}_+ \to \mathbb{R}$ should be generated by convolution with a one-parameter family of kernels $L\colon \mathbb{Z} \times \mathbb{R}_+ \to \mathbb{R}$, i.e., let $L(x;\ 0) = f(x)$ and

$$L(x;\ t) = \sum_{n=-\infty}^{\infty} T(n;\ t) f(x-n). \tag{3.40}$$

This form of the smoothing formula reflects the requirements about *linear shift-invariant smoothing* and a *continuous scale parameter*. The amount of structure in a signal must not increase with scale. This means that for any $t_2 > t_1$ the number of local extrema in $L(x;\ t_2)$ must not exceed the number of local extrema in $L(x;\ t_1)$. In particular, by setting t_1 to zero it follows that each $T(\cdot;\ t)$ must be a scale-space kernel.

To simplify the analysis, impose a *semi-group* requirement

$$T(\cdot;\ s) * T(\cdot;\ t) = T(\cdot;\ s+t) \tag{3.41}$$

on the family of kernels. This property means that any representation $L(\cdot;\ t_2)$ at a coarse level t_2 can be computed from the representation $L(\cdot;\ t_1)$ at a finer level t_1 ($t_2 > t_1$) by convolution with a kernel *from the* one-parameter *family*. In summary,

$$\begin{aligned} L(\cdot;\ t_2) &= \{\text{definition}\} = T(\cdot;\ t_2) * f = \\ &= \{\text{semi-group}\} = (T(\cdot;\ t_2 - t_1) * T(\cdot;\ t_1)) * f \\ &= \{\text{associativity}\} = T(\cdot;\ t_2 - t_1) * (T(\cdot;\ t_1) * f) \\ &= \{\text{definition}\} = T(\cdot;\ t_2 - t_1) * L(\cdot;\ t_1). \end{aligned} \tag{3.42}$$

Since each $T(\cdot;\ t)$ is required to be a scale-space kernel, the semi-group property ensures that the scale-space property holds between any two levels of scale. It also means that all scale levels will be treated in a uniform manner.

It will now be shown that the conditions mentioned, combined with a *normalization* criterion

$$\sum_{n=-\infty}^{\infty} T(n;\ t) = 1, \tag{3.43}$$

and a *symmetry* constraint $T(-n;\ t) = T(n;\ t)$ determine the family of kernels up to a positive scaling parameter[8] α. One obtains,

$$T(n;\ t) = e^{-\alpha t} I_n(\alpha t), \tag{3.44}$$

[8]For simplicity, the parameter α, which only affects the scaling of the scale parameter, will be set to 1 after this section.

where I_n are the modified Bessel functions of integer order. These functions with real arguments are related to the ordinary Bessel functions J_n of integer order with purely imaginary arguments by

$$I_n(t) = I_{-n}(t) = (-i)^n J_n(it) \quad (n \geq 0,\ t > 0). \tag{3.45}$$

THEOREM 3.21. (SCALE-SPACE FOR DISCRETE SIGNALS; NECESSITY AND SUFFICIENCY) *Given any one-dimensional signal* $f\colon \mathbb{Z} \to \mathbb{R}$ *let* $L\colon \mathbb{Z} \times \mathbb{R}_+ \to \mathbb{R}$ *be a one-parameter family of functions defined by* $L(x;\ 0) = f(x)$ $(x \in \mathbb{Z})$ *and*

$$L(x;\ t) = \sum_{n=-\infty}^{\infty} T(n;\ t) f(x-n) \tag{3.46}$$

$(x \in \mathbb{Z},\ t \in \mathbb{R}_+)$, *where* $T\colon \mathbb{Z} \times \mathbb{R}_+ \to \mathbb{R}$ *is a one-parameter family of symmetric functions satisfying the semi-group property* $T(\cdot;\ s) * T(\cdot;\ t) = T(\cdot;\ s+t)$ *and the normalization criterion* $\sum_{n=-\infty}^{\infty} T(n;\ t) = 1$. *For all signals* f, *it is required that if* $t_2 > t_1$ *then the number of local extrema (zero-crossings) in* $L(x;\ t_2)$ *must not exceed the number of local extrema (zero-crossings) in* $L(x;\ t_1)$. *Then, necessarily (and sufficiently),*

$$T(n;\ t) = e^{-\alpha t} I_n(\alpha t) \tag{3.47}$$

for some non-negative real α, *where* I_n *are the modified Bessel functions of integer order.*

Proof. As mentioned above, every kernel $T(n;\ t)$ must be a scale-space kernel. A theorem by Karlin (1968) states that the only semi-group of normalized Pólya frequency sequences has a generating function of the form

$$\varphi(z) = e^{t(az^{-1}+bz)} \tag{3.48}$$

where $t > 0$ and $a, b \geq 0$. This result, which forms the basis of the proof can be easily understood from theorem 3.16. If a family $h(\cdot;\ t)$ possesses the semi-group property $h(\cdot;\ s) * h(\cdot;\ t) = h(\cdot;\ s+t)$ then its generating function must necessarily obey the relation $\varphi_{h(\cdot;\ s)}\varphi_{h(\cdot;\ t)} = \varphi_{h(\cdot;\ s+t)}$ for all non-negative s and t. This excludes the factors z^k, $(1+\alpha_i z)$, $(1+\delta_i z^{-1})$, $(1-\beta_i z)$ and $(1-\gamma_i z^{-1})$ from (3.32). What remains are the constant and the exponential factors. The argument of the exponential factor must also be linear in t in order to fulfill the adding property of the scale parameters of the kernels under convolution.

Due to the symmetry the generating function must satisfy $\varphi_h(z^{-1}) = \varphi_h(z)$, which in our case leads to $a = b$. For simplicity, let $a = b = \frac{\alpha}{2}$, and we get the generating function for the modified Bessel functions of integer order, see Abramowitz and Stegun (1964)

$$\varphi_t(z) = e^{\frac{\alpha t}{2}(z^{-1}+z)} = \sum_{n=-\infty}^{\infty} I_n(\alpha t)\, z^n. \tag{3.49}$$

A normalized kernel is obtained if $T\colon \mathbb{Z} \times \mathbb{R}_+ \to \mathbb{R}$ is defined by $T(n;\ t) = e^{-\alpha t} I_n(\alpha t)$. Set z to 1 in the generating function (3.49). Then, it follows that $\sum_{n=-\infty}^{\infty} I_n(\alpha t) = e^{\alpha t}$, which means that $\sum_{n=-\infty}^{\infty} T(n;\ t) = 1$. The semi-group property is trivially preserved after normalization. □

This theorem, which is one of the main results of this chapter, provides an explicit and controlled method to preserve structure in the spatial domain when smoothing a discrete signal to coarser level of scales. The kernel $T(n;\ t) = e^{-\alpha t} I_n(\alpha t)$ possesses similar properties in the discrete case as those who make the ordinary Gaussian kernel special in the continuous case. Therefore it is natural to refer to it as *the discrete analogue of the Gaussian kernel;* see also Norman (1960).

DEFINITION 3.22. (DISCRETE ANALOGUE OF THE GAUSSIAN KERNEL) *The kernel $T\colon \mathbb{Z} \times \mathbb{R}_+ \to \mathbb{R}$ given by $T(n;\ t) = e^{-\alpha t} I_n(\alpha t)$ is called the discrete analogue of the Gaussian kernel, or shorter, the discrete Gaussian.*

3.4.2.2. Properties of the discrete analogue of the Gaussian kernel

Now, some elementary properties of this kernel will be pointed out. In the special case when $t = 0$ it holds that

$$I_n(0) = I_{-n}(0) = \delta(n) = \begin{cases} 1 & \text{if } n = 0, \\ 0 & \text{otherwise,} \end{cases} \tag{3.50}$$

which means that $T(\cdot;\ 0) = \delta(\cdot)$ and the convolution expression (3.46) with T according to (3.47) is valid for $t = 0$ as well. Observe that when the scale parameter tends to zero, the continuous Gaussian kernel tends to the continuous delta function while the discrete analogue of the Gaussian kernel instead tends to the discrete delta function.

For large t on the other hand, it holds that the discrete analogue of the Gaussian kernel approaches the continuous Gaussian. This can be understood by studying an asymptotic expression for the modified Bessel functions for large t, see Abramowitz and Stegun (1964) (9.7.1)

$$I_n(t) = \frac{e^t}{\sqrt{2\pi t}}\left(1 - \frac{4n^2-1}{8t} + O\left(\frac{1}{t^2}\right)\right), \tag{3.51}$$

which shows that

$$T(n;\ t) - g(n;\ t) = e^{-t} I_n(t) - \frac{1}{\sqrt{2\pi t}} e^{-n^2/2t} = \frac{1}{\sqrt{2\pi t}} (\frac{1}{8t} + O(\frac{1}{t^2})).$$

If the relation (3.49) is multiplied with the factor e^{-t} and if z is replaced with $e^{-i\theta}$, then the result is an analytical expression for the Fourier transform of $T(n;\ t)$.

PROPOSITION 3.23. (FOURIER TRANSFORM OF DISCRETE GAUSSIAN) *The Fourier transform of the kernel $T(n;\ t) = e^{-\alpha t} I_n(\alpha t)$ is*

$$\psi_T(\theta) = \sum_{n=-\infty}^{\infty} T(n;\ t) e^{-in\theta} = e^{\alpha t(\cos\theta - 1)} \tag{3.52}$$

For completeness, we give the variance of this kernel as well

PROPOSITION 3.24. (VARIANCE OF THE DISCRETE GAUSSIAN KERNEL) *The variance of the kernel $T(n;\ t) = e^{-\alpha t} I_n(\alpha t)$ is*

$$\sum_{n=-\infty}^{\infty} n^2 T(n;\ t) = t \tag{3.53}$$

Proof. This can be easily shown from a recurrence relation for the modified Bessel functions (Abramowitz and Stegun 1964)

$$I_{n-1}(t) - I_{n+1}(t) = \frac{2n}{t} I_n(t), \tag{3.54}$$

and the normalization condition. It holds that

$$\begin{aligned}
\sum_{n=-\infty}^{\infty} n^2 e^{-t} I_n(t) &= \sum_{n=-\infty}^{\infty} n^2 e^{-t} \frac{t}{2n} (I_{n-1}(t) - I_{n+1}(t)) \\
&= \frac{te^{-t}}{2} \left(\sum_{n=-\infty}^{\infty} n I_{n-1}(t) - \sum_{n=-\infty}^{\infty} n I_{n+1}(t) \right) \\
&= \frac{te^{-t}}{2} \left(\sum_{m=-\infty}^{\infty} (m+1) I_m(t) - \sum_{m=-\infty}^{\infty} (m-1) I_m(t) \right) \\
&= \frac{te^{-t}}{2} \sum_{m=-\infty}^{\infty} (m+1-m+1) I_m(t) = t \sum_{m=-\infty}^{\infty} T(m;\ t).
\end{aligned}$$

□

Compare with the variance of the continuous Gaussian kernel, which is $\sigma^2 = t$. All moments of odd order are of course zero due to symmetry.

3.5. Axiomatic construction of continuous scale-space

If similar arguments are applied in the continuous case, then the continuous Gaussian kernel is obtained. To give a background to the analysis, some important theorems from the theory of variation-diminishing convolution transformations for continuous signals will first be briefly reviewed. Then, those results will be used for stating an equivalent formulation of the scale-space for continuous signals.

3.5.1. Background

Let $S^-(f)$ denotes the number of sign changes in a function f defined by

$$S^-(f) = \sup V^-(f(t_1), f(t_2), \ldots, f(t_m)), \tag{3.55}$$

where the supremum is extended over all sets $t_1 < t_2 < \cdots < t_m$ ($t_i \in \mathbb{R}$), m is arbitrary but finite, and $V^-(\bar{x})$ denotes the number of sign changes in a vector $\bar{x}$ defined in section 3.3.1. The transformation

$$f_{out}(\eta) = \int_{\xi=-\infty}^{\infty} f_{in}(\eta - \xi)\, dG(\xi), \tag{3.56}$$

where G is a distribution function, is said to be variation-diminishing if

$$S^-(f_{out}) \leq S^-(f_{in}) \tag{3.57}$$

holds for all continuous and bounded f_{in}. The continuous correspondence to Pólya frequency sequences is called Pólya frequency functions. Also this concept is defined in terms of total positivity and shift invariance (Karlin 1968).

> TOTAL POSITIVITY (CONTINUOUS CASE): *A continuous kernel $K(x, y)\colon \mathbb{R} \times \mathbb{R} \to \mathbb{R}$ is said to be totally positive (TP_∞) if all minors, of every order r from $1, 2$ to infinity, are non-negative, i.e. if there for all choices of $x_1 < x_2 < \cdots < x_r$ and $y_1 < y_2 < \cdots < y_r$ from $\mathbb{R}$ holds that*
>
> $$K\begin{pmatrix} x_1, x_2, \ldots, x_r \\ y_1, y_2, \ldots, y_r \end{pmatrix} \geq 0 \tag{3.58}$$
>
> $$x_1 < x_2 < \cdots < x_r, \quad y_1 < y_2 < \cdots < y_r, \quad r = 1, 2, \ldots, \infty.$$
>
> PÓLYA FREQUENCY FUNCTIONS: *A function $k\colon \mathbb{R} \to \mathbb{R}$ is said to be a Pólya frequency function if the function $K\colon \mathbb{R}\times\mathbb{R} \to \mathbb{R}$ given by $K(x, y) = k(x - y)$ is totally positive.*

The variation-diminishing property of continuous convolution transformations on the form (3.56) can be completely characterized in terms of Pólya frequency functions. The following results are due to Schoenberg (1950), see also Hirschmann and Widder (1955), or Karlin (1968).

CLASSIFICATION OF CONTINUOUS VARIATION-DIMINISHING TRANSFORMATIONS I: *The transformation* (3.56) *is variation-diminishing if and only if G is either, up to a sign change, a cumulative Pólya frequency function*

$$G(t) = \epsilon \int_{u=-\infty}^{t} k(u)\, du, \tag{3.59}$$

where $\epsilon = \pm 1$ and $k(u)$ is a Pólya frequency function, or else G is a step function with only one jump.

CLASSIFICATION OF CONTINUOUS VARIATION-DIMINISHING TRANSFORMATIONS II: *The transformation* (3.56) *is variation-diminishing if and only if G has a bilateral Laplace-Stieltjes transform of the form*

$$\int_{\xi=-\infty}^{\infty} e^{-s\xi}\, dG(\xi) = C\, e^{\gamma s^2 + \delta s} \prod_{i=1}^{\infty} \frac{e^{a_i s}}{1 + a_i s} \quad (-c < Re(s) < c) \tag{3.60}$$

for some $c > 0$, where $C \neq 0$, $\gamma \geq 0$, δ and a_i are real, and $\sum_{i=1}^{\infty} a_1^2$ is convergent.

Interpreted in the spatial domain, these results imply that for continuous signals there are four primitive types of linear and shift-invariant smoothing transformations; convolution with the *Gaussian kernel,*

$$h(\xi) = e^{-\gamma \xi^2}, \tag{3.61}$$

convolution with the *truncated exponential functions,*

$$h(\xi) = \begin{cases} e^{-|\lambda|\xi} & \xi \geq 0, \\ 0 & \xi < 0, \end{cases} \qquad h(\xi) = \begin{cases} e^{|\lambda|\xi} & \xi \leq 0, \\ 0 & \xi > 0, \end{cases} \tag{3.62}$$

as well as trivial *translation* and *rescaling.* Moreover, it means that a shift-invariant linear transformation is a smoothing operation if and only if it can be decomposed into these primitive operations.

3.5.2. Continuous scale-space with continuous scale parameter

These results show that the Pólya frequency functions are the natural functions to start from when defining a scale-space representation for continuous signals, or equivalently, that *the Pólya frequency functions are the continuous scale-space kernels.* If again a semi-group requirement and a symmetry constraint are imposed on these kernels the Gaussian kernel will remain as the only candidate.

THEOREM 3.25. (SCALE-SPACE FOR CONTINUOUS SIGNALS; NECESSITY AND SUFFICIENCY) *Given any one-dimensional continuous signal* $f: \mathbb{R} \to \mathbb{R}$*, let* $L: \mathbb{R} \times \mathbb{R}_+ \to \mathbb{R}$ *be a one-parameter family of functions defined by* $L(\cdot;\ 0) = f(\cdot)$ *and*

$$L(x;\ t) = \int_{\xi=-\infty}^{\infty} g(\xi;\ t)\, f(x-\xi)\, d\xi \tag{3.63}$$

($x \in \mathbb{R}, t > 0$)*, where* $g: \mathbb{R} \times \mathbb{R}_+ \to \mathbb{R}$ *is a one-parameter family of symmetric functions satisfying the semi-group property* $g(\cdot;\ s) * g(\cdot;\ t) = g(\cdot;\ s+t)$ *and the normalization criterion*

$$\int_{\xi=-\infty}^{\infty} g(\xi;\ t)\, d\xi = 1. \tag{3.64}$$

For all signals f *it is required that if* $t_2 > t_1$ *then the number of local extrema*[9] *(zero-crossings) in* $L(x;\ t_2)$ *must not exceed the number of local extrema (zero-crossings) in* $L(x;\ t_1)$*. Suppose also that* $g(\xi;\ t)$ *is Borel-measurable as a function of* t*. Then, necessarily (and sufficiently),*

$$g(\xi;\ t) = \frac{1}{\sqrt{2\pi\alpha t}}\, e^{-\xi^2/2\alpha t} \tag{3.65}$$

for some non-negative real α*.*

Proof. According to the above treatment every kernel $g(\cdot;\ t)$ must be a continuous scale-space kernel, that is, a Pólya frequency function. A theorem by Karlin (1968) shows that these conditions uniquely define the Gaussian family of kernels.

> CLASSIFICATION OF SEMI-GROUPS OF PÓLYA FREQUENCY FUNCTIONS: Let $g: \mathbb{R} \times \mathbb{R}_+ \to \mathbb{R}$ denote a one-parameter family of Pólya frequency functions integrable on the real axis and satisfying the semi-group property
>
> $$g(\cdot;\ t_1) * g(\cdot;\ t_2) = g(\cdot;\ t_1 + t_2) \tag{3.66}$$
>
> Suppose also that $g(x;\ t))$ is Borel-measurable as a function of t. Then, necessarily (where $x \in \mathbb{R}$, $t \in \mathbb{R}_+$, $\delta \in \mathbb{R}$)
>
> $$g(x;\ t) = \frac{1}{\sqrt{2\pi\alpha t}}\, e^{-(x-\delta t)^2/2\alpha t}. \tag{3.67}$$

[9] In the continuous case, the variation-diminishing property is normally expressed in terms of zero-crossings. Thus, this formulation is valid only if the differentiation operator commutes with the convolution operator. If problems occur, the primary formulation should be interpreted in terms of zero-crossings instead.

Because of the symmetry constraint the constant δ must be zero. The constant α only affects the scaling of the scale parameter. Hence, it can be set to one without loss of generality. □

Consequently, this theorem provides an alternative formulation of the one-dimensional scale-space theory for continuous signals, leading to the same result as the work by Koenderink (1984) and Babaud *et al.* (1986), as well as further support for the firm belief that theorem 3.21 states the canonical way to define a scale-space for discrete signals. The assumption of Borel-measurability means no important restriction. It is well-known that all continuous functions are Borel-measurable.

3.6. Numerical approximations of continuous scale-space

Now, some numerical approximations will be considered, which are close at hand for the convolution integral (1.3) and the diffusion equation (1.5). Using the classification results derived in previous sections, it will be investigated if the corresponding transformations possess scale-space properties in the discrete sense. One aim is to analyze the previously commonly adapted approach where the filter coefficients are set to sampled values of the Gaussian kernel. It will be shown that some undesired effects occur, mainly due to the fact that the semi-group property does not hold after discretization. It will also be shown that the transformation obtained by convolution with the presented discrete analogue of the Gaussian kernel is equivalent to the solution of a semi-discretized version of the diffusion equation. This result as well as some other interconnections between the scale-space formulations for continuous and discrete signals provide further motivation for the selection of the discrete analogue of the Gaussian kernel as the canonical discrete scale-space kernel.

The rendering is of necessity somewhat technical, and the details can be skipped by the hasty reader without loss of continuity.

3.6.1. The sampled Gaussian kernel

Presumably, the most obvious way to approximate the convolution integral

$$L(x;t) = \int_{\xi=-\infty}^{\infty} \frac{1}{\sqrt{2\pi t}}\, e^{-\xi^2/2t}\, f(x-\xi)\, d\xi \qquad (3.68)$$

numerically is by the rectangle rule of integration. Provided that no truncation of the infinite integration interval is performed, this leads to the approximation

$$\tilde{L}(x;\ t) = \sum_{n=-\infty}^{\infty} \frac{1}{\sqrt{2\pi t}}\, e^{-n^2/2t}\, f(x-n), \qquad (3.69)$$

which shows discrete convolution with the sampled Gaussian kernel. It will be demonstrated that this representation may lead to undesirable effects. From the definitions of PF_∞-functions and PF_∞-sequences in terms of minors it is clear that

LEMMA 3.26. (A SAMPLED PF_∞-FUNCTION IS A PF_∞-SEQUENCE) *Uniform sampling of a continuous scale-space kernel gives a discrete scale-space kernel.*

Therefore, since the Gaussian kernel is a PF_∞ function, it follows that the transformation from the zero level $L(\cdot;\ 0)$ to a higher level never increases the number of local extrema (zero-crossings). However, it will be shown that the transformation from an arbitrary low level $\tilde{L}(\cdot;\ t_1)$ to an arbitrary higher level $\tilde{L}(\cdot;\ t_2)$ is in general *not* a scale-space transformation. Thus, it is not guaranteed that the amount of structure will decrease monotonically with scale. More precisely,

PROPOSITION 3.27. (SCALE-SPACE PROPERTIES OF THE SAMPLED GAUSSIAN KERNEL) *The transformation from a low level $t_1 \geq 0$ to an arbitrary higher level $t_2 > t_1$ in the representation* (3.69), *generated by discrete convolution with the sampled Gaussian kernel, is a scale-space transformation if and only if either t_1 is zero or the ratio t_2/t_1 is an odd integer.*

Proof. Assume that the "scale-space" for a discrete signal is constructed by convolution with the sampled Gaussian kernel, i.e. given a discrete signal $f\colon \mathbb{Z} \to \mathbb{R}$ define the family of functions $\tilde{L}\colon \mathbb{Z} \times \mathbb{R}_+ \to \mathbb{R}$ by $\tilde{L}(x;\ 0) = f(x)$ $(x \in \mathbb{Z})$ and for $t > 0$ by

$$\tilde{L}(x;\ t) = \sum_{n=-\infty}^{\infty} g(n;\ t)\, f(x-n). \tag{3.70}$$

where

$$g(n;\ t) = \frac{1}{\sqrt{2\pi t}}\, e^{-n^2/2t}. \tag{3.71}$$

We will make use of an expression for the generating function for the discrete kernel corresponding to the sampled Gaussian. For simplicity, let $q_t = e^{-\frac{1}{2t}}$. It can be shown (Mumford 1983) that

$$\begin{aligned}\varphi_t(z) &= \sum_{n=-\infty}^{\infty} g(n;\ t)\ z^n = \frac{1}{\sqrt{2\pi t}} \sum_{n=-\infty}^{\infty} q_t^{n^2}\ z^n \\ &= C_t \prod_{n=0}^{\infty} (1 + q_t^{2n+1} z)(1 + q_t^{2n+1} z^{-1}),\end{aligned}$$

where

$$C_t = \frac{1}{\sqrt{2\pi t}} \prod_{n=1}^{\infty} (1 - q_t^{2n}). \tag{3.72}$$

Comparison with the complete characterization of the generating function of a discrete scale-space kernel (3.34) in theorem 3.16 shows that the sampled Gaussian kernel is a discrete scale-space kernel. This constitutes another proof of the property that for any signal f the number of local extrema in $\tilde{L}(x;\ t)$ $(t > 0)$ does not exceed the number of local extrema in f. However, it will now be shown that this scale-space property does *not* hold between two *arbitrary* levels.

Let t_1 and t_2 be two levels $(t_2 > t_1 > 0)$ of the representation (3.70), and let φ_{in} be the generating function of the input signal. Then, the generating functions of $\tilde{L}(x;\ t_1)$ and $\tilde{L}(x;\ t_2)$ are

$$\varphi_{\tilde{L}_1}(z) = \varphi_{t_1}(z)\, \varphi_{in}(z), \qquad \varphi_{\tilde{L}_2}(z) = \varphi_{t_2}(z)\, \varphi_{in}(z). \tag{3.73}$$

Let φ_{diff} describe the transformation from $\tilde{L}(x;\ t_1)$ to $\tilde{L}(x;\ t_2)$. Thus,

$$\varphi_{\tilde{L}_2}(z) = \varphi_{diff}(z)\, \varphi_{\tilde{L}_1}(z). \tag{3.74}$$

Combination of (3.73), (3.74) and (3.72) gives

$$\varphi_{diff}(z) = \frac{\varphi_{\tilde{L}_2}(z)}{\varphi_{\tilde{L}_1}(z)} = \frac{C_{t_2}}{C_{t_1}} \cdot \frac{\prod_{m=0}^{\infty}(1 + q_{t_2}^{2m+1} z)(1 + q_{t_2}^{2m+1} z^{-1})}{\prod_{n=0}^{\infty}(1 + q_{t_1}^{2n+1} z)(1 + q_{t_1}^{2n+1} z^{-1})}. \tag{3.75}$$

According to the complete characterization of scale-space kernels it follows that the corresponding kernel is a scale-space kernel if and only

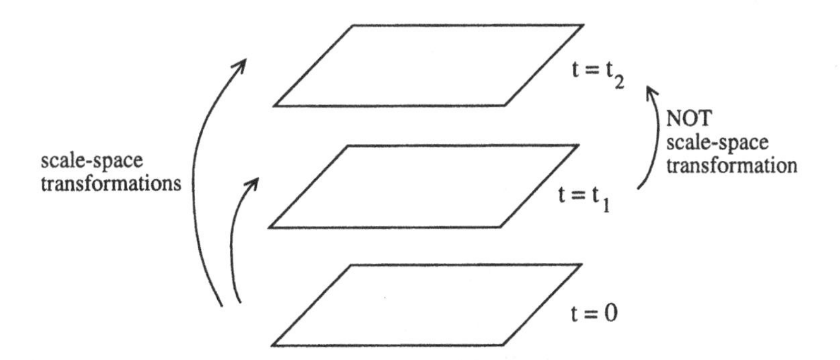

Figure 3.10. In the "scale-space representation" produced by discrete convolution with the sampled Gaussian kernel, the transformation from the zero level to any coarser level of scale is always a scale-space transformation. However, the transformation between two arbitrary levels is in general not a scale-space transformation.

if (3.75) can be written on the form (3.32). Then, for each factor $(1 + q_{t_1}^{2n+1} z^{\pm 1})$ in the denominator there must exist a corresponding factor in the numerator $(1 + q_{t_2}^{2m+1} z^{\pm 1})$, i.e for each n there must exist an m such that

$$q_{t_1}^{2n+1} = q_{t_2}^{2m+1}. \tag{3.76}$$

Insertion of $q_{t_i} = e^{-1/2t_i}$ and reduction gives the necessary and sufficient requirement

$$2m = \frac{t_2}{t_1}(2n+1) - 1. \tag{3.77}$$

It is clear that this relation cannot hold for all $n \in \mathbb{Z}$ if t_1 and t_2 are chosen arbitrarily. The transformation from $\tilde{L}(x;\, t_1)$ to $\tilde{L}(x;\, t_2)$ $(t_2 > t_1)$ is a scale-space transformation if and only if the ratio t_2/t_1 is an odd integer. □

The result constitutes an example of the fact that properties derived in the continuous case might be violated after discretization. The main reason why the scale-space property fails to hold between two arbitrary levels is because *the semi-group property of the Gaussian kernel is not preserved after discretization.*[10]

3.6.2. Discretized diffusion equation

The scale-space family generated by (3.40) and (3.44) can be interpreted in terms of a discretized version of the diffusion equation.

THEOREM 3.28. (DIFFUSION FORMULATION OF DISCRETE SCALE-SPACE) *Given a discrete signal $f\colon \mathbb{Z} \to \mathbb{R}$ in l_1 let $L\colon \mathbb{Z} \times \mathbb{R}_+ \to \mathbb{R}$ be the discrete scale-space representation given by*

$$L(x;\, t) = \sum_{n=-\infty}^{\infty} T(n;\, t)\, f(x-n), \tag{3.78}$$

where $T\colon \mathbb{Z} \times \mathbb{R}_+ \to \mathbb{R}$ is the discrete analogue of the Gaussian kernel. Then, L satisfies the solution of the system of ordinary differential equations ($x \in \mathbb{Z}$)

$$\partial_t L(x;\, t) = \tfrac{1}{2}(L(x+1;\, t) - 2L(x;\, t) + L(x-1;\, t)) \tag{3.79}$$

with initial conditions $L(x;\, 0) = f(x)$, i.e., the system of differential equations obtained if the diffusion equation (1.5) is discretized in space, but solved analytically in time.

[10]This means that if a representation at a level $t_2 > 0$ is computed via an intermediate level t_1 $(0 < t_1 < t_2)$ by application of the approximation formula (3.70) in two steps, then computation does yield the same result as if it would have been computed directly from the original signal.

Proof. From the relation

$$2I'_n(t) = I_{n-1}(t) + I_{n+1}(t) \tag{3.80}$$

for modified Bessel functions (Abramowitz and Stegun 1964), it can be easily shown that $T(n;\ t) = e^{-t}I_n(t)$ satisfies:

$$\begin{aligned}\partial_t T(n;\ t) &= \partial_t(e^{-t}\,I_n(t)) = e^{-t}I'_n(t) - e^{-t}I_n(t)\\ &= \frac{e^{-t}}{2}(I_{n-1}(t) + I_{n+1}(t)) - e^{-t}I_n(t)\\ &= \tfrac{1}{2}(T(n-1;\ t) - 2T(n;\ t) + T(n+1;\ t)),\end{aligned} \tag{3.81}$$

which in turn means that

$$\begin{aligned}\partial_t L(x;\ t) &= \partial_t \sum_{n=-\infty}^{\infty} T(n;\ t)\, f(x-n) = \sum_{n=-\infty}^{\infty} \partial_t T(n;\ t)\, f(x-n)\\ &= \sum_{n=-\infty}^{\infty} \tfrac{1}{2}\,(T(n-1;\ t) - 2T(n;\ t) + T(n+1;\ t)) f(x-n)\\ &= \tfrac{1}{2}(L(x-1;\ t) - 2L(x;\ t) + L(x+1;\ t)).\end{aligned}$$

The regularity condition on f justifies the change of order between differentiation and infinite summation. □

This provides another motivation for the selection of $T(n;\ t) = e^{-t}I_n(t)$ as the canonical discrete scale-space kernel. If (3.79) is further discretized in scale using Euler's method, this gives the iteration formula

$$L_i^{k+1} = \tfrac{1}{2}\,\Delta t\, L_{i+1}^k + (1-\Delta t)\, L_i^k + \tfrac{1}{2}\,\Delta t\, L_{i-1}^k, \tag{3.82}$$

where the subscripts denote the spatial coordinates and the superscripts represent the iteration indices. Equivalently, one iteration with this formula can be described as discrete convolution with the three-kernel

$$\left(\tfrac{1}{2}\,\Delta t,\quad 1-\Delta t,\quad \tfrac{1}{2}\Delta t\right). \tag{3.83}$$

Proposition 3.13 states that this kernel is a scale-space kernel if and only if

$$\Delta t \leq \tfrac{1}{2}, \tag{3.84}$$

which is a stronger condition on Δt than induced by the stability criterion for Euler's forward method (Dahlquist 1974; Strang 1986). From corollary 3.20 it follows that all symmetric scale-space kernels with finite support can be derived from kernels of this latter form. Hence, they provide a possible set of primitive kernels for the scale-space with a discrete scale parameter discussed in section 3.4.1.

PROPOSITION 3.29. (DIFFUSION EQUATION AND DISCRETE SCALE-SPACE KERNELS) *All symmetric discrete scale-space kernels with finite support arise from repeated application of the discretization of the diffusion equation* (3.82), *using if necessary different* $\Delta t_k \in [0, \frac{1}{2}]$.

In many applications the scale step in multi-scale representations with discrete scale parameter is selected such that $\Delta t = \frac{1}{2}$. Note, however, that for any $0 \leq \Delta t \leq \frac{1}{2}$ the kernel given by (3.83) is a discrete scale-space kernel. Hence, it enables a finer sampling in scale also for the scale-space with discrete scale parameter.

It is not too difficult to derive the analytical solution to the system of scale-continuous equations (3.79). Assume that we want to compute the scale-space representation for a fixed value of t. We can use the discretization (3.82) with n steps in the scale-direction such that the step size $\Delta t = t/n$ satisfies (3.84). Since each iteration step consists of a linear convolution, the final solution can equivalently be obtained by convolution with the composed kernel

$$K_{composed} = *_{i=1}^{n} K_{step}. \tag{3.85}$$

Let us derive an asymptotic expression for its generating function. The generating function for the transformation corresponding to one iteration with the formula (3.82) is

$$\varphi_{step}(z) = \tfrac{1}{2}\,\Delta t\, z^{-1} + (1 - \Delta t) + \tfrac{1}{2}\,\Delta t\, z, \tag{3.86}$$

and the generating function of the composed kernel describing the transformation from the scale zero to scale t is

$$\begin{aligned} \varphi_{composed,n}(z) &= (\varphi_{step}(z))^n = (\tfrac{1}{2}\,\Delta t\, z^{-1} + (1 - \Delta t) + \tfrac{1}{2}\,\Delta t\, z)^n \\ &= \left(1 + \frac{t}{n}(\frac{z^{-1}}{2} - 1 + \frac{z^1}{2})\right)^n. \end{aligned} \tag{3.87}$$

Since $\lim_{n\to\infty}(1 + \alpha_n/n)^n = e^{\alpha}$ if $\lim_{n\to\infty} \alpha_n = \alpha$, it follows that

$$\lim_{n\to\infty} \varphi_{composed,n}(z) = e^{-t(1-z^{-1})/2}\, e^{-t(1-z)/2} = e^{-t}\, e^{t(z^{-1}+z)/2}. \tag{3.88}$$

This expression can be recognized as the generating function of the family of discrete kernels that we arrived at when we constructed the discrete scale-space in section 3.4.2. e^{-t} is the normalization factor.

Consequently, this result provides a constructive proof of the property that the transformation obtained by convolution with the discrete analogue of the Gaussian is equivalent[11] to the analytical solution of the system of equations obtained by discretizing the diffusion equation on a fixed equidistant grid in space.

[11]The conclusion is valid only if the solution to the discretization (3.82) converges to the solution of the continuous equations (3.79) when $\Delta t \to 0$. This holds, for example, if $f \in l_1$ or $f \in l_2$.

PROPOSITION 3.30. (REPEATED AVERAGING AND THE DIFFUSION EQUATION) *The discrete scale-space generated by convolution with the discrete Gaussian kernel* (3.47), *or equivalently as the solution to the semi-discretized version of the diffusion equation* (3.79), *describes the limit case of repeated iteration of the recurrence relation* (3.82) *when the scale step* Δt *tends to zero.*

This is not surprising bearing theorem 3.25 in mind. These essence of this treatment is that when the scale-space theory is to be applied to discrete signals, only what is necessary should be discretized, namely along the spatial coordinate. The continuous scale parameter can be left untouched.

3.6.3. Integrated Gaussian kernel

Another way to discretize the convolution integral (3.68) is by integrating the continuous Gaussian kernel over each pixel support region. This method can be regarded as giving "a more true approximation"[12] than the method with the sampled Gaussian, especially at fine scales (compare also with section 5.5). The resulting approximation formula corresponds to discrete convolution with the kernel given by

$$c_i = \int_{i-\frac{1}{2}}^{i+\frac{1}{2}} g(\xi;\ t)\, d\xi. \tag{3.89}$$

This choice of filter coefficients is equivalent to the continuous formulation (3.68) if the continuous signal f is a piecewise constant function, which is equal to the discrete pixel value over each pixel support region. Another possibility is to let f in (3.68) be defined by linear interpolation between the discrete values. This leads to

$$c_i = \int_{i-1}^{i} (\xi - i + 1)\, g(\xi;\ t)\, d\xi + \int_{i}^{i+1} (i - \xi + 1)\, g(\xi;\ t)\, d\xi. \tag{3.90}$$

According to a theorem by Karlin (1968) it holds that that a kernel, given by the difference operator applied to uniformly sampled values of an integrated Pólya frequency function, is a Pólya frequency sequence.

[12] This issue actually comes down to philosophical questions in the image formation process. What do the recorded pixel values actually represent? Often they are implicitly regarded as sampled values of the underlying physical light intensity in the real world. In reality, this is certainly not true, but under that assumption the formula (3.69) should be a proper discretization. (Except for the fact that the grid is not dense enough to resolve the rapid variations in the integrand.) Presumably, a more correct statement is that the pixel values should be regarded as the result of first applying a continuous convolution operator to the physical light intensity, and then as a second step sampling that output uniformly. The integration formula defined by (3.89) is an example of the latter model. In that case the kernel function is assumed to be one within the entire pixel support region and zero outside. Probably, a bell-shaped kernel would be more realistic.

UNIFORM SAMPLING OF INTEGRATED PF_∞ FUNCTIONS: *Let $f(x)$ be a PF_∞ sequence and form*

$$g(x) = \int_{\xi=-\infty}^{\infty} f(\xi)\, d\xi. \tag{3.91}$$

Then, $(\Delta g)(n) = g(n+1) - g(n)$ constitutes a PF_∞ sequence.

This means that the transformation from the original signal ($t = 0$) to an arbitrary level of scale ($t_1 > 0$) is always a scale-space transformation. However, a semigroup property cannot be expected to hold exactly, and we will probably arrive at similar scale-space problems as with the sampled Gaussian kernel when considering transformations between arbitrary scale levels. It is left as an open problem whether or not the second kernel (3.90) is a scale-space kernel.

PROPOSITION 3.31. (SCALE-SPACE PROPERTIES OF THE INTEGRATED GAUSSIAN KERNEL) *The transformation from the zero level to a coarser level in the representation generated by discrete convolution with the integrated Gaussian kernel, given by* (3.89), *is a discrete scale-space transformation.*

3.7. Summary and discussion

The aim of this treatment has been to investigate discrete aspects of one-dimensional scale-space theory. Linear and shift-invariant transformations have been studied, and a requirement has been stated on scale-space kernels: the number of local extrema in a convolved signal must not exceed the number of local extrema in the original signal.

As an immediate consequence of this, the sequence of filter coefficients sequence must be non-negative and unimodal in the spatial domain. For symmetric kernels the same requirements hold for the Fourier transform.

It has been shown that the interesting kernels can be completely classified in terms of total positivity—all shift-invariant discrete scale-space kernels are equivalent to normalized Pólya frequency sequences. The generating function of such a sequence/kernel possesses a very simple characterization

$$\varphi(z) = c\, z^k\, e^{(q_{-1}z^{-1} + q_1 z)} \prod_{i=1}^{\infty} \frac{(1+\alpha_i z)(1+\delta_i z^{-1})}{(1-\beta_i z)(1-\gamma_i z^{-1})},$$

implying that there are only three non-trivial types of primitive smoothing transformations: repeated averaging, recursive smoothing, and diffusion smoothing. In particular, it was shown that all discrete scale-space kernels of finite support arise from generalized binomial smoothing.

Then, a continuous scale parameter was introduced, and it was shown that the only reasonable way to define a scale-space for discrete signals is by convolution with the one-parameter family of kernels

$$T(n;\ t) = e^{-t} I_n(t),$$

where I_n are the modified Bessel functions of integer order. When similar arguments were applied in the continuous case, we were uniquely lead to the Gaussian kernel. The kernel T does also have the attractive property that it is equivalent to the limit case of a certain discretization of the diffusion equation

$$\partial_t L(x;\ t) = \tfrac{1}{2} \nabla_3^2 L(x;\ t) = \tfrac{1}{2} (L(x+1;\ t) - 2L(x;\ t) + L(x-1;\ t)).$$

The idea of a continuous scale parameter even for discrete signals is of considerable importance, since it permits arbitrary degrees of smoothing, i.e., we are no longer restricted to specific predetermined levels of scale. Due to the semi-group property, the scale-space condition holds between any two levels of representation. It was shown that the commonly used technique, in which the "scale-space" is constructed by convolution with the sampled Gaussian kernel, might lead to undesirable effects, since in general the transformation from an arbitrary fine level to a randomly selected coarser level is not a scale-space transformation.

Finally, some other aspects of the presented theory will be mentioned that have not been mentioned elsewhere.

3.7.1. *Ideal low-pass filters and block average filters*

The unimodality requirement on discrete scale-space kernels implies that an "ideal low-pass filter" is not a smoothing kernel in this sense because of the ringing phenomena in the spatial domain. This means that the first pre-filtering step that is often carried out in digital signal processing (to guarantee band-limited signals) actually violates the scale-space conditions. Neither does a block average filter possess scale-space properties, unless its width is either 1 or 2. This can be easily understood from the ringing phenomena and the non-negative values introduced in the frequency domain.

3.7.2. *Positivity and unimodality is necessary but not sufficient*

Note that the positivity and unimodality requirements for discrete scale-space kernels are necessary but not sufficient requirements. In other words, there exist kernels, which are non-negative and unimodal both in the spatial and the frequency domain but are not discrete scale-space kernels. This can be easily shown, for example, by considering a symmetric five-kernel having a generating function with only complex roots.

OBSERVATION 3.32. (POSITIVITY AND UNIMODALITY NOT SUFFICIENT) *The positivity and unimodality requirements in the spatial and the frequency domain are necessary but not sufficient conditions for a one-dimensional discrete kernel $\mathbb{Z} \to \mathbb{R}$ to be a discrete scale-space kernel.*

Proof. Consider the kernel with the generating function

$$\begin{aligned}\varphi(z) &= (z+2+2i)(z+2-2i)(z+\tfrac{1}{4}+\tfrac{1}{4}i)(z+\tfrac{1}{4}-\tfrac{1}{4}i)z^{-2} \\ &= z^2+\tfrac{36}{8}z+\tfrac{81}{8}+\tfrac{36}{8}z^{-1}+z^{-2}.\end{aligned} \tag{3.92}$$

It is easy to verify that the Fourier transform

$$\psi(\theta) = \tfrac{1}{8}\,(81+72\cos\theta+16\cos 2\theta) \tag{3.93}$$

is non-negative and unimodal. From the characterization of discrete scale-space kernels in section 3.3.2 it follows that this kernel cannot possess scale-space properties, since its generating function has non-real roots. □

3.7.3. Recursive filters

According to the classification of discrete scale-space kernels, it follows that the recursive filters suggested by Deriche (1985, 1987) possess discrete scale-space properties if and only if they can be implemented as a series of first-order smoothing filters, i.e., if and only if their generating functions

$$H_{a,b}(z) = \frac{b_0+b_1z^{-1}+\cdots+b_{n-1}^{-(n-1)}}{1+a_1z^{-1}+\cdots+a_n^{-n}} \tag{3.94}$$

can be factorized to the form

$$\varphi_K(z) = c\prod_{k=1}^{n}\frac{1+\delta_k z^{-1}}{1-\gamma_k z^{-1}}, \tag{3.95}$$

where $c > 0$, $\gamma_k, \delta_k \geq 0$ and $\gamma_k < 1$, compare with (3.34).

3.8. Conclusion: Scale-space for 1-D discrete signals

The results from this one-dimensional treatment all point in the same direction: *The natural way to apply the scale-space theory to discrete signals is by discretizing the diffusion equation—not the convolution integral.*

4

Scale-space for N-D discrete signals

The extension of the previous one-dimensional scale-space theory to two and higher dimensions is not obvious, since it is possible to show that there are no non-trivial kernels on $\mathbb{R}^2$ or $\mathbb{Z}^2$ with the property that they never introduce new local extrema. Lifshitz and Pizer (1987) have presented an illuminating counter-example (quoted freely):

> Imagine a two-dimensional image function consisting of two hills, one of them somewhat higher than the other one (see figure 4.1). Assume that they are smooth, wide, rather bell-shaped surfaces situated some distance apart, clearly separated by a deep valley running between them. Connect the two tops by a narrow sloping ridge without any local extrema, so that the top point of the lower hill no longer is a local maximum. Let this configuration be the input image. When the operator corresponding to the diffusion equation is applied to the geometry, the ridge will erode much faster than the hills. After a while it has eroded so much that the lower hill appears as a local maximum again. Thus, a new local extremum has been created.

The same argument can be carried out in the discrete case. Of course, the notion of connectivity has to be considered when defining local extrema. But this question is only of a formal nature. Given an arbitrary non-trivial convolution kernel, it is always possible to create a counter-example where the number of local extrema can increase, provided that the peaks are sufficiently far apart and the valley between them is sufficiently deep.

Anyway, we should not be too surprised. The decomposition of the sense in the above example is intuitively quite reasonable. The narrow ridge is a fine-scale phenomenon and should therefore disappear before the coarse-scale peaks. In this case, it is the measure on structure rather than the smoothing method that is the decisive factor.

From the counter-example it is clear that a level curve may split into two during scale-space smoothing. Hence, there are no non-trivial ker-

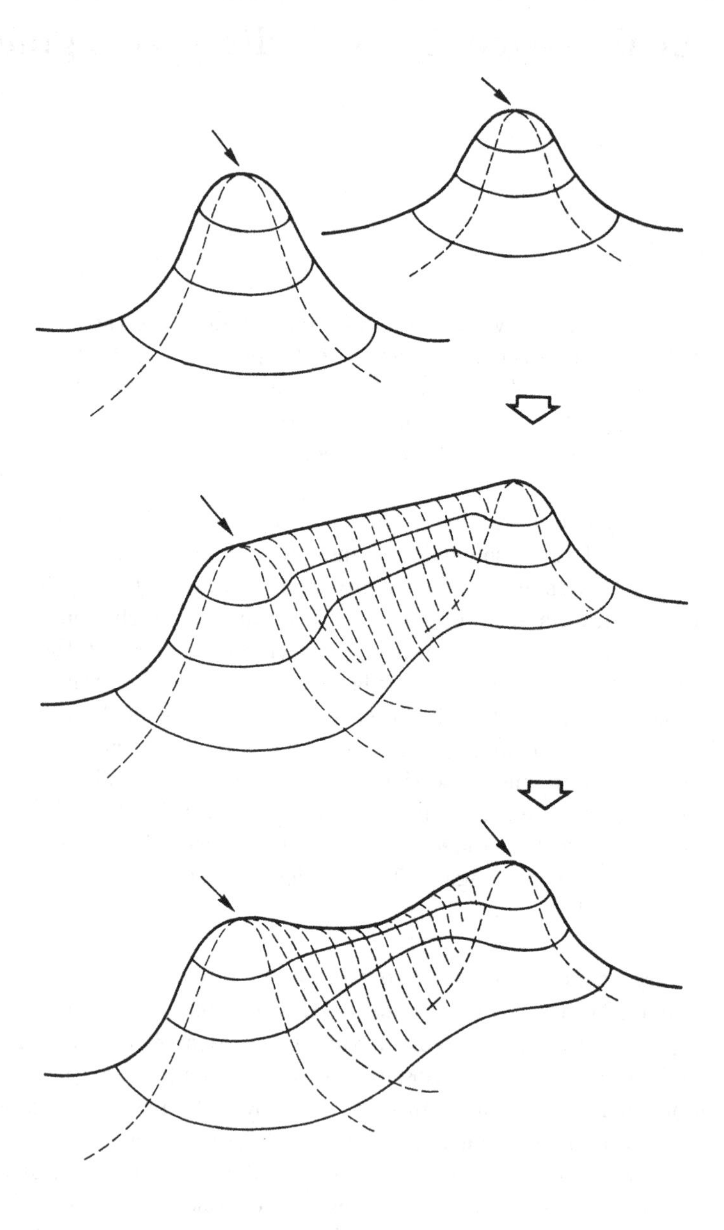

Figure 4.1. New local extrema can be created by the diffusion equation in the two-dimensional case

nels never increasing the number of zero-crossing curves either.[1] Also other types of features, such as elliptic regions, may be created with increasing scale (Yuille 1988). The property that new local extrema (zero-crossings/elliptic regions) can be created by linear smoothing is inherent and inescapable in two and higher dimensions. Therefore, when extending the theory to higher dimensions, we should not be locked to the previously given definition of discrete scale-space kernel. In one dimension, the number of local extrema is a natural measure of structure, on which a theory can be founded—in higher dimensions obviously not. Instead, the previously given treatment should be understood in a wider sense, as providing a deep understanding of which one-dimensional linear transformations can be regarded as smoothing transformations. It also shows that the only reasonable way to convert the one-dimensional scale-space theory from continuous signals to discrete signals is by discretizing the diffusion equation.

In this chapter a scale-space theory in higher dimensions is developed based on somewhat different axioms, which, however, in one dimension turn out to be equivalent to the previously stated formulation.

4.1. Scale-space axioms in higher dimensions

Koenderink (1984) derives the scale-space for two-dimensional continuous images from three assumptions; causality, homogeneity, and isotropy. The main idea is that it should be possible to trace every grey-level at a coarse scale to a corresponding grey-level at a finer scale. In other words, no new level curves should be created when the scale parameter increases. Using differential geometry he shows that these requirements uniquely lead to the diffusion equation.

It is of course impossible to apply these ideas directly in the discrete case since there are no direct correspondences to level curves or differential geometry for discrete signals. Neither can the scaling argument by Florack *et al.* (1992) be carried out in a discrete situation; a perfectly scale invariant operator cannot be defined on a discrete grid, which has a certain preferred scale given by the distance between adjacent grid points.

An alternative way of expressing the causality requirement, however, is by requiring that *if for some scale level* t_0 *a point* x_0 *is a local maximum for the scale-space representation at that level (regarded as a function of the space coordinates only) then its value must not increase when the scale parameter increases.* Analogously, if a point is a local minimum then its value must not decrease when the scale parameter increases.

It is clear that this formulation is equivalent to the formulation in

[1] However, new zero-crossing curves of the Laplacian, not arising from splits of previously existing zero-crossing curves of the Laplacian, cannot be created due to the causality property.

terms of level curves for continuous images, since if the grey-level value at a local maximum (minimum) increases (decreases), a new level curve is created. Conversely, if a new level curve is created, some local maximum (minimum) must have increased (decreased). An intuitive description of this requirement is that it prevents local extrema from being enhanced and from "popping up out of nowhere." In fact, it is closely related to the maximum principle for parabolic differential equations (Widder 1975).

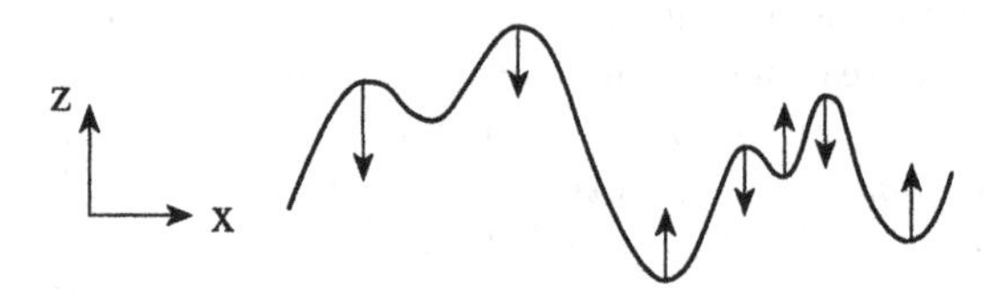

Figure 4.2. The non-enhancement condition of local extrema means that the grey-level value of a local maximum must not increase with scale and that the grey-level of a local minimum must not decrease.

In next section it is shown that this condition combined with a continuous scale parameter means a strong restriction on the smoothing method also in the discrete case, and again it will lead to a discretized version of the diffusion equation. In a special case, the scale-space representation is reduced to the family of functions generated by separated convolution with the discrete analogue of the Gaussian kernel, $T(n;\ t)$.

4.1.1. Basic definitions

Given a point $x \in \mathbb{Z}^N$ denote its neighbourhood of connected points by

$$N(x) = \{\xi \in \mathbb{Z}^N\colon\ (\| x - \xi \|_\infty \leq 1) \wedge (\xi \neq x)\}. \tag{4.1}$$

This corresponds to what is known as eight-connectivity in the two-dimensional case. The corresponding set including the central point x is written $N_+(x)$. Define (weak) extremum points as follows:

DEFINITION 4.1. (DISCRETE LOCAL MAXIMUM)
A point $x \in \mathbb{Z}^N$ is said to be a (weak) local maximum of a function $g\colon \mathbb{Z}^N \to \mathbb{R}$ if $g(x) \geq g(\xi)$ for all $\xi \in N(x)$.

DEFINITION 4.2. (DISCRETE LOCAL MINIMUM)
A point $x \in \mathbb{Z}^N$ is said to be a (weak) local minimum of a function $g\colon \mathbb{Z}^N \to \mathbb{R}$ if $g(x) \leq g(\xi)$ for all $\xi \in N(x)$.

The following operators are natural discrete correspondences to the Laplacian operator (below the notation $f_{-1,0,1}$ stands for $f(x_1 - 1, x_2, x_3 + 1)$);

the three-point operator in one dimension

$$(\nabla_3^2 f)_0 = f_{-1} - 2f_0 + f_1,$$

the five-point operator ∇_5^2 and the cross operator $\nabla_{\times^2}^2$ in two dimensions

$$(\nabla_5^2 f)_{0,0} = f_{-1,0} + f_{+1,0} + f_{0,-1} + f_{0,+1} - 4f_{0,0},$$
$$(\nabla_{\times^2}^2 f)_{0,0} = \tfrac{1}{2}\,(f_{-1,-1} + f_{-1,+1} + f_{+1,-1} + f_{+1,+1} - 4f_{0,0}),$$

as well as the corresponding operators ∇_7^2, $\nabla_{+^3}^2$, and $\nabla_{\times^3}^2$ in three dimensions respectively:

$$\begin{aligned}
(\nabla_7^2 f)_{0,0,0} &= f_{-1,0,0} + f_{+1,0,0} + f_{0,-1,0} \\
&\quad + f_{0,+1,0} + f_{0,0,-1} + f_{0,0,+1} - 6f_{0,0,0}, \\
(\nabla_{+^3}^2 f)_{0,0,0} &= \tfrac{1}{4}(f_{-1,-1,0} + f_{-1,+1,0} + f_{+1,-1,0} + f_{+1,+1,0} \\
&\quad + f_{-1,0,-1} + f_{-1,0,+1} + f_{+1,0,-1} + f_{+1,0,+1} \\
&\quad + f_{0,-1,-1} + f_{0,-1,+1} + f_{0,+1,-1} + f_{0,+1,+1} - 12f_{0,0,0}), \\
(\nabla_{\times^3}^2 f)_{0,0,0} &= \tfrac{1}{4}(f_{-1,-1,-1} + f_{-1,-1,+1} + f_{-1,+1,-1} + f_{-1,+1,+1} \\
&\quad + f_{+1,-1,-1} + f_{+1,-1,+1} + f_{+1,+1,-1} + f_{+1,+1,+1} - 8f_{0,0,0}).
\end{aligned}$$

$$\begin{pmatrix} 1 & -2 & 1 \end{pmatrix} \qquad \begin{pmatrix} & 1 & \\ 1 & -4 & 1 \\ & 1 & \end{pmatrix} \qquad \begin{pmatrix} \frac{1}{2} & & \frac{1}{2} \\ & -2 & \\ \frac{1}{2} & & \frac{1}{2} \end{pmatrix}$$

(a) (b) (c)

Figure 4.3. Computational molecules for (a) the three-point operator ∇_3^2, (b) the five-point operator ∇_5^2, (c) the cross operator $\nabla_\times^2$. (Throughout this treatment a unit step size $h = 1$ is used.)

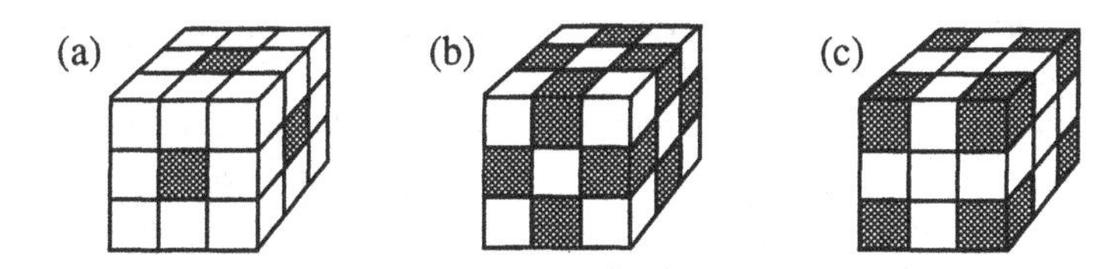

Figure 4.4. Computational molecules for three different discrete approximations to the Laplacian operator in the three-dimensional case; (a) the seven-point operator ∇_7^2, (b) the cross-operator $\nabla_{+^3}^2$, and (c) the diagonal cross-operator $\nabla_{\times^3}^2$. Except for the central point, ∇_7^2 has non-zero entries only in $\pm e_i$, $\nabla_{+^3}^2$ only in $\pm e_i \pm e_j$, and $\nabla_{\times^3}^2$ only in $\pm e_i \pm e_j \pm e_k$ $(i, j, k \in [1..3])$.

4.2. Axiomatic discrete scale-space formulation

Given that the task is to state an axiomatic formulation of the first stages of visual processing, *the visual front-end*, a list of desired properties may be long:

> linearity, translational invariance, rotational symmetry, mirror symmetry, semi-group, causality, positivity, unimodality, continuity, differentiability, normalization to one, nice *scaling* behaviour, locality, rapidly decreasing for large x and t, existence of an *infinitesimal generator* (explained below), invariance with respect to certain grey-level transformations, etc.

Such a list will, however, contain redundancies, as does this one. Here, a (minimal) subset of these properties is taken as axioms. In fact, it can be shown that all the other above-mentioned properties follow from the selected subset.

The scale-space representation for higher-dimensional signals can be constructed analogously to the one-dimensional case. To start with, postulate that the scale-space should be generated by convolution with a one-parameter family of kernels, i.e., $L(x;\ 0) = f(x)$ and for $t > 0$

$$L(x;\ t) = \sum_{\xi \in \mathbb{Z}^N} T(\xi;\ t)\, f(x - \xi). \tag{4.2}$$

This form of the smoothing formula corresponds to natural requirements about *linear shift-invariant smoothing* and the existence of a *continuous scale parameter.* It is natural to require that all coordinate directions should be handled identically. Therefore all kernels should be *symmetric.* Impose also a *semi-group* condition on the family T. This means that all scale levels will be treated similarly, that is, the smoothing operation does not depend on the scale value, and the transformation from a lower scale level to a higher scale level is always given by convolution with a kernel from the family (compare with (3.42)).

As smoothing criterion the *non-enhancement* requirement for local extrema is taken. It is convenient to express it as a condition of the derivative of the scale-space family with respect to the scale parameter. To ensure a proper statement of this condition, where differentiability is guaranteed, it is necessary to state a series of preliminary definitions leading to the desired formulation.

4.2.1. Definitions

Let us summarize this (minimal) set of basic properties a family should satisfy to be a candidate family for generating a (linear) scale-space.

DEFINITION 4.3. (PRE-SCALE-SPACE FAMILY OF KERNELS)
A one-parameter family of kernels $T \colon \mathbb{Z}^N \times \mathbb{R}_+ \to \mathbb{R}$ is said to be a pre-scale-space family of kernels if it satisfies

- $T(\cdot;\ 0) = \delta(\cdot)$,
- *the semi-group property* $T(\cdot;\ s) * T(\cdot;\ t) = T(\cdot;\ s+t)$,
- *the symmetry properties* $T(-x_1, x_2, \ldots, x_N;\ t) = T(x_1, x_2, \ldots, x_N;\ t)$ *and* $T(P_k^N(x_1, x_2, \ldots, x_N);\ t) = T(x_1, x_2, \ldots, x_N;\ t)$ *for all* $x = (x_1, x_2, \ldots, x_N) \in \mathbb{Z}^n$, *all* $t \in \mathbb{R}_+$, *and all possible permutations* P_k^N *of* N *elements, and*
- *the continuity requirement* $\| T(\cdot;\ t) - \delta(\cdot) \|_1 \to 0$ *when* $t \downarrow 0$.

DEFINITION 4.4. (PRE-SCALE-SPACE REPRESENTATION)
Let $f \colon \mathbb{Z}^N \to \mathbb{R}$ be a discrete signal and let $T \colon \mathbb{Z}^N \times \mathbb{R}_+ \to \mathbb{R}$ be a pre-scale-space family of kernels. Then, the one-parameter family of signals $L \colon \mathbb{Z}^N \times \mathbb{R}_+ \to \mathbb{R}$ given by (4.2) is said to be the pre-scale-space representation of f generated by T.

Provided that the input signal f is sufficiently regular, these conditions on the family of kernels T guarantee that the representation L is differentiable and satisfies a system of linear differential equations.

LEMMA 4.5. (A PRE-SCALE-SPACE REPRESENTATION IS DIFFERENTIABLE)
Let $L \colon \mathbb{Z}^N \times \mathbb{R}_+ \to \mathbb{R}$ be the pre-scale-space representation of a signal $f \colon \mathbb{Z}^N \to \mathbb{R}$ in l_1. Then L satisfies the differential equation

$$\partial_t L = \mathcal{A} L \tag{4.3}$$

for some linear and shift-invariant operator $\mathcal{A}$.

Proof. If f is sufficiently regular, e.g., if $f \in l_1$, define a family of operators $\{\mathcal{T}_t, t > 0\}$, here from from l_1 to l_1, by $\mathcal{T}_t f = T(\cdot;\ t) * f$. Due to the conditions imposed on the kernels it will satisfy the relation

$$\lim_{t \to t_0} \| (\mathcal{T}_t - \mathcal{T}_{t_0}) f \|_1 = \lim_{t \to t_0} \| (\mathcal{T}_{t-t_0} - \mathcal{I})(\mathcal{T}_{t_0} f) \|_1 = 0, \tag{4.4}$$

where $\mathcal{I}$ is the identity operator. Such a family is called a strongly continuous semigroup of operators (see Hille and Phillips 1957: p. 58–59).

A semi-group is often characterized by its *infinitesimal generator* $\mathcal{A}$ defined by

$$\mathcal{A} f = \lim_{h \downarrow 0} \frac{\mathcal{T}_h f - f}{h}. \tag{4.5}$$

The set of elements f for which $\mathcal{A}$ exists is denoted $\mathcal{D}(\mathcal{A})$. This set is not empty and never reduces to the zero element. Actually, it is even dense in l_1 (see Hille and Phillips (1957: p. 307)). If this operator exists then

$$\lim_{h \downarrow 0} \frac{L(\cdot, \cdot;\ t+h) - L(\cdot, \cdot;\ t)}{h} = \lim_{h \downarrow 0} \frac{\mathcal{T}_{t+h} f - \mathcal{T}_t f}{h}$$
$$= \lim_{h \downarrow 0} \frac{\mathcal{T}_h(\mathcal{T}_t f) - (\mathcal{T}_t f)}{h} = \mathcal{A}(\mathcal{T}_t f) = \mathcal{A} L(\cdot;\ t). \quad (4.6)$$

According to a theorem by Hille and Phillips (1957: p. 308) strong continuity implies $\partial_t(\mathcal{T}_t f) = \mathcal{A}\mathcal{T}_t f = \mathcal{T}_t \mathcal{A} f$ for all $f \in \mathcal{D}(\mathcal{A})$. Hence, the scale-space family L must obey the differential equation $\partial_t L = \mathcal{A} L$ for some linear operator $\mathcal{A}$. Since L is generated from f by a convolution operation it follows that $\mathcal{A}$ must be shift-invariant. □

This property makes it possible to formulate the previously indicated scale-space property in terms of derivatives of the scale-space representation with respect to the scale parameter. As in the maximum principle, the grey-level value in every local maximum point must not increase, while the grey-level value in every local minimum point must not decrease.

DEFINITION 4.6. (PRE-SCALE-SPACE PROPERTY: NON-ENHANCEMENT)
A differentiable one-parameter family of signals $L\colon \mathbb{Z}^N \times \mathbb{R}_+ \to \mathbb{R}$ *is said to possess pre-scale-space properties, or equivalently not to enhance local extrema, if for every value of the scale parameter* $t_0 \in \mathbb{R}_+$ *it holds that if* $x_0 \in \mathbb{Z}^N$ *is a local extremum point for the mapping* $x \mapsto L(x;\ t_0)$ *then the derivative of* L *with respect to* t *in this point satisfies*

$$\partial_t L(x_0;\ t_0) \leq 0 \quad \text{if } x_0 \text{ is a local maximum point,} \quad (4.7)$$

$$\partial_t L(x_0;\ t_0) \geq 0 \quad \text{if } x_0 \text{ is a local minimum point.} \quad (4.8)$$

Now it can be stated that a pre-scale-space family of kernels is a scale-space family of kernels if it satisfies this property for *any* input signal.

DEFINITION 4.7. (SCALE-SPACE FAMILY OF KERNELS)
A one-parameter family of pre-scale-space kernels $T\colon \mathbb{Z}^N \times \mathbb{R}_+ \to \mathbb{R}$ *is said to be a scale-space family of kernels if for any signal* $f\colon \mathbb{Z}^N \to \mathbb{R} \in l_1$ *the pre-scale-space representation of* f *generated by* T *possesses pre-scale-space properties, i.e., if* for any *signal local extrema are never enhanced.*

DEFINITION 4.8. (SCALE-SPACE REPRESENTATION)
A pre-scale-space representation $L\colon \mathbb{Z}^N \times \mathbb{R}_+ \to \mathbb{R}$ *of a signal* $f\colon \mathbb{Z}^N \to \mathbb{R}$ *generated by a family of kernels* $T\colon \mathbb{Z}^N \times \mathbb{R}_+ \to \mathbb{R}$*, which are scale-space kernels, is said to be a scale-space representation of* f.

In the next section it is shown how these requirements strongly restrict the possible class of kernels and scale-space representations. For example, they will lead to a number of restrictions on the operator $\mathcal{A}$ in lemma 4.5:

DEFINITION 4.9. (INFINITESIMAL SCALE-SPACE GENERATOR)
A shift-invariant linear operator $\mathcal{A}$ from l_1 to l_1

$$(\mathcal{A}L)(x;\ t) = \sum_{\xi \in \mathbb{Z}^N} a_\xi L(x - \xi;\ t), \tag{4.9}$$

is said to be an infinitesimal scale-space generator, denoted $\mathcal{A}_{ScSp}$, if the coefficients $a_\xi \in R$ satisfy

- *the locality condition $a_\xi = 0$ if $\xi \notin N_+(0)$,*
- *the positivity constraint $a_\xi \geq 0$ if $\xi \neq 0$,*
- *the zero sum condition $\sum_{\xi \in \mathbb{Z}^N} a_\xi = 0$, as well as*
- *the symmetry requirements*

$$a_{(-\xi_1, \xi_2, \ldots, \xi_N)} = a_{(\xi_1, \xi_2, \ldots, \xi_N)} \quad \text{and}$$
$$a_{P_k^N(\xi_1, \xi_2, \ldots, \xi_N)} = a_{(\xi_1, \xi_2, \ldots, \xi_N)}$$

for all $\xi = (\xi_1, \xi_2, \ldots, \xi_N) \in \mathbb{Z}^N$ and all possible permutations P_k^N of N elements.

4.2.2. Necessity

It will first be shown that these conditions necessarily imply that the family L satisfies a semi-discretized version of the diffusion equation.

THEOREM 4.10. (SCALE-SPACE FOR DISCRETE SIGNALS: NECESSITY)
A scale-space representation $L\colon \mathbb{Z}^N \times \mathbb{R}_+ \to \mathbb{R}$ of a signal $f\colon \mathbb{Z}^N \to \mathbb{R}$ satisfies the differential equation

$$\partial_t L = \mathcal{A}_{ScSp} L \tag{4.10}$$

with initial condition $L(\cdot;\ 0) = f(\cdot)$ for some infinitesimal scale-space generator $\mathcal{A}_{ScSp}$. In one, two and three dimensions respectively (4.10) reduces to

$$\partial_t L = \alpha_1 \nabla_3^2 L, \tag{4.11}$$
$$\partial_t L = \alpha_1 \nabla_5^2 L + \alpha_2 \nabla_{\times^2}^2 L, \tag{4.12}$$
$$\partial_t L = \alpha_1 \nabla_7^2 L + \alpha_2 \nabla_{+^3}^2 L + \alpha_3 \nabla_{\times^3}^2 L, \tag{4.13}$$

for some constants $\alpha_1 \geq 0$, $\alpha_2 \geq 0$ and $\alpha_3 \geq 0$.

Proof. The proof consists of two parts. The first part has already been presented in lemma 4.5, where it was shown that the requirements on the kernels imply that the family L obeys a linear differential equation. Because of the shift invariance $\mathcal{A}L$ can be written in the form (4.9). In the second part counterexamples are constructed from various simple test functions in order to delimit the class of possible operators.

The extremum point conditions (4.7), (4.8) combined with definitions 4.7-4.8 mean that $\mathcal{A}$ must be *local*, i.e., that $a_\xi = 0$ if $\xi \notin N_+(0)$. This is easily understood by studying the following counterexample: First, assume that $a_{\xi_0} > 0$ for some $\xi_0 \notin N_+(0)$ and define a function $f_1 \colon \mathbb{Z}^N \to \mathbb{R}$ by

$$f_1(x) = \begin{cases} \varepsilon > 0 & \text{if } x = 0, \\ 0 & \text{if } x \in N(0), \\ 1 & \text{if } x = \xi_0, \text{ and} \\ 0 & \text{otherwise.} \end{cases} \tag{4.14}$$

Obviously, 0 is a local maximum point for f_1. From (4.3) and (4.9) one obtains $\partial_t L(0;\ 0) = \epsilon a_0 + a_{\xi_0}$. It is clear that this value can be positive provided that ε is chosen small enough. Hence, L cannot satisfy (4.7). Similarly, it can also be shown that $a_{\xi_0} < 0$ leads to a violation of the non-enhancement property (4.8) (let $\varepsilon < 0$). Consequently, a_ξ must be zero if $\xi \notin N_+(0)$.

Moreover, the *symmetry* conditions imply that permuted and reflected coefficients must be equal, i.e., $a_{(-\xi_1,\xi_2,\ldots,\xi_N)} = a_{(\xi_1,\xi_2,\ldots,\xi_N)}$ and $a_{P_k^N(\xi_1,\xi_2,\ldots,\xi_N)} = a_{(\xi_1,\xi_2,\ldots,\xi_N)}$ for all $\xi = (\xi_1, \xi_2, \ldots, \xi_N) \in \mathbb{Z}^N$ and all possible permutations P_k^N of N elements. For example, the two-dimensional version of (4.9) reads

$$\partial_t L = \begin{pmatrix} a & b & a \\ b & c & b \\ a & b & a \end{pmatrix} L \tag{4.15}$$

for some a, b and c. Then, consider the function given by

$$f_2(x, y) = \begin{cases} 1 & \text{if } x \in N_+(0), \text{ and} \\ 0 & \text{otherwise.} \end{cases} \tag{4.16}$$

With the given (weak) definitions of local extremum points it is clear that 0 is both a local maximum point and a local minimum point. Hence $\partial_t L(0;\ 0)$ must be zero, and the *coefficients sum to zero*

$$\sum_{\xi \in \mathbb{Z}^N} a_\xi = 0, \tag{4.17}$$

which in two dimensions reduces to $4a + 4b + c = 0$ in (4.15). Obviously, (4.9) can be written

$$\partial_t L = (\mathcal{A}L)(x;\ t) = \sum_{\xi \in N(0)} a_\xi (L(x-\xi;\ t) - L(x;\ t)), \tag{4.18}$$

and the two-dimensional special case (4.15) reduces to

$$\partial_t L = \alpha_1 \begin{pmatrix} & 1 & \\ 1 & -4 & 1 \\ & 1 & \end{pmatrix} L + \alpha_2 \begin{pmatrix} \frac{1}{2} & & \frac{1}{2} \\ & -2 & \\ \frac{1}{2} & & \frac{1}{2} \end{pmatrix} L = \alpha_1 \nabla_5^2 L + \alpha_2 \nabla_{\times^2}^2 L. \tag{4.19}$$

Finally, by considering the test function

$$f_3(x, y) = \begin{cases} \epsilon > 0 & \text{if } x = 0, \\ -1 & \text{if } x = \tilde{\xi}, \text{ and} \\ 0 & \text{otherwise}, \end{cases} \tag{4.20}$$

for some $\tilde{\xi}$ in $N(0)$ one easily realizes that a_ξ must be *non-negative* if $\xi \in N(0)$. It follows that $\alpha_1 \geq 0$ and $\alpha_2 \geq 0$ in (4.19), which proves (4.12). (4.11) and (4.13) follow from similar straightforward considerations. The initial condition $L(\cdot;\ 0) = f$ is a direct consequence of the definition of pre-scale-space kernel. □

4.2.3. Sufficiency

The reverse statement of theorem 4.10 is also true.

THEOREM 4.11. (SCALE-SPACE FOR DISCRETE SIGNALS: SUFFICIENCY) *Let $f\colon \mathbb{Z}^N \to \mathbb{R}$ be a discrete signal in l_1, let $\mathcal{A}_{ScSp}$ be an infinitesimal scale-space generator, and let $L\colon \mathbb{Z}^N \times \mathbb{R}_+ \to \mathbb{R}$ be the representation generated by the solution to the differential equation*

$$\partial_t L = \mathcal{A}_{ScSp} L$$

with initial condition $L(\cdot;\ 0) = f(\cdot)$. Then, L is a scale-space representation of f.

Proof. It follows almost trivially that L possesses pre-scale-space properties, i.e., that L does not enhance local extrema, if the differential equation is rewritten in the form

$$\partial_t L = (\mathcal{A}L)(x;\ t) = \sum_{\xi \in N(0)} a_\xi (L(x-\xi;\ t) - L(x;\ t)). \tag{4.21}$$

If at some scale level t a point x is a local maximum point then all differences $L(x-\xi;\ t) - L(x;\ t)$ are non-positive, which means that

$\partial_t L(x;\ t) \leq 0$ provided that $a_\xi \geq 0$. Similarly, if a point is a local minimum point then the differences are all non-negative and $\partial_t L(x;\ t) \geq 0$.

What remains to be verified is that L actually satisfies the requirements for being a pre-scale-space representation. Since L is generated by a linear differential equation, it follows that L can be written as the convolution of f with some kernel T, i.e., $L(\cdot;\ t) = T(\cdot;\ t) * f$. The requirements of pre-scale-space kernels can be shown to hold by letting the input signal f be the discrete delta function. The semi-group property of the kernels follows from the fact that the coefficients ξ are constant, and the solution at a time $s+t$ hence can be computed from the solution at an earlier time s by letting the time increase by t. The symmetry properties of the kernel are obvious from the symmetry of the differential equation. The continuity at the origin follows directly from the differentiability of the representation. □

These results show that *a one-parameter family of discrete signals is a scale-space representation if and only if it satisfies the differential equation* (4.10) *for some infinitesimal scale-space generator.*

4.3. Parameter determination

For simplicity, from now on mainly two-dimensional signals will be considered. If (4.12) is rewritten in the form

$$\partial_t L = C\left((1-\gamma)\nabla_5^2 L + \gamma \nabla_{\times^2}^2 L\right) = C\,\nabla_\gamma^2 L, \tag{4.22}$$

the interpretation of the parameter C is just a trivial rescaling of the scale parameter. Thus, without loss of generality C may be set to $\frac{1}{2}$ to get the same scaling constant as in the one-dimensional case. What is left to investigate is how the remaining degree of freedom in the parameter $\gamma \in [0, 1]$ affects the scale-space representation.

If $\gamma = 1$ then a undesirable situation appears. Since the cross-operator only links diagonal points, the system of ordinary differential equations given by (4.22) can be split into two *uncoupled* systems, one operating on the points with even coordinate sum $x_1 + x_2$ and the other operating on the points with odd coordinate sum. It is clear that this is really an unwanted behaviour, since even after a substantial amount of "blurring," for certain types of input signals the "smoothed" grey-level landscape may still have a rather saw-toothed shape.

4.3.1. Derivation of the Fourier transform

Further arguments showing that γ must not be too large can be obtained by studying the Fourier transform of the corresponding scale-space family of kernels.

PROPOSITION 4.12. (FOURIER TRANSFORM OF DISCRETE SCALE-SPACE) *Let $L\colon \mathbb{Z}^2 \times \mathbb{R}_+ \to \mathbb{R}$ be the scale-space representation of a discrete signal $f\colon \mathbb{Z}^2 \to \mathbb{R}$ generated by (4.22) with initial condition $L(\cdot;\ 0) = f(\cdot)$. Assume that $f \in l_1$. Then the generating function of the kernel describing the transformation from the original signal to the representation at a certain scale t is given by*

$$\begin{aligned}\varphi_T(z, w) &= \sum_{(m,n)\in\mathbb{Z}^2} T(m, n;\ t)\, z^m\, w^n \\ &= \exp(-(2-\gamma) \\ &\qquad + (1-\gamma)\,(z^{-1} + z + w^{-1} + w)/2 \\ &\qquad + \gamma\,(z^{-1}w^{-1} + z^{-1}w + zw^{-1} + zw)/4).\end{aligned} \tag{4.23}$$

Its Fourier transform is

$$\begin{aligned}\varphi_T(z, w) &= \psi_T(e^{-iu}, e^{-iv}) \\ &= \exp(-(2-\gamma)t \\ &\qquad + (1-\gamma)(\cos u + \cos v)t \\ &\qquad + (\gamma \cos u \cos u)t).\end{aligned} \tag{4.24}$$

Proof. Discretizing (4.22) further in scale using Euler's explicit method with scale step Δt, gives an iteration formula of the form

$$\begin{aligned}L_{i,j}^{k+1} &= (1-(2-\gamma)\,\Delta t)\, L_{i,j}^{k} \\ &\quad + \tfrac{1}{2}\,(1-\gamma)\,\Delta t\,(L_{i-1,j}^{k} + L_{i+1,j}^{k} + L_{i,j-1}^{k} + L_{i,j+1}^{k}) \\ &\quad + \tfrac{1}{4}\,\gamma\,\Delta t\,(L_{i-1,j-1}^{k} + L_{i-1,j+1}^{k} + L_{i+1,j-1}^{k} + L_{i+1,j+1}^{k}),\end{aligned} \tag{4.25}$$

where the subscripts i and j denote the spatial coordinates x and y respectively, and the superscript k denotes the iteration index. The generating function describing one iteration with this transformation is

$$\begin{aligned}\varphi_{step}(z, w) &= (1-(2-\gamma)\Delta t) \\ &\quad + \tfrac{1}{2}\,(1-\gamma)\,\Delta t\,(z^{-1} + z + w^{-1} + w) \\ &\quad + \tfrac{1}{4}\,\gamma\,\Delta t\,(z^{-1}w^{-1} + z^{-1}w + zw^{-1} + zw).\end{aligned} \tag{4.26}$$

Assume that the scale-space representation at a scale level t is computed using n iterations with a scale step $\Delta t = \frac{t}{n}$. Then, the generating function describing the composed transformation can be written

$$\varphi_{composed,n}(z, w) = (\varphi_{step}(z, w))^n. \tag{4.27}$$

After substitution of Δt for $\frac{t}{n}$ and using

$$\lim_{n\to\infty} (1 + \frac{\alpha_n}{n})^n = e^{\alpha} \quad \text{if} \quad \lim_{n\to\infty} \alpha_n = \alpha, \tag{4.28}$$

it follows that $\varphi_{composed,n}(z)$ tends to $\varphi_T(z, w)$ according to (4.23) when $n \to \infty$, provided that the discretization (4.25) converges to the actual solution of (4.22). □

4.3.2. *Unimodality in the Fourier domain*

It is easy to verify that the Fourier transform is unimodal if and only if $\gamma \leq \frac{1}{2}$.

PROPOSITION 4.13. (UNIMODALITY OF THE FOURIER TRANSFORM; 2D) *The Fourier transform (4.24) of the kernel describing the transformation from the original signal to the smoothed representation at a coarser level of scale is unimodal if and only if $\gamma \leq \frac{1}{2}$.*

Proof. Differentiation of (4.24) gives

$$\begin{aligned} \partial_u \psi &= -\psi(u,v) \sin u \,(1 - \gamma(1 + \cos v))\, t, \\ \partial_v \psi &= -\psi(u,v) \sin v \,(1 - \gamma(1 + \cos u))\, t. \end{aligned} \tag{4.29}$$

The Fourier transform decreases with $|u|$ and $|v|$ for all u and v in $[-\pi, \pi]$ if and only if the factors $(1 - \gamma(1 + \cos v))$ and $(1 - \gamma(1 + \cos u))$ are non-negative for all u and v, i.e., if and only if $\gamma \leq \frac{1}{2}$. Then, any directional derivative away from the origin is negative. □

4.3.3. *Separability*

The transformation kernel is separable if and only if its Fourier transform is separable, that is, if and only if $\psi_T(u,v)$ can be written on the form $U_T(u)V_T(v)$ for some functions U_T and V_T. From (4.24) it is realized that this separation is possible if and only if $\gamma = 0$. Hence,

PROPOSITION 4.14. (SEPARABILITY OF 2D DISCRETE SCALE-SPACE) *The convolution kernel associated with the scale-space representation defined by $L(x, y;\ 0) = f(x, y)$ and*

$$\partial_t L = \tfrac{1}{2}\left((1-\gamma)\, \nabla_5^2 L + \gamma\, \nabla_{\times^2}^2\right) L \tag{4.30}$$

is separable if and only if $\gamma = 0$. Then, L is given by

$$L(x, y;\ t) = \sum_{m=-\infty}^{\infty} T(m;\ t) \sum_{n=-\infty}^{\infty} T(n;\ t)\, f(x-m, y-n) \quad (t > 0), \tag{4.31}$$

where $T(n;\ t) = e^{-t} I_n(t)$ and I_n are the modified Bessel functions of integer order.

Proof. The Fourier transform $\psi_T(u,v)$ can be written in the form $U_T(u)V_T(v)$ for some functions U_T and V_T if and only if the term with $\cos u \cos v$ can be eliminated from the argument of the exponential function, i.e., if and only if γ is zero. In that case the Fourier transform reduces to

$$\begin{aligned} \psi_T(u,v) &= \exp((-2 + \cos u + \cos v)t) \\ &= \exp((-1 + \cos u)t)\, \exp((-1 + \cos v)t), \end{aligned} \tag{4.32}$$

which corresponds to separated smoothing with the one-dimensional discrete analogue of the Gaussian kernel along each coordinate direction.

It can also be verified directly that (4.31) satisfies (4.30). Consider the possible scale-space representation of an N-dimensional signal generated by separable convolution with the one-dimensional discrete analogue of the Gaussian kernel; i.e., given $f\colon \mathbb{Z}^N \to \mathbb{R}$ define $L\colon \mathbb{Z}^N \times \mathbb{R}_+ \to \mathbb{R}$ by

$$L(x;\ t) = \sum_{x \in \mathbb{Z}^N} T_N(\xi;\ t) f(x - \xi) \quad (t > 0), \tag{4.33}$$

where $T_N : \mathbb{Z}^N \times \mathbb{R}_+ \to \mathbb{R}$ is given by

$$T_N(\xi;\ t) = \prod_{i=1}^{N} T_1(\xi_i;\ t), \tag{4.34}$$

$\xi = (\xi_1, \ldots, \xi_N)$, and $T_1 : \mathbb{Z} \times \mathbb{R}_+ \to \mathbb{R}$ is the discrete analogue of the Gaussian kernel, $T_1(n;\ t) = e^{-t} I_n(t)$. It will be shown that this representation satisfies a semi-discretized version of the two-dimensional diffusion equation

$$\partial_t L = \tfrac{1}{2} \nabla^2_{2N+1} L, \tag{4.35}$$

where

$$(\nabla^2_{2N+1} L)(x;\ t) = \sum_{i=1}^{N} L(x + e_i;\ t) - 2L(x;\ t) + L(x - e_i;\ t), \tag{4.36}$$

and e_i denotes the unit vector in the ith coordinate direction. Consider

$$(\partial_t T_N)(\xi;\ t) = \sum_{i=1}^{N} (\partial_t T_1)(\xi_i;\ t) \prod_{j \neq i} T_1(\xi_j;\ t). \tag{4.37}$$

Since T_1 satisfies (3.79), this expression can be written

$$(\partial_t T_N)(\xi;\ t) = \sum_{i=1}^{N} \tfrac{1}{2} (T_1(\xi_i - 1;\ t) - 2T_1(\xi_i;\ t) + T_1(\xi_i + 1;\ t)) \prod_{j \neq i} T_1(\xi_j;\ t),$$

which is obviously equivalent to

$$\partial_t T_N = \tfrac{1}{2} \nabla^2_{2N+1} T_N. \tag{4.38}$$

The same relation holds for L provided that the differentiation and infinite summation operators commute. □

In other words, in the separable case the resulting higher-dimensional discrete scale-space corresponds to repeated application of the one-dimensional scale-space concept along each coordinate direction.

$$\begin{pmatrix} \frac{1}{16} & \frac{1}{8} & \frac{1}{16} \\ \frac{1}{8} & \frac{1}{4} & \frac{1}{8} \\ \frac{1}{16} & \frac{1}{8} & \frac{1}{16} \end{pmatrix} \qquad \begin{pmatrix} \frac{1}{8} & \frac{2}{8} & \frac{1}{8} \\ \frac{2}{8} & -\frac{12}{8} & \frac{2}{8} \\ \frac{1}{8} & \frac{2}{8} & \frac{1}{8} \end{pmatrix}$$

(a) (b)

$$\begin{pmatrix} \frac{1}{36} & \frac{1}{9} & \frac{1}{36} \\ \frac{1}{9} & \frac{4}{9} & \frac{1}{9} \\ \frac{1}{36} & \frac{1}{9} & \frac{1}{36} \end{pmatrix} \qquad \begin{pmatrix} \frac{1}{6} & \frac{4}{6} & \frac{1}{6} \\ \frac{4}{6} & -\frac{20}{3} & \frac{4}{6} \\ \frac{1}{6} & \frac{4}{6} & \frac{1}{6} \end{pmatrix}$$

(c) (d)

Figure 4.5. Computational molecules corresponding to; (a) discrete iteration with $\Delta t = \gamma = \frac{1}{2}$, (b) the Laplacian operator when $\gamma = \frac{1}{2}$, (c) discrete iteration with $\Delta t = \gamma = \frac{1}{3}$, and (d) the Laplacian operator when $\gamma = \frac{1}{3}$.

4.3.4. Discrete iterations

The discretization of (4.22) in (4.25) using Euler's explicit method with scale step Δt corresponds to iterating with a kernel with the computational molecule

$$\begin{pmatrix} \frac{1}{4}\gamma\,\Delta t & \frac{1}{2}(1-\gamma)\,\Delta t & \frac{1}{4}\gamma\,\Delta t \\ \frac{1}{2}(1-\gamma)\,\Delta t & 1-(2-\gamma)\Delta t & \frac{1}{2}(1-\gamma)\,\Delta t \\ \frac{1}{4}\gamma\,\Delta t & \frac{1}{2}(1-\gamma)\,\Delta t & \frac{1}{4}\gamma\,\Delta t \end{pmatrix}. \tag{4.39}$$

Clearly, this kernel is unimodal if and only if $\gamma \leq \frac{2}{3}$. It is separable if and only if $\gamma = \Delta t$ (see below). In that case, the corresponding one-dimensional kernel is a discrete scale-space kernel in the sense of definition 3.1 if and only if $\Delta t \leq \frac{1}{2}$ (see proposition 3.14). This gives a further indication that γ should not exceed the value $\frac{1}{2}$.

PROPOSITION 4.15. (SEPARABILITY OF THE ITERATION KERNEL)
The iteration kernel (4.39), *corresponding to discrete forward iteration with Euler's explicit method, is separable if and only if* $\gamma = \Delta t$. *In that case, the corresponding one-dimensional kernel is a discrete scale-space kernel if and only if* $0 \leq \gamma \leq \frac{1}{2}$.

Proof. Since the kernel is symmetric and the coefficients sum to one, the kernel is separable if and only if it can be written as a kernel

$$(a,\ 1-2a,\ a) \tag{4.40}$$

convolved with itself, i.e., if and only if there exists an $a \geq 0$ such that

$$\begin{aligned} a^2 &= \tfrac{1}{4}\gamma\,\Delta t, \\ a(1-a) &= \tfrac{1}{2}(1-\gamma)\,\Delta t, \\ (1-a)^2 &= 1-(2-\gamma)\,\Delta t. \end{aligned} \tag{4.41}$$

The first equation has one non-negative root $a = \frac{1}{2}\sqrt{\gamma\Delta t}$. Insertion into the second equation gives two conditions for Δt; either $\Delta t = 0$ or $\Delta t = \gamma$. One verifies that these roots satisfy the third equation. The kernel $(a,\ 1-2a,\ a)$ is a discrete scale-space kernel if and only if $a \leq \frac{1}{2}$ (see (3.14); compare also with theorem 3.16). □

The boundary case $\gamma = \Delta t = \frac{1}{2}$ gives the iteration kernel in figure 4.5(a) corresponding to separated convolution with the one-dimensional binomial kernel $(\frac{1}{4}, \frac{1}{2}, \frac{1}{4})$ frequently used in pyramid generation (Crowley and Stern 1984).

4.3.5. Spatial isotropy

Another aspect that might affect the selection of γ is spatial isotropy. It is not clear that rotational invariance is a primary quality to be aimed at in the discrete case, since then one is locked to a fixed square grid. It is also far from obvious as to what is meant by spatial isotropy in a discrete situation. Possibly, it is better to talk about the lack of spatial isotropy, spatial anisotropy, or rotational asymmetry. However, since the Fourier transform is a continuous function of u and v, one can regard its variation as a function of the polar angle, given a fixed value of the radius, as one measure of this property. By expressing $\psi_T(u, v)$ in polar coordinates $u = \omega\cos\phi$, $v = \omega\sin\phi$ and examining the resulting expression,

$$\psi_T(\omega\cos\phi, \omega\sin\phi) = \exp(h(\omega\cos\phi, \omega\sin\phi)t), \tag{4.42}$$

where

$$\begin{aligned} h(\omega\cos\phi, \omega\sin\phi) = &-(2-\gamma) \\ &+(1-\gamma)(\cos(\omega\cos\phi)+\cos(\omega\sin\phi)) \\ &+\gamma\cos(\omega\cos\phi)\cos(\omega\sin\phi), \end{aligned} \tag{4.43}$$

one realizes that the value of γ that gives the smallest angular variation for a fixed value of ω, depends on ω. Hence, with this formulation, the "rotational invariance" is scale dependent. At coarse scales one obtains:

PROPOSITION 4.16. (ROTATIONAL INVARIANCE IN FOURIER DOMAIN) *The value of γ that gives the least rotational asymmetry for large scale phenomena in the solution to the differential equation* (4.22) *is* $\gamma = \frac{1}{3}$.

Proof. The Taylor expansion of h for small values of ω is (Lindeberg 1991: appendix A.2.3)

$$h(\omega\cos\phi, \omega\sin\phi) = -\tfrac{1}{2}\,\omega^2 + \tfrac{1}{24}\,(1 + (6\gamma - 2)\cos^2\phi\sin^2\phi)\,\omega^4 + O(\omega^6),$$

where the $O(\omega^6)$ term depends on both ϕ and γ. Observe that if $\gamma = \frac{1}{3}$ then the ϕ-dependence decreases with ω as ω^6 instead of as ω^4 □

This means that $\gamma = \frac{1}{3}$ *asymptotically*, i.e., with increasing spatial scale, gives the most isotropic smoothing effect on coarse-scale events. The reason spatial isotropy is desired at coarse scales rather than at fine scales is that the grid effects become smaller for coarse-scale phenomena, which in turn makes it more meaningful to talk about rotational invariance. This selection of γ corresponds to approximating the Laplacian operator with the "the nine-point operator" (see figure 4.5(d)). In numerical analysis this operator is known as the rotationally most symmetric 3×3 discrete approximation to the Laplacian operator (Dahlquist *et al.* 1974).

Note that when the separability is violated by using a non-zero value of γ, the discrete scale-space representation can anyway be computed efficiently in the Fourier domain using (4.24).

4.4. Summary and discussion

The proper way to apply the scale-space theory to discrete signals is apparently by discretizing the diffusion equation. Starting from a requirement that local extrema must not be enhanced when the scale parameter is increased continuously, it has been shown that within the class of linear transformations a necessary and sufficient condition for a one-parameter family of representations $L\colon \mathbb{Z}^N \times \mathbb{R}_+ \to \mathbb{R}$ to be a scale-space family of a discrete signal $f\colon \mathbb{Z}^N \to \mathbb{R}$ is that it satisfies the differential equation

$$\partial_t L = \mathcal{A}_{ScSp} L, \tag{4.44}$$

with initial condition $L(\cdot;\ 0) = f(\cdot)$ for some infinitesimal scale-space generator $\mathcal{A}_{ScSp}$. In one, two, and three dimensions it can equivalently be stated that a family is a scale-space family if and only if for some linear reparametrization of the scale parameter t and for some $\gamma_i \in [0, 1]$, it satisfies

$$\partial_t L = \tfrac{1}{2}\,\nabla_3^2 L, \tag{4.45}$$

$$\partial_t L = \tfrac{1}{2}\,((1 - \gamma_1)\,\nabla_5^2 L + \gamma_1\,\nabla_{\times^2}^2 L) \tag{4.46}$$

$$\partial_t L = \tfrac{1}{2}\,((1 - \gamma_1 - \gamma_2)\,\nabla_7^2 L + \gamma_1\,\nabla_{+^3}^2 L + \gamma_2\,\nabla_{\times^3}^2 L). \tag{4.47}$$

The essence of (4.44)–(4.47) is that these equations correspond to discretizations of *first-order* differential operators in *scale*, and *second-order* differential operators in *space*.

The effect of using different values of γ_1 in the two-dimensional case has been analyzed in detail. Nevertheless, the question about definite selection is left open. Unimodality considerations indicate that γ must not exceed $\frac{1}{2}$, while $\gamma = \frac{1}{3}$ gives the least degree of rotational asymmetry in the Fourier domain.

The family of scale-space kernels is separable if and only if $\gamma = 0$. In this case the scale-space family is given by convolution with the one-dimensional discrete analogue of the Gaussian kernel along each dimension. For this parameter setting the closed-form expressions for several derived entities simplify (see chapter 5). Observe also that $\gamma = 0$ arises a necessary consequence if the neighbourhood concept (defined in section 4.1.1) is redefined as

$$N(x) = \{\xi \in \mathbb{Z}^N : (\| x - \xi \|_1 \leq 1) \wedge (\xi \neq x)\} \tag{4.48}$$

(corresponding to what is known as four-connectivity in the two-dimensional case), since then necessarily $\alpha_i = 0$ $(i > 1)$ in (4.12) and (4.13). Similar results hold in higher dimensions. A possible disadvantage with choosing $\gamma = 0$ is that it emphasizes the role of the coordinate axes as being special directions.

Finally, it should be remarked that if a linear and shift-invariant operator $\mathcal{L}$, commuting with the smoothing operator $T*$, is applied to the scale-space representation L of a signal f, then $\mathcal{L}L$ will be a scale-space representation of $\mathcal{L}f$. One consequence of this is that multi-scale discrete derivative approximations defined by linear filtering of the smoothed signal preserve the scale-space properties. This property, which provides a natural way to discretize the multi-scale N-jet representation proposed by Koenderink and van Doorn (1987, 1992), is developed in chapter 5.

4.5. Possible extensions

The treatment so far has been restricted to signals defined on infinite and uniformly sampled square grids using uniform smoothing of all grid points. Next, a number of ways to generalize these notions are outlined.

4.5.1. Anisotropic smoothing

Perona and Malik (1990) proposed anisotropic smoothing as a way to reduce the shape distortions arising in edge detection by smoothing across object boundaries (see also Nordström 1990). The suggested methodology is to modify the diffusion coefficients to favour intraregion smoothing over interregion smoothing.

Using the maximum principle they show that the resulting anisotropic scale-space representation possesses a suppression property for local extrema similar to that used in Koenderink's (1984) continuous scale-space formulation and this discrete treatment. From the proofs of theorems 4.10 and 4.11 it is obvious that the discrete scale-space concept can easily be extended to such anisotropic diffusion by letting the coefficients in the operator $\mathcal{A}_{ScSp}$ depend upon the input signal. By this, the locality, positivity, and zero sum conditions will be preserved, while the symmetry requirements must be relaxed. Introducing such an anisotropic diffusion equation, however, violates the convolution form of smoothing as well as the semi-group property. Therefore, when proving the necessity of the representation a certain form of the smoothing formula may have to be assumed (for example, of the form (4.3) with the filter coefficients depending upon the input signal). Note that, if the translational invariance and the symmetry with respect to coordinate interchanges are relaxed in (4.44), this equation corresponds to the (spatial) discretization of the (second-order) diffusion equation with *variable conductance*, $c(x;\ t)$,

$$(\partial_t L)(x;\ t) = \nabla(c(x;\ t)\,\nabla L(x;\ t)). \tag{4.49}$$

Throughout this work uniform smoothing has been used at the cost of possible smoothing across object boundaries. The motivation behind this choice has been the main interest in using scale-space for *detecting* image structures. Therefore, in the absence of any prior information, it is natural that the first processing steps should be as *uncommitted* as possible. The approach taken has been to first detect candidate regions of interest, and then, once candidates have been detected as regions, improve their localization. Possibly, variable conductance could be useful in the second step of this process. Another natural application is to avoid the negative effects of smoothing thin or elongated structures.

There are, however, some problems that need to be further analyzed. Modifying the diffusion coefficients requires some kind of *a priori* information concerning which structures in the image are to be smoothed and which are not. In the Perona and Malik method, a tuning function has to be determined, giving the diffusion coefficient as a function of the gradient magnitude. When the scale parameter t tends to infinity, the solution to the anisotropic diffusion equation tends to a function with various sharp edges. Hence, choosing a tuning function somehow implies an implicit assumption about a "final segmentation" of the image. It is not clear that such a concept exists or can be rigorously defined.

4.5.2. Finite data

A practical problem always arising in linear filtering is what to do with pixels near the image boundary for which a part of the filter mask stretches outside the available image.

The most conservative outlook is, of course, to regard the output as undefined as soon as a computation requires image data outside the available domain. This is, however, hardly desirable for scale-space smoothing, since the (untruncated) convolution masks have infinite support, while the peripheral coefficients decrease towards zero very rapidly. A variety of ad hoc methods have been proposed to deal with this; extension methods, subtraction of steady-state components, solving the diffusion equation on a limited domain with (say, adiabatic) boundary conditions, etc. However, no such technique can overcome the problem with missing data. In some simple situations ad hoc extensions may do. But this requires some kind of *a priori* information about the contents of the image.

Inevitably, the peripheral image values of a smoothed finite image will be less reliable than the central ones. Instead, if accurate values really are required near the image boundary, then the vision system should try to acquire additional data such that the convolution operation becomes well-defined up to the prescribed accuracy. This is easily achieved within the active vision paradigm simply by moving the camera so that more values become available in a sufficiently large neighbourhood of the object of interest. The task of analyzing an object manifesting itself at a certain scale requires input data in a region around the object. The width of this frame depends both on the current level of scale and the prescribed accuracy of the analysis.

Of course, a genuinely finite approach is also possible. In this presentation this subject has not been developed, since the associated problems are somehow artificial and difficult to handle in a consistent manner, although the non-enhancement property can be easily formulated for finite data and although in the one-dimensional case the concepts of sign-regularity and semi-groups of totally positive matrices (Karlin 1968) in principle provide possible tools for dealing with this issue. One way to avoid both the infiniteness and the boundary problems is by using a spherical camera. Then, the ordinary planar camera geometry appears as an approximate description for foveal vision, that is, small solid angles in the central field of vision.

4.5.3. Other types of grids

The assumption of a square grid is not a necessary restriction. The same type of treatment can be carried out on, for example, a hexagonal grid with the semi-group property preserved, and also on a grid corresponding to non-uniform spatial sampling provided that the diffusion coefficients are modified accordingly. In the latter case some *a priori* form of the smoothing formula may have to be adopted when proving the necessity of the representation. An interesting case to consider might actually be the non-uniformly sampled spherical camera.

4.5.4. Further work

Finally, it should be pointed out that there is one main issue that has not been considered here, namely *scale-dependent spatial sampling.* This issue is important in order to improve the computational efficiency both when computing the representation and for algorithms working on the data. The scale-space concept outlined here uses the same spatial resolution at all levels of scale. The pyramid representations, on the other hand, imply a fixed relation between scale and resolution beyond which refinements are not possible.

Since the smoothed images at coarser scales become progressively more redundant, it seems plausible that some kind of subsampling can be done at the coarser scales without too much loss of information. It would be interesting to carry out an analysis about how much information is lost by such an operation, and to what extent a subsampling operator can be introduced in this representation while still maintaining the theoretical properties associated with having a continuous scale parameter, and without introducing any severe discontinuities along the scale direction that would be a potential source to numerical difficulties for algorithms working on the output from the representation.

5

Discrete derivative approximations with scale-space properties

A commonly occurring problem in computer vision concerns how to compute derivative approximations from discrete data. This problem arises, for example, when computing image descriptors such as features or differential invariants from image data and when relating image properties to phenomena in the outside world. Since differential geometry is a natural framework for describing geometric relations, formulations in terms of derivatives can be expected to arise in a large number of vision problems.

It is, however, well-known that derivative computation is not a well-posed problem. Derivative estimators are known to enhance the noise, and so frequently the argument is made that "some sort of smoothing is necessary".

Ultimately, the task of defining derivatives from real-world data boils down to a fundamental and *inherent measurement problem*, namely that objects in the world and features in images, in contrast to ideal mathematical entities, like "point" or "line," exist only as meaningful entities over certain finite ranges of scale.

5.1. Numerical approximation of derivatives

Mathematically, a partial derivative of a continuous function $f\colon \mathbb{R}^N \to \mathbb{R}$ with respect to a variable $x_i \in \mathbb{R}$ is defined as a limit value

$$\partial_{x_i} f(x_0) = \lim_{h \to 0} \frac{f(x_0 + he_i) - f(x_0)}{h}, \tag{5.1}$$

where $x_0 \in \mathbb{R}^N$ denotes the point at which the derivative is computed, and $e_i \in \mathbb{R}^N$ is a unit vector along the ith coordinate direction.

In numerical algorithms, derivatives are usually approximated by difference operators. One of the most common discrete approximations is the central difference operator

$$\partial_{x_i} f(x_0) \approx \delta_{x_i} f(x_0) = \frac{f(x_0 + he_i) - f(x_0 - he_i)}{2h}. \tag{5.2}$$

Depending on the choice of step size $h \in \mathbb{R}$, discrete approximations of different accuracy will be obtained. In traditional numerical analysis one is usually concerned with the simulation of mathematical models, which are is assumed to be exact, and the errors are usually estimated from the first term(s) in a Taylor expansion of the discretization error

$$\partial_{x_i} f(x_0) - \delta_{x_i} f(x_0) = \tfrac{1}{6} f_{x_i^3}(x_0)\, h^2 + \mathcal{O}(h^4). \tag{5.3}$$

A decreasing value of h will, in general, decrease the error (until round-off errors due to finite precision become important), and provided that the step length is selected small enough, the truncation error due to the missing higher-order terms in the Taylor expansion can usually be neglected.

If the task is to compute derivative approximations from measured data, such as in computer vision, the situation is different. Given a discrete set of measured image data, there will always be a finite inner scale of the image beyond which the step length cannot be decreased.[1] More important, the data can be expected to be corrupted by a substantial amount of noise and depending on the choice of step length, a difference operator used for derivative approximations will respond to structures in the image at different scales. A detailed discussion about this problem is given in section 1.3.3.

5.2. Scale-space derivatives

The scale-space representation provides a well-defined way to operationalize the notion of scale in derivative computations. A partial derivative

$$L_{x^\alpha} = \partial_{x^\alpha} L = \partial_{x_1^{\alpha_1} \dots x_N^{\alpha_N}} L \tag{5.4}$$

of the scale-space representation can be written

$$L_{x^\alpha}(x;\, t) = \int_{\xi \in \mathbb{R}^N} g_{x^\alpha}(\xi;\, t)\, f(x - \xi)\, d\xi, \tag{5.5}$$

and is always determined by an aperture function. Hence, scale-space derivatives provide a qualitative link between continuous derivatives and discrete approximations. The weighted difference computed by convolving the original signal with a derivative of the Gaussian may be seen as a generalization of a difference operator. In the limit case, when the scale parameter tends to zero, the scale-space derivative approaches to true derivative of the function.

In this respect, there is a high degree of similarity with Schwartz distribution theory (1951). In the theory of generalized functions, two functions f_1 and f_2 are treated as equivalent if

$$\lim_{t \to 0} (f_1 - f_2) * \phi(\cdot;\, t) = 0 \tag{5.6}$$

[1]When analysing pre-recorded images, the actual sampling density imposes a strong limitation. In an active vision system, on the other hand, increasing freedom is obtained by the ability to acquire new images with increased resolution. A simple example of this will be explored in section 11.3.

holds given a one-parameter family of test functions ϕ.[2] A derivative $\tilde{f}_{x^\alpha}$ of a generalized function f is in turn defined, basically, as the limit of convolving the function with the derivatives of the family of test functions

$$\tilde{f}_{x^\alpha} = \lim_{t \to 0} \phi_{x^\alpha}(\cdot;\ t) * f \tag{5.7}$$

Observe the analogy with scale-space representation,

$$\tilde{f}_{x^\alpha} = \lim_{t \to 0} L_{x^\alpha}, \tag{5.8}$$

although in the latter case it is neither needed nor desired to explicitly compute the limit case. Contrary, a major aim with scale-space analysis is to provide an internally consistent way to define a multi-scale calculus on any given signal, where not only the limit case $t \to 0$ is considered.

A strong regularizing property of the scale-space representation, is that under very mild conditions on the function f, the scale-space derivatives (5.5) will be well-defined entities, although the original signal may not be differentiable of any order. For example, if f is bounded by some polynomial, e.g., if there exist some constants $C_1, C_2 \in \mathbb{R}_+$ such that

$$|f(x)| \leq C_1 \, (1 + x^T x)^{C_2}, \tag{5.9}$$

then the integral is guaranteed to converge for any $t > 0$. Thus, the scale-space representation can for every $t > 0$ be treated as infinitely differentiable.

5.2.1. Algebraic properties of the multi-scale N-jet representation

Since for Gaussian smoothing the derivative operator commutes with the smoothing operator, the "smoothed derivatives" obtained in this way satisfy

$$\partial_{x^\alpha}(g * f) = (\partial_{x^\alpha} g) * f = g * (\partial_{x^\alpha} f), \tag{5.10}$$

implying that there are in principle three equivalent[3] ways to compute them; (i) by differentiating the smoothed signal, (ii) by convolving the signal with the differentiated smoothing kernel, (iii) by smoothing the differentiated signal. Moreover, the spatial derivatives satisfy the diffusion equation

$$\partial_t L_{x^\alpha} = \tfrac{1}{2} \nabla^2 L_{x^\alpha} \tag{5.11}$$

and inherit the cascade smoothing property of Gaussian smoothing

$$L_{x^\alpha}(\cdot;\ t_2) = g(\cdot;\ t_2 - t_1) * L_{x^\alpha}(\cdot;\ t_1) \quad (t_2 > t_1 \geq 0) \tag{5.12}$$

[2] A common choice of test functions is, in fact, the family of Gaussian kernels $\tau = g$.
[3] The last equality in (5.10) is valid only if f is sufficiently regular.

associated with the semi-group property of the Gaussian kernel

$$g(\cdot;\ t_2) = g(\cdot;\ t_2 - t_1) * g(\cdot;\ t_1) \quad (t_2 > t_1 \geq 0). \tag{5.13}$$

The latter result is a special case of the more general statement

$$g_{x^\alpha}(\cdot;\ t_1) * g_{x^\beta}(\cdot;\ t_2) = g_{x^{\alpha+\beta}}(\cdot;\ t_2 + t_1), \tag{5.14}$$

whose validity follows directly from the commutative property of convolution and differentiation.

This type of multi-scale representation based on smoothed derivatives, or *Gaussian derivatives*, has been proposed by Koenderink and van Doorn (1987, 1992) as a plausible model for the local processing in the receptive fields in a visual system.

By combining the output from these operators at any specific scale, smoothed differential geometric descriptors can be defined at that scale. If such descriptors are defined at *all* scales, the result is multi-scale differential geometric representation of the signal. As we shall see examples of in chapter 6 and chapter 14, this type of framework is useful for a variety of early vision tasks.

5.2.2. Discrete analogue of the multi-scale N-jet representation

A problem that arises when to apply this theory to image data concerns how these operations should be discretized when they are to be implemented in a machine vision system. The Gaussian derivative model is expressed for continuous signals, while realistic signals obtained from standard cameras are *discrete.* Although, as mentioned earlier, a standard discretization of the continuous equations may be expected to give a behaviour, which in some sense is "relatively close" to the behaviour in the continuous case (especially at coarse scales where the grid effects can be expected to be smaller), it is not guaranteed that the original scale-space conditions, however formulated, will be preserved after the discretization. Another important question concerns how sensitive these operations will be to noise, in particular when derivatives of high order are to be computed from noisy measured data.

The goal of this chapter is to describe how the earlier developed discrete scale-space theory can be generalized to derivatives, and to present a *discrete analogue of the multi-scale Gaussian derivative representation.* We will treat the case with discrete signals defined on an infinite and uniformly sampled square grid, and derive a set of discrete operators, which in a certain sense represents the canonical discretization of the previously stated continuous expressions. By replacing

(i) the continuous signal by a discrete signal,

(ii) the convolution with the continuous Gaussian kernel g by discrete convolution with the discrete analogue of the Gaussian kernel T,

(iii) the derivative operators ∂_{x^α} with a set of difference operators δ_{x^α},

it will be shown how a multi-scale representation of discrete derivative approximations can be defined, so that discrete analogues of (5.5) and (5.10)–(5.13) *hold exactly after discretization.*

The representation to be proposed has theoretical advantages compared to traditional discretizations based on different versions of the sampled or integrated Gaussian kernel, and discretizations carried out in the frequency domain, in the sense that it preserves the commutative properties—operators that *commute* before the discretization, commute also after the discretization. An important computational implication of this is that the derivative approximations can be computed *directly* from the smoothed grey-level values at different scales, without any need for re-doing the smoothing part of the operation, which is usually the computationally most expensive part when computing smoothed derivative approximations. Another positive side effect is that a large number of *normalization conditions* concerning the sums of the filter coefficients are transferred to the discrete domain.

As a further support for the presented methodology, experimental results of using these operations for a few different visual tasks will then be presented in chapter 6. A straightforward edge detection scheme will be described, which is similar to Canny's method (1986), but does not need any direct estimates of the gradient direction. Instead zero-crossings are detected in a certain polynomial expression in terms of derivatives up to order two, and tests are made on the sign of another polynomial expression in terms of derivatives up to order three (in order to eliminate "false edges"). Qualitatively the results obtained are similar those of Canny, although the proposed scheme is given by a conceptually much simpler framework, and in addition has the advantage that sub-pixel accuracy is obtained automatically. It will also be illustrated how a junction detector can be straightforwardly implemented by detecting extrema in another polynomial expression in terms of derivatives up to order two. Later in chapter 14 it will be shown how a shape-from-texture method can be expressed in terms of similar operations.

5.3. Discrete approximation of scale-space derivatives

If we are to construct discrete analogues of derivatives at multiple scales, which are to possess scale-space properties in a discrete sense, what properties are desirable? Let us start by observing that convolution operators, in general, commute.[4] In this specific case, it means that the scale-space smoothing operator commutes with any sufficiently regular difference operator. Applied to the scale-space representation of discrete signals (defined in chapter 4), we have that:

OBSERVATION 5.1. (DISCRETE DERIVATIVE APPROXIMATIONS)
Given a discrete signal $f\colon \mathbb{Z}^N \to \mathbb{R}$ in l_1 and an infinitesimal scale-space generator $\mathcal{A}_{ScSp}$, let $L\colon \mathbb{Z}^N \times \mathbb{R}_+ \to \mathbb{R}$ be the discrete scale-space representation of f generated from

$$\partial_t L = \mathcal{A}_{ScSp} L \tag{5.15}$$

with initial condition $L(\cdot;\ t) = f$, and let $\mathcal{D}_{x^\alpha}$ be a linear and shift-invariant operator from l_1 to l_1 corresponding to discrete convolution with a kernel of finite support.[5] *Then, the derivative approximation operator $\mathcal{D}_{x^\alpha}$ commutes with the scale-space smoothing operator $T(\cdot;\ t)*$ defined by*

$$L(\cdot;\ t) = T(\cdot;\ t) * f, \tag{5.16}$$

in other words,

$$\mathcal{D}_{x^\alpha} L = \mathcal{D}_{x^\alpha}(T * f) = (\mathcal{D}_{x^\alpha} T) * f = T * (\mathcal{D}_{x^\alpha} f). \tag{5.17}$$

Moreover, the discrete derivative approximation obeys the cascade smoothing property

$$(\mathcal{D}_{x^\alpha} L)(\cdot;\ t_2) = T(\cdot;\ t_2 - t_1) * (\mathcal{D}_{x^\alpha} L)(\cdot;\ t_1) \quad (t_2 > t_1 \geq 0) \tag{5.18}$$

and satisfies the semi-discretized version of the diffusion equation

$$\partial_t(\mathcal{D}_{x^\alpha} L) = \mathcal{A}_{ScSp}(\mathcal{D}_{x^\alpha} L). \tag{5.19}$$

In particular, the discrete derivative approximation fulfills the following non-enhancement property of local extrema; if $x_0 \in \mathbb{Z}^N$ is a local extremum point for the mapping $x \mapsto (\mathcal{D}_{x^\alpha} L)(x;\ t_0)$, then the derivative of $\mathcal{D}_{x^\alpha} L$ with respect to t in this point satisfies

$$\partial_t(\mathcal{D}_{x^\alpha} L)(x_0;\ t_0) \leq 0 \quad \text{if } x_0 \text{ is a local maximum point,} \tag{5.20}$$

$$\partial_t(\mathcal{D}_{x^\alpha} L)(x_0;\ t_0) \geq 0 \quad \text{if } x_0 \text{ is a local minimum point.} \tag{5.21}$$

To summarize, if a smoothed derivative L_{x^α} is defined as the result of applying a linear and shift-invariant operator $\mathcal{D}_{x^\alpha}$ to the smoothed signal L, i.e.,

$$L_{x^\alpha} = \mathcal{D}_{x^\alpha} L, \tag{5.22}$$

then L_{x^α} inherits the scale-space properties in the discrete case.

[4] Assuming that the functions and kernels involved are sufficiently regular.

[5] The conditions concerning finite support convolution kernel and $f \in l_1$ can be weakened. However, the generality of this statement is sufficient for our purpose.

Proof. The validity of (5.17) follows directly from the commutative property of convolution transformations, as does (5.18) if the lemma is combined with the cascade smoothing property of the discrete scale-space. The validity of (5.19) can be derived by using $\mathcal{D}_{x^\alpha} L = T * (\mathcal{D}_{x^\alpha} f)$ from (5.17).

Finally, (5.20) and (5.21) are direct consequences of the fact that due to (5.19) it holds that $\mathcal{D}_{x^\alpha} L$ is a scale-space representation of $\mathcal{D}_{x^\alpha} f$; see theorem 4.11 for a direct proof. □

In other words, *if a discrete derivative approximation is defined as the result of applying a convolution operator to the smoothed signal, then it will possess all the scale-space properties listed in the introduction*, i.e., equations (5.5)–(5.13). Obviously, the derivative approximation should also be selected such that it in a numerical sense approximates the continuous derivative. A natural minimum requirement to pose is that the discrete operator $\mathcal{D}_{x^\alpha}$ should constitute a *consistent*[6] approximation of the continuous derivative operator.

5.3.1. Necessity

The necessity of this type of representation can be derived by postulating the following structure on the scale-space of derivative approximation operators, which is similar to, but not equal to the structure postulated on the smoothing operation in the derivation of the traditional (zero-order) discrete scale-space representation in section 4.2.1.

Because of simplicity of presentation, the treatment will be restricted to two-dimensional signals. The approach is, however, valid in arbitrary dimensions. Nevertheless, certain details will be somewhat technical. The hasty reader may without loss of continuity proceed directly to corollary 5.9, where a summary is given.

DEFINITION 5.2. (PRE-SCALE-SPACE FAMILY OF DERIVATIVE APPROXIMATION KERNELS) *A one-parameter family of kernels $D_{x^\alpha}\colon \mathbb{Z}^2 \times \mathbb{R}_+ \to \mathbb{R}$ is said to be a pre-scale-space family of α-derivative approximation kernels if*

- *$D_{x^\alpha}(\cdot;\ 0)$ is a finite support kernel corresponding to a consistent discrete approximation of the derivative operator ∂_{x^α}, and*
- *D_{x^α} satisfies the cascade smoothing property*

$$D_{x^\alpha}(\cdot;\ t_2) = T(\cdot;\ t_2 - t_1) * D_{x^\alpha}(\cdot;\ t_1) \quad (t_2 \geq t_1 \geq 0) \quad (5.23)$$

[6] A discrete derivative approximation, $\delta_{x^\alpha}^{(h)}$ depending on a step length h, is said to be *consistent* if (under reasonable assumptions on the signal $L\colon \mathbb{R}^2 \to \mathbb{R}$) the truncation error tends to zero as the step length tends to zero, i.e., if $\lim_{h \downarrow 0} (\delta_{x^\alpha}^{(h)} L)(x_0) = (\partial_{x^\alpha} L)(x_0)$. In our considerations h is omitted from the notation, since the grid spacing is throughout assumed to be equal to one.

for some family of kernels $T\colon \mathbb{Z}^2 \times \mathbb{R}_+ \to \mathbb{R}$ *in* l_1*, which in turn obeys*

- *the symmetry properties*[7] $T(-x; y\ t) = T(x, y;\ t)$ *and* $T(y, x;\ t) = T(x, y;\ t)$ *for all* $x = (x, y) \in \mathbb{Z}^2$*, and*
- *the continuity requirement* $\| T(\cdot;\ t) - \delta(\cdot) \|_1 \to 0$ *when* $t \downarrow 0$.

Remark: The structure imposed on these kernels is almost identical to the structure on the pre-scale-space kernels postulated in definition 4.3. The only difference is that the pre-scale-space kernels have to be a semi-group, while it is sufficient that the derivative approximation kernels $\mathcal{D}_{x^\alpha}$ satisfy a cascade smoothing property.

DEFINITION 5.3. (PRE-SCALE-SPACE REPRESENTATION OF DERIVATIVE APPROXIMATIONS) *Let* $f\colon \mathbb{Z}^2 \to \mathbb{R}$ *be a discrete signal in* l_1 *and let* $D_{x^\alpha}\colon \mathbb{Z}^2 \times \mathbb{R}_+ \to \mathbb{R}$ *be a pre-scale-space family of* α*-derivative approximation kernels. Then, the one-parameter family of signals* $L_{x^\alpha}\colon \mathbb{Z}^2 \times \mathbb{R}_+ \to \mathbb{R}$ *given by*

$$L_{x^\alpha}(x;\ t) = \sum_{\xi \in \mathbb{Z}^2} D_{x^\alpha}(\xi;\ t) f(x - \xi) \tag{5.24}$$

is said to be the pre-scale-space representation of α*-derivative approximations of* f *generated by* D_{x^α}.

As mentioned previously, the linear type of smoothing is a consequence of the principle that the first stages of visual processing, the *visual front-end*, should be as *uncommitted* as possible and make no actual "irreversible decisions." More technically, the linearity can also be motivated by requiring the discrete derivative approximations to obey similar linearity properties as the continuous derivatives.

The convolution type of smoothing and the symmetry requirements on T correspond to the assumption that in the absence of any information about what the image can be expected to contain, all spatial points should be treated in the same way, i.e., the smoothing should be *spatially shift invariant* and *spatially isotropic*.

The cascade form of smoothing and the continuity with respect to the continuous scale parameter reflect the properties that any coarse scale representation should be computable from any fine scale representation, and that all scale levels should be treated in a similar manner. In other words, there should be *no preferred scale*.

In analogy with lemma 4.5, these requirements imply that L_{x^α} is *differentiable* with respect to the scale parameter.

[7] $T(x, -y;\ t) = T(x, y;\ t)$ is implied from the two other properties.

LEMMA 5.4. (DIFFERENTIABILITY OF DERIVATIVE APPROXIMATIONS) *Let $L_{x^\alpha}\colon \mathbb{Z}^2 \times \mathbb{R}_+ \to \mathbb{R}$ be a pre-scale-space representation of α-derivative approximations to a signal $f\colon \mathbb{Z}^2 \to \mathbb{R}$ in l_1. Then, L_{x^α} satisfies the differential equation*

$$\partial_t L_{x^\alpha} = \mathcal{A} L_{x^\alpha} \tag{5.25}$$

for some linear and shift-invariant operator $\mathcal{A}$.

Proof. Due to the cascade smoothing property of D_{x^α} we have that

$$T(\cdot;\ t_2) * D_{x^\alpha}(\cdot;\ 0) * f' = T(\cdot;\ t_2 - t_1) * T(\cdot;\ t_1) * D_{x^\alpha}(\cdot;\ 0) * f' \tag{5.26}$$

and

$$(f' * D_{x^\alpha}(\cdot;\ 0)) * (T(\cdot;\ t_2) - T(\cdot;\ t_2 - t_1) * T(\cdot;\ t_1)) = 0 \tag{5.27}$$

hold for any $f'\colon \mathbb{Z}^2 \to \mathbb{R}$ and any $t_2 \geq t_1 \geq 0$. Hence, $T(\cdot;\ t_2) - T(\cdot;\ t_2 - t_1) * T(\cdot;\ t_1)$ will always be in the null space of $D_{x^\alpha}(\cdot;\ 0)$, and we can with respect to the effect on L_{x^α} of D_{x^α} without loss of generality assume that T obeys the semi-group property $T(\cdot;\ t_2) = T(\cdot;\ t_2 - t_1) * T(\cdot;\ t_1)$. This means that L_{x^α} is a *pre-scale-space representation* of $D_{x^\alpha}(\cdot;\ 0) * f$ (see definition 4.4). According to lemma 4.5 it follows that L_{x^α} satisfies (5.25). □

In analogy with definition 4.7 we can now state that a pre-scale-space family of derivative approximation kernels is to be regarded as a scale-space family of derivative approximation kernels if *for any input signal* it satisfies the pre-scale-space property stated in definition 4.6.

DEFINITION 5.5. (SCALE-SPACE FAMILY OF DERIVATIVE APPROXIMATION KERNELS) *A one-parameter family of pre-scale-space α-derivative approximation kernels $D_{x^\alpha}\colon \mathbb{Z}^2 \times \mathbb{R}_+ \to \mathbb{R}$ is said to be a scale-space family of α-derivative approximation kernels if for any signal $f\colon \mathbb{Z}^2 \to \mathbb{R} \in l_1$ the pre-scale-space representation of α-derivative approximations to f generated by D_{x^α} obeys the non-enhancement property stated in definition 4.6. i.e., if for any signal $f \in l_1$ local extrema in L_{x^α} are never enhanced.*

DEFINITION 5.6. (SCALE-SPACE REPRESENTATION OF DERIVATIVE APPROXIMATIONS) *Let $f\colon \mathbb{Z}^2 \to \mathbb{R}$ be a discrete signal in l_1 and let $D_{x^\alpha}\colon \mathbb{Z}^2 \times \mathbb{R}_+ \to \mathbb{R}$ be a family of scale-space α-derivative approximations kernels. Then, the pre-scale-space representation of α-derivative approximations $L_{x^\alpha}\colon \mathbb{Z}^2 \times \mathbb{R}_+ \to \mathbb{R}$ of f is said to be a scale-space representation of α-derivative approximations to f.*

From these definitions it can be shown that the structure of the scale-space representation is determined up to two arbitrary constants, and that L_{x^α} must satisfy a semi-discretized version of the diffusion equation.

THEOREM 5.7. (DISCRETE DERIVATIVE APPROXIMATIONS: NECESSITY)

A scale-space representation of α-derivative approximations $L_{x^\alpha}\colon \mathbb{Z}^2 \times \mathbb{R}_+ \to \mathbb{R}$ to a signal $f\colon \mathbb{Z}^2 \to \mathbb{R}$ satisfies the differential equation

$$\partial_t L_{x^\alpha} = \alpha \nabla_5^2 L_{x^\alpha} + \beta \nabla_\times^2 L_{x^\alpha} \tag{5.28}$$

with initial condition $L_{x^\alpha}(\cdot;\ 0) = D_{x^\alpha}(\cdot;\ 0) * f(\cdot)$ for some constants $\alpha \geq 0$ and $\beta \geq 0$ and some finite support kernel D_{x^α}.

Proof. See (Lindeberg 1993: appendix A.2). □

THEOREM 5.8. (DISCRETE DERIVATIVE APPROXIMATIONS: SUFFICIENCY) *Let $f\colon \mathbb{Z}^2 \to \mathbb{R}$ be a discrete signal in l_1, and let $L_{x^\alpha}\colon \mathbb{Z}^2 \times \mathbb{R}_+ \to \mathbb{R}$ be the representation generated by the solution to differential equation*

$$\partial_t L_{x^\alpha} = \alpha \nabla_5^2 L_{x^\alpha} + \beta \nabla_\times^2 L_{x^\alpha} \tag{5.29}$$

with initial condition $L_{x^\alpha}(\cdot;\ 0) = D_{x^\alpha}(\cdot;\ 0) * f(\cdot)$ for some fixed $\alpha \geq 0$ and $\beta \geq 0$ and some finite support kernel $D_{x^\alpha}(\cdot;\ 0)$, corresponding to a consistent approximation to the derivative operator ∂_{x^α}. Then, L_{x^α} is a scale-space representation of α-derivative approximations to f.

Proof. See (Lindeberg 1993: appendix A.3). □

By reparametrizing $\alpha = C(1-\gamma)$ and $\beta = C\gamma$ (where $\gamma \in [0,1]$), and by (linearly) transforming the scale parameter t such that $C = 1/2$, it follows without loss of generality that the necessity and sufficiency results can be summarized in the following way:

COROLLARY 5.9. (DERIVATIVE APPROXIMATIONS PRESERVING SCALE-SPACE PROPERTIES; 2D) *Within the class of linear transformations of convolution type that obey the cascade smoothing property, a multi-scale representation of discrete derivative approximations L_{x^α} of a signal f, satisfying*

$$L_{x^\alpha}(\cdot;\ 0) = \mathcal{D}_{x^\alpha} * f \tag{5.30}$$

for some finite support convolution operator $\mathcal{D}_{x^\alpha}$, possesses scale-space properties in the discrete sense if and only if it is defined as the result of applying the operator $\mathcal{D}_{x^\alpha}$ to the scale-space representation of f at any scale, i.e. if an only if L_{x^α} is defined as

$$L_{x^\alpha} = \mathcal{D}_{x^\alpha} L = \mathcal{D}_{x^\alpha}(T_\gamma * f) \tag{5.31}$$

where T_γ denotes the discrete analogue of the Gaussian kernel defined as the kernel describing the solution to (5.32) for some fixed $\gamma \in [0, 1]$.

Equivalently, the derivative approximation possesses discrete scale-space properties if and only if it, for some fixed $\gamma \in [0, 1]$ and for some linear transformation of the scale parameter ($t = \alpha' t'$ where $\alpha' > 0$), satisfies the semi-discretized version of the diffusion equation

$$\partial_t L_{x^\alpha} = \tfrac{1}{2} \nabla_\gamma^2 L_{x^\alpha} \tag{5.32}$$

with initial condition $L_{x^\alpha}(\cdot;\ 0) = D_{x^\alpha}(\cdot;\ 0) * f$.

The result has been expressed in a general form in order to indicate that similar results hold in the one-dimensional case as well as in higher dimensions. (For example, in the one-dimensional case the operator ∇_γ^2 is replaced by $\nabla_3^2 L$). Now, what remains, is to *define*[8] how such derivative approximations are to be computed within the given class of operations.

5.3.2. One-dimensional signals

In the one-dimensional case it is natural to define the discrete correspondence to the derivative operator ∂_x as the first-order difference operator δ_x. This gives

$$(\mathcal{D}_x L)(x;\ t) = (\delta_x L)(x;\ t) = \tfrac{1}{2}(L(x+1;\ t) - L(x-1;\ t)) \tag{5.33}$$

and the striking similarity between the discrete and continuous relations,

$$(\delta_x T)(x;\ t) = -\frac{x}{t} T(x;\ t), \quad (\partial_x G)(x;\ t) = -\frac{x}{t} G(x;\ t). \tag{5.34}$$

Similarly, it is natural to define the discrete correspondence of the second-order derivative operator ∂_{xx} as the second-order difference operator δ_{xx} given by

$$(\mathcal{D}_{xx} L)(x;\ t) = (\delta_{xx} L)(x;\ t) = L(x+1;\ t) - 2L(x;\ t) + L(x-1;\ t) \tag{5.35}$$

From the diffusion equation it follows that the following relations are satisfied,

$$(\delta_{xx} T)(x;\ t) = 2(\partial_t T)(x;\ t), \quad (\partial_{xx} g)(x;\ t) = 2(\partial_t g)(x;\ t) \tag{5.36}$$

Note, however, the clear problem in this discrete case[9]

$$\delta_x \delta_x \neq \delta_{xx}. \tag{5.37}$$

---

[8] Note that it has nowhere in the proofs been made use of the fact that D_{x^α} is a derivative approximation operator. Corresponding results hold if D_{x^α} is replaced by an *arbitrary linear operator.*

[9] The second difference operator can, of course, also be defined as $\mathcal{D}_{xx} = \delta_x \delta_x$. Then, however, $\mathcal{D}_{xx} \neq \nabla_3^2$. Another possibility, is to use both the forward difference

Difference operators of higher order can be defined in an analogous manner

$$\mathcal{D}_{x^{2m}} = \delta_{x^{2m}} = (\delta_{xx})^m, \quad \mathcal{D}_{x^{2m+1}} = \delta_{x^{2m+1}} = \delta_x \delta_{x^{2m}}, \tag{5.38}$$

which means that the derivative approximations of different order are related by

$$L_{x^{m+2}} = \delta_{xx} L_{x^m}, \quad L_{x^{2m+1}} = \delta_x L_{x^{2m}}. \tag{5.39}$$

5.3.3. *Two-dimensional signals*

For two-dimensional signals it is natural to let the definitions of the derivative approximations depend on the value of γ.

5.3.3.1. *Separable case*

In the separable case $\gamma_{sep} = 0$, it is natural to inherit the definitions from the one-dimensional case

$$\mathcal{D}_{x^m y^n} = \delta_{x^m y^n} = \delta_{x^m} \delta_{y^n}, \tag{5.40}$$

where the operator δ_{y^n} should be interpreted as a similar difference operator in the y-direction as δ_{x^m} is in the x-direction. This gives

$$\nabla_\gamma^2 = \nabla_5^2 = \delta_{xx} + \delta_{yy}. \tag{5.41}$$

If $T^{(x)}, T^{(y)} \colon \mathbb{Z}^2 \times \mathbb{R}_+ \to \mathbb{R}$ are defined as the two-dimensional kernels corresponding to convolution with the one-dimensional discrete analogue of the Gaussian kernel $T \colon \mathbb{Z} \times \mathbb{R}_+ \to \mathbb{R}$ along the x- and y-directions respectively, then the effect of the derivative approximation method can be written

$$L_\alpha = \delta_\alpha(T_{\gamma=0} * f) = \delta_{x^m} \delta_{y^n} (T^{(x)} * T^{(y)} * f) = \\ (\delta_{x^m} T^{(x)}) * (\delta_{y^n} T^{(y)}) * f, \tag{5.42}$$

which implies that a two-dimensional derivative approximation L_α of order $|\alpha| = m + n$ exactly corresponds to the result of applying a one-dimensional derivative approximation kernel $\delta_{x^m} T^{(x)}$ of order m along the x-direction and a one-dimensional derivative approximation kernel $\delta_{y^n} T^{(y)}$ of order n along the y-direction.

---

operator, $(\delta_{x+} L)(x;\ t) = L(x+1;\ t) - L(x;\ t)$, and the backward difference operator, $(\delta_{x-} L)(x;\ t) = L(x;\ t) - L(x-1;\ t)$, and, e.g, define $\tilde{\mathcal{D}}_x = \delta_{x+}$ and $\tilde{\mathcal{D}}_{xx} = \delta_{x-}\delta_{x+}$. By this, $\tilde{\mathcal{D}}_{xx}$ will correspond to (a translate of) $\tilde{\mathcal{D}}_x^2$. Then, however, the odd-order derivatives are no longer estimated at the grid points, and (5.34) no longer holds. Nevertheless, the commutative algebraic structure with respect to smoothing and derivative approximations is preserved independent of this choice.

5.3.3.2. *Rotationally symmetric case*

The case $\gamma_{symm} = \frac{1}{3}$ corresponds to approximating the continuous Laplacian with the discrete nine-point operator illustrated in figure 4.5(d). Assuming that the second derivative approximation operators should be symmetric and satisfy[10] $\tilde{\delta}_{xx} + \tilde{\delta}_{yy} = \nabla_9^2$, it is natural to assume that for some values of a and b, $\tilde{\delta}_{xx}$ and $\tilde{\delta}_{yy}$ can be represented by 3×3 computational molecules of the form

$$\tilde{\delta}_{xx} = \begin{pmatrix} a & -2a & a \\ b & -2b & b \\ a & -2a & a \end{pmatrix}, \quad \tilde{\delta}_{yy} = \begin{pmatrix} a & b & a \\ -2a & -2b & -2a \\ a & b & a \end{pmatrix}. \tag{5.43}$$

The condition $\tilde{\delta}_{xx} + \tilde{\delta}_{yy} = \nabla_9^2$ then gives

$$\tilde{\delta}_{xx} = \begin{pmatrix} \frac{1}{12} & -\frac{1}{6} & \frac{1}{12} \\ \frac{5}{6} & -\frac{5}{3} & \frac{5}{6} \\ \frac{1}{12} & -\frac{1}{6} & \frac{1}{12} \end{pmatrix}, \quad \tilde{\delta}_{yy} = \begin{pmatrix} \frac{1}{12} & \frac{5}{6} & \frac{1}{12} \\ -\frac{1}{6} & -\frac{5}{3} & -\frac{1}{6} \\ \frac{1}{12} & \frac{5}{6} & \frac{1}{12} \end{pmatrix}. \tag{5.44}$$

We leave the question open about how to define the other operators $\mathcal{D}_x$, $\mathcal{D}_y$ and $\mathcal{D}_{xy}$.

5.3.4. *Other possible choices*

Concerning the choice of these operators, it should be remarked that these (in principle, arbitrary) selections[11] were based on the relations to standard discrete operators used in numerical analysis (Dahlquist *et al.* 1974). Other design criteria may lead to other operators, see, e.g., Haralick (1984), Meer and Weiss (1992), and Vieville and Faugeras (1992). Nevertheless, the *algebraic* scale-space properties are preserved whatever linear operators are used.

5.4. Computational implications

An immediate consequence of the proposed scale-space representation of discrete derivative approximations, is that the derivative approximations can be computed *directly* from the smoothed grey-level values at different scales, and that this will (up to numerical truncation and rounding errors) give *exactly* the same result as convolving the signal with the discrete analogue of the Gaussian derivative kernel. This has a clear advantage in terms of computational efficiency, since the derivative approximations operators, $\mathcal{D}_{x^\alpha}$, usually have a small support region and contain a small number of non-zero filter coefficients. Hence, there is absolutely

[10] This condition is a necessary requirement for $\partial_t L = \frac{1}{2}(\tilde{\delta}_{xx} + \tilde{\delta}_{yy})L$ to hold.

[11] Similar operators have also been used in *pyramid* representations (Burt and Adelson 1983; Crowley and Stern 1984).

no need for *re-doing* the smoothing part of the transformation, as is the case if *several* derivative approximations are computed by convolution with smoothed derivative filters, e.g., some discrete approximations to the Gaussian derivatives.[12] This issue is of particular importance when computing multi-scale differential geometric descriptors of high derivation order. It also leads to a conceptually very simple computational architecture; see figure 5.1.

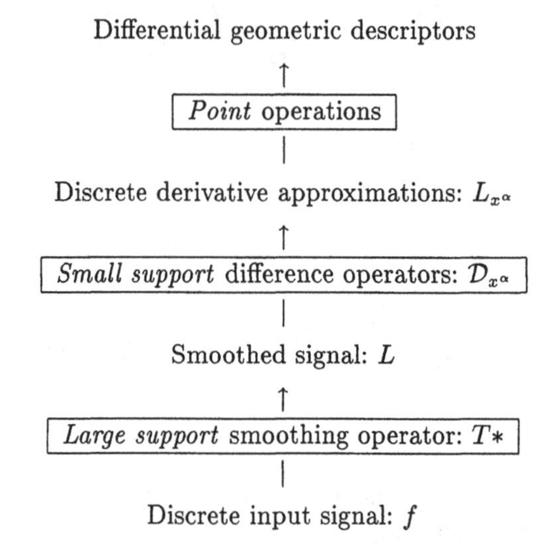

Figure 5.1. Schematic overview of the different types of computations required for computing multi-scale derivative approximations and discrete approximations to differential geometric descriptors using the proposed framework.

5.4.1. Normalization of the filter coefficients

Below, a number of continuous relations are listed together with their discrete correspondences. When the scale parameter tends to zero, the continuous and discrete Gaussian kernels tend to the continuous and discrete delta functions respectively, δ_{cont} and δ_{disc};

$$\lim_{t \downarrow 0} T(x;\ t) = \delta_{disc}(x), \quad \lim_{t \downarrow 0} g(x;\ t) = \delta_{cont}(x). \tag{5.45}$$

Concerning the normalization of the filter coefficients, it holds that

$$\sum_{\xi=-\infty}^{\infty} T(\xi;\ t) = 1, \quad \int_{\xi=-\infty}^{\infty} g(\xi;\ t)\, d\xi = 1, \tag{5.46}$$

[12] The same problem arises also if the computations are performed in the Fourier domain, since at least one inverse FFT transformation will be needed for each derivative approximation.

which means that in the discrete case the sum of the smoothing coefficients is exactly one. Similarly, the sum of the filter coefficients in a derivative approximation kernel is exactly zero,

$$\sum_{\xi=-\infty}^{\infty} (\delta_{x^\alpha} T)(\xi;\ t) = 0, \qquad \int_{\xi=-\infty}^{\infty} (\partial_{x^\alpha} g)(\xi;\ t)\, d\xi = 0. \tag{5.47}$$

for any integer $n \geq 1$. A trivial consequence of this is that the sum of the filter coefficients in the discrete Laplacian of the Gaussian is exactly zero, and there is no need for "modifying the filter coefficients" as has been the case in previous implementations of this operator.

In some situations it is useful to normalize the kernels used for derivative computations so that the integral of positive parts remains constant over scales. Such kernels are useful in edge detection (Korn 1988; Zhang and Bergholm 1991), more generally for automatic scale selection in feature detection (see chapter 13), and for shape-from-texture (see chapter 14). Then, the following relations are useful;

$$\int_{\xi=0}^{\infty} (\partial_x g)(\xi;\ t)\, d\xi = -g(0;\ t) = \frac{1}{\sqrt{2\pi t}}, \quad \sum_{\xi=0}^{\infty} (\delta_x T)(\xi;\ t) = -T(0;\ t). \tag{5.48}$$

In practice, to give the equivalent effect of normalizing the kernels such that the sum of the positive values is always equal to one, it is, of course, sufficient to divide $\delta_x L$ and $\delta_y L$ by $T(0;\ t)$. Note that with increasing t, this correction factor asymptotically approaches the corresponding normalization factor in the continuous case $1/g(0;\ t) = \sqrt{2\pi t}$, while at $t = 0$ the discrete normalization factor it is exactly one—in contrast to the continuous normalization factor which then is zero; see also section 5.4.3.

5.4.2. Comparisons with other discretizations

As a comparison with other possible approaches, observe that if continuously equivalent expressions are discretized in a straightforward way, say by approximating the convolution integral using the rectangle rule of integration, and by approximating the derivative operator ∂_{x^α} with the difference operator δ_{x^α}, then the discretized results will depend upon which actual expression in selected. Consider, for example, the three equivalent expressions in (5.10), where,

(i) the discretization of the left expression corresponds to discrete convolution with the *sampled Gaussian* kernel followed by the application of a *difference* operator,

(ii) the discretization of the central expression corresponds to discrete convolution with the *sampled derivative* of the Gaussian kernel, and

(iii) the discretization of the right expression corresponds to the application of the central *difference* operator to the signal followed by discrete convolution with the *sampled Gaussian* kernel.

It is clear that the equivalence is violated; (i) and (iii) describe equivalent operations, while (i) and (ii) do not. Considering the particular case of the Laplacian of the Gaussian, $\nabla^2 g$, it is well-known that this kernel is not separable. When performing the computations in the spatial domain, the fact that g satisfies the diffusion equation, $\nabla^2 g = 2\,\partial_t g$, is sometimes used for reducing the computational work; by computing $g * f$ at two adjacent scales, forming the difference, and then dividing by the scale difference. In the literature, this method is usually referred to as the difference of Gaussians (DOG) approach (Marr and Hildreth 1980). Note, however, that when the scale difference tends to zero, the result of this operation is not guaranteed to converge to the actual result, of say convolving the original signal with the sampled Laplacian of the Gaussian; not even if the calculations (on the spatially sampled data) are represented with infinite precision in the grey-level domain.

For the proposed discrete framework on the other hand, the discrete analogues of these entities are *exactly equivalent*; see (5.18) and (5.19). The main reason why the "discrete scale-space representations" generated from different versions of the sampled Gaussian kernel do not possess discrete scale-space properties, is that the commutativity is violated when using this discretization method—the discrete correspondences to operators that commute before the discretization do not *commute after the discretization.*

5.4.3. Discrete modelling of feature detectors

The proposed discrete kernels are also suitable for discrete modelling of feature detectors. As an example of this, consider the *diffuse step edges* studied by Zhang and Bergholm (1991). In the continuous case, the intensity profile perpendicular to such a (straight and unit height) edge may be modelled by an integrated Gaussian

$$\Phi(x;\ t_0) = \int_{\xi=-\infty}^{x} g(\xi;\ t_0)\, d\xi, \tag{5.49}$$

where t_0 describes the degree of diffuseness. In a scale-space representation L_{t_0} of $\Phi(\cdot;\ t_0)$, the variation over scales of the gradient magnitude, computed at the origin and normalized as described in section 5.4.1, is

$$(\partial_x L_{t_0})(0;\ t) \sim \frac{g(0;\ t_0+t)}{g(0;\ t)} = \frac{\sqrt{t}}{\sqrt{t_0+t}}. \tag{5.50}$$

If a corresponding *discrete diffuse step edge* is defined by

$$\tilde{\Phi}(x;\ t_0) = \sum_{\xi=-\infty}^{x} T(\xi;\ t_0), \tag{5.51}$$

and if the gradient is approximated by the backward difference operator δ_{x^-}, then the analytic expression for the corresponding discrete analogue of the gradient magnitude will be algebraically similar to that in the continuous case,

$$(\delta_{x^-} L_{t_0})(0;\ t) \sim \frac{T(0;\ t_0+t)}{T(0;\ t)}. \tag{5.52}$$

Moreover, it is easy to show from (3.51) that when t and t_0 increase, $(\delta_{x^-} L_{t_0})(0;\ t)$ approaches $\sqrt{t}/\sqrt{t_0+t}$, which agrees with the continuous expression for $(\partial_x L_{t_0})(0;\ t)$ obtained from (5.50).

5.5. Kernel graphs

Figures 5.2 and 5.3 illustrate the differences and similarities between the proposed discrete kernels and the derivatives of the continuous Gaussian kernel. These figures show the graphs of

$$T_{x^n} = \delta_{x^n} T \quad \text{and} \quad g_{x^n} = \partial_{x^n} g \tag{5.53}$$

for a few order of derivatives/differences and at two different scales. These kernels describe the equivalent effect of computing smoothed derivatives in the one-dimensional case as well as the separable two-dimensional case; see (5.42). For comparison, the equivalent discrete kernels corresponding to sampling the derivatives of the Gaussian kernel are displayed. It can be seen that the difference is largest at fine scales, and that it decreases as the kernels approach each other at coarser scales.

Figure 5.4 and 5.5 show corresponding two-dimensional kernels represented by grey-level values. Figure 5.6 and 5.7 also show examples of equivalent directional derivatives corresponding to pointwise linear combinations of the components of L_{x^β} using the well-known expression for the nth-order directional derivative $\partial_{\bar{\beta}}^n$ of a function L in any direction β,

$$\partial_{\bar{\beta}}^n L = (\cos\beta\, \partial_x + \sin\beta\, \partial_y)^n L. \tag{5.54}$$

In the terminology of Freeman and Adelson (1990), and Perona (1992), these kernels are trivially "steerable" (as is the directional derivative of any continuously differentiable function).

Figure 5.8 gives an illustration of what might happen if the sampling problems at fine scales are not properly treated. It shows a situation where the slope at the origin of the sampled (fifth-order) derivative of the Gaussian kernel is reversed. If this kernel is used for derivative estimation, then even the sign of the derivative can be wrong. For the corresponding discrete kernel, however, the qualitative behaviour is correct.

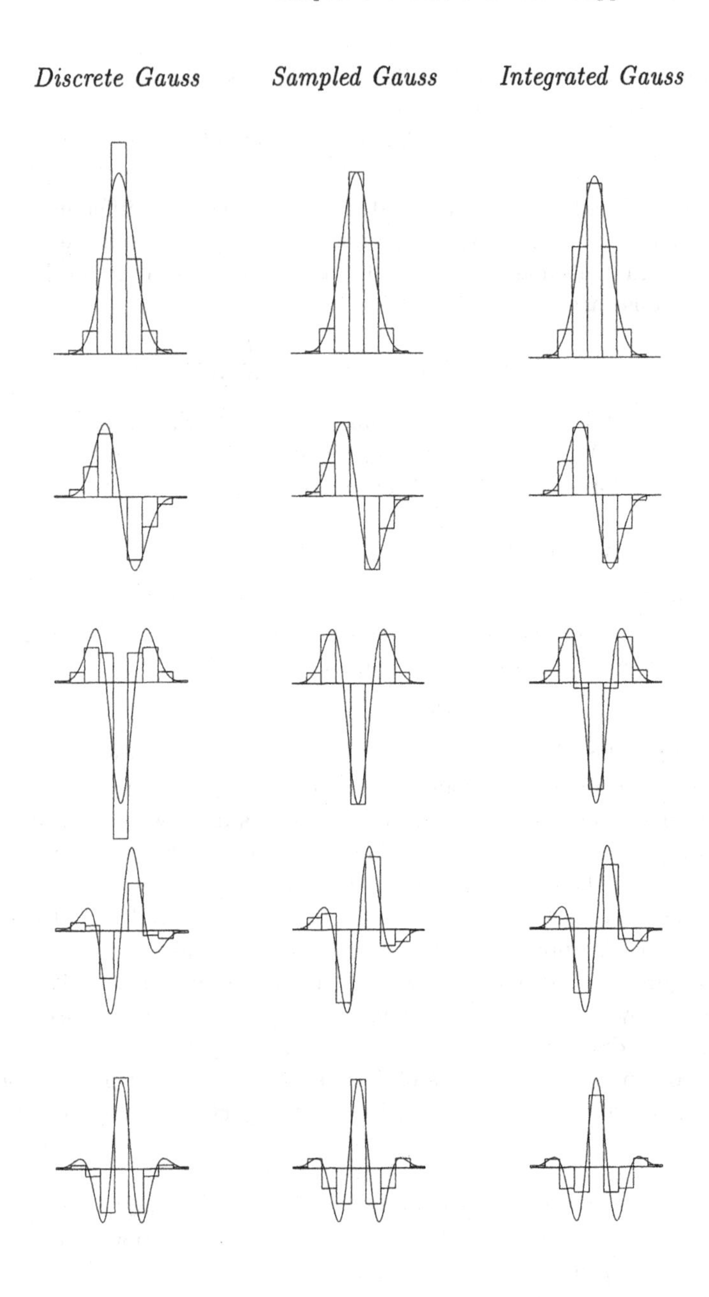

Figure 5.2. Comparisons between the discrete analogue of the Gaussian kernel and the sampled Gaussian kernel at scale level $t = 1.0$. The columns show from left to right; the discrete Gaussian, the sampled Gaussian, and and integrated Gaussian. The derivative/difference order increases from top to bottom; the upper row shows the raw smoothing kernel; then follow the first-, second-, third-, and fourth-order derivative/difference kernels. The block diagrams indicate the discrete kernels and the smooth curve the continuous Gaussian.

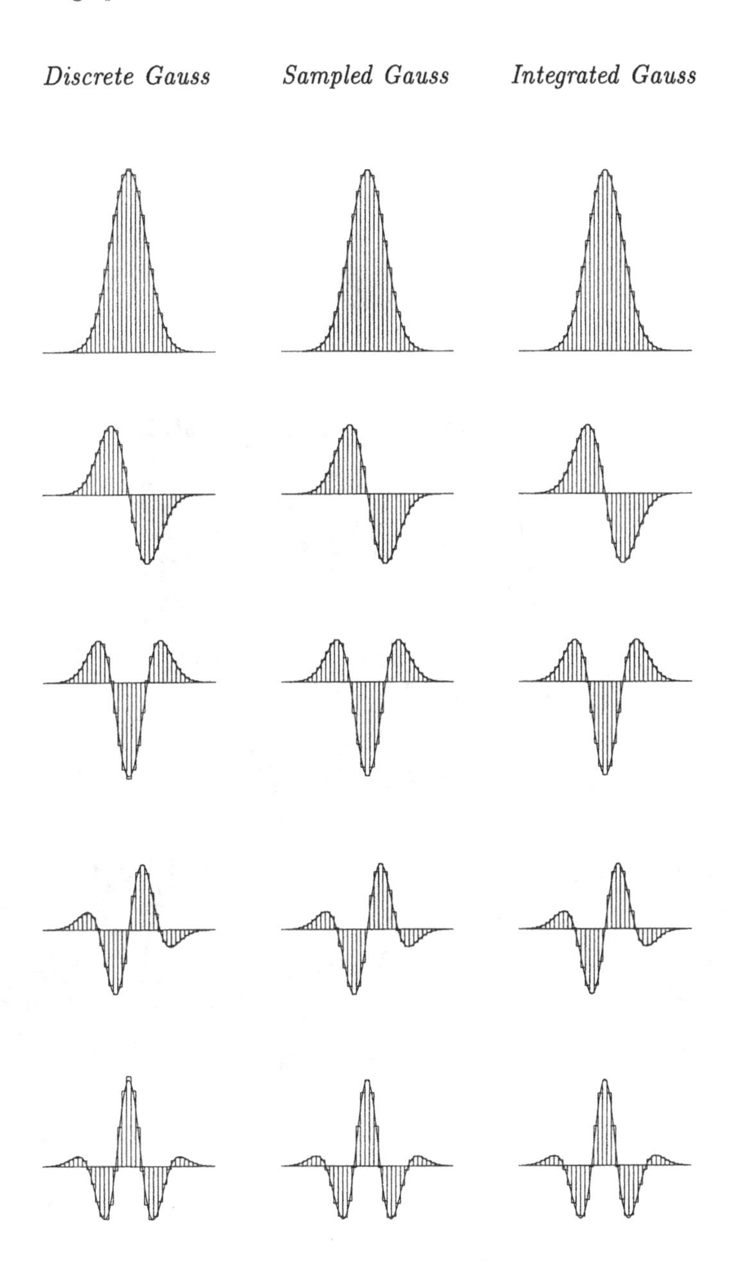

Figure 5.3. Comparisons between the discrete analogue of the Gaussian kernel and the sampled Gaussian kernel at scale level $t = 16.0$. The columns show from left to right; the discrete Gaussian, the sampled Gaussian, and and integrated Gaussian. The derivative/difference order increases from top to bottom; the upper row shows the raw smoothing kernel; then follow the first-, second-, third-, and fourth-order derivative/difference kernels. The block diagrams indicate the discrete kernels and the smooth curve the continuous Gaussian.

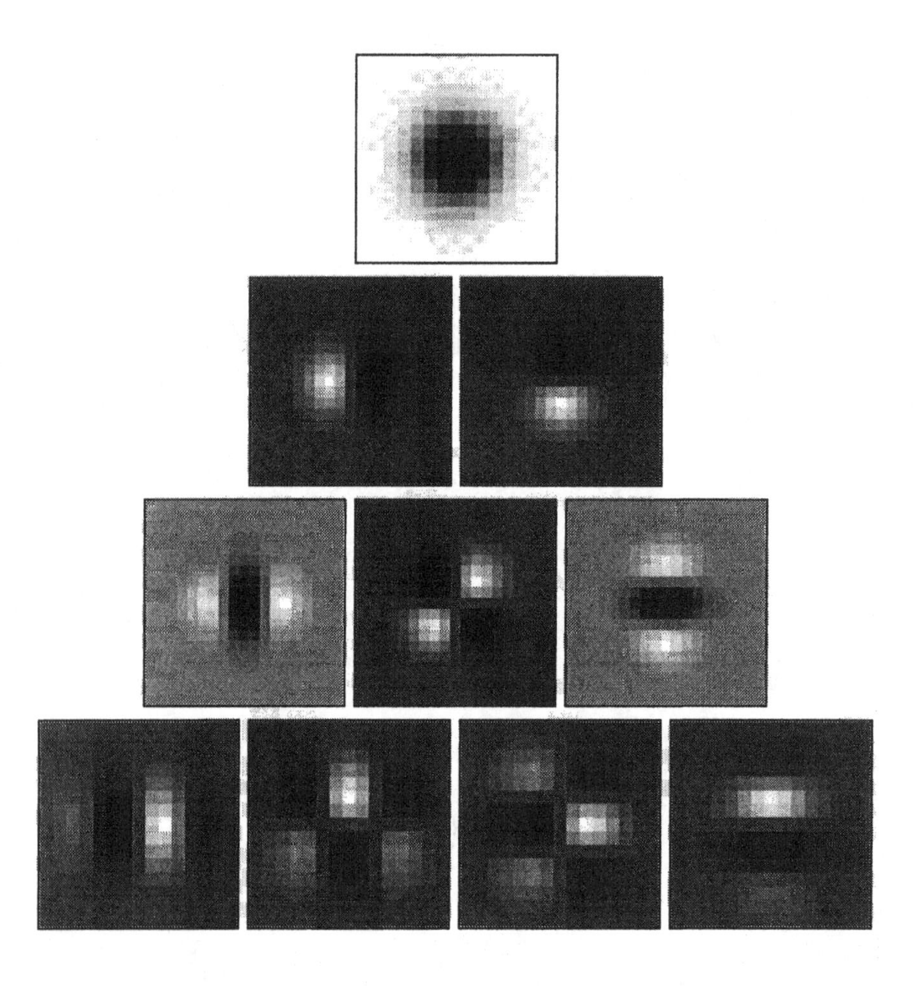

Figure 5.4. Grey-level illustrations of the equivalent two-dimensional discrete derivative approximation kernels up to order three (in the separable case corresponding to $\gamma = 0$). (row 1) (a) Zero-order smoothing kernel, T, (inverted). (row 2) (b–c) First-order derivative approximation kernels, $\delta_x T$ and $\delta_y T$. (row 3) (d–f) Second-order derivative approximation kernels $\delta_{xx} T$, $\delta_{xy} T$, $\delta_{yy} T$. (row 4) (g–j) Third-order derivative approximation kernels $\delta_{xxx} T$, $\delta_{xxy} T$, $\delta_{xyy} T$, $\delta_{yyy} T$. (Scale level $t = 4.0$, image size 15×15 pixels).

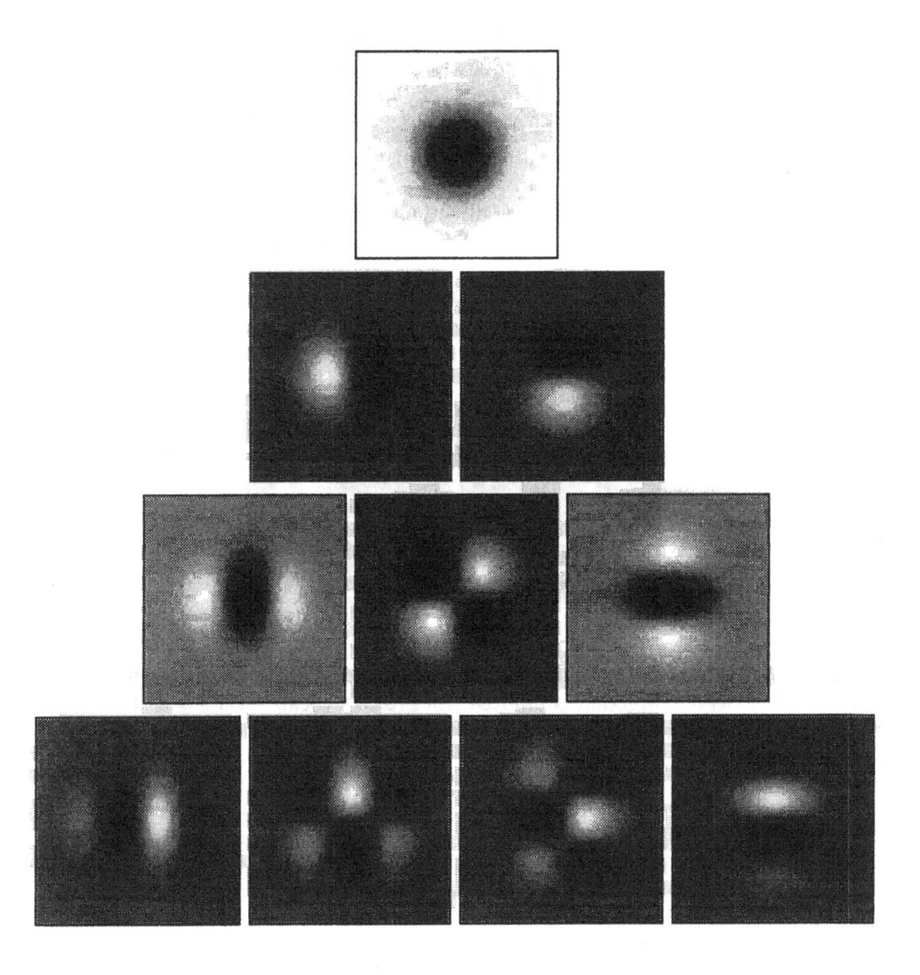

Figure 5.5. Grey-level illustrations of the equivalent two-dimensional discrete derivative approximation kernels up to order three (in the separable case corresponding to $\gamma = 0$). (row 1) (a) Zero-order smoothing kernel, T, (inverted). (row 2) (b–c) First-order derivative approximation kernels, $\delta_x T$ and $\delta_y T$. (row 3) (d–f) Second-order derivative approximation kernels $\delta_{xx} T$, $\delta_{xy} T$, $\delta_{yy} T$. (row 4) (g–j) Third-order derivative approximation kernels $\delta_{xxx} T$, $\delta_{xxy} T$, $\delta_{xyy} T$, $\delta_{yyy} T$. (Scale level $t = 64.0$, image size 127×127 pixels).

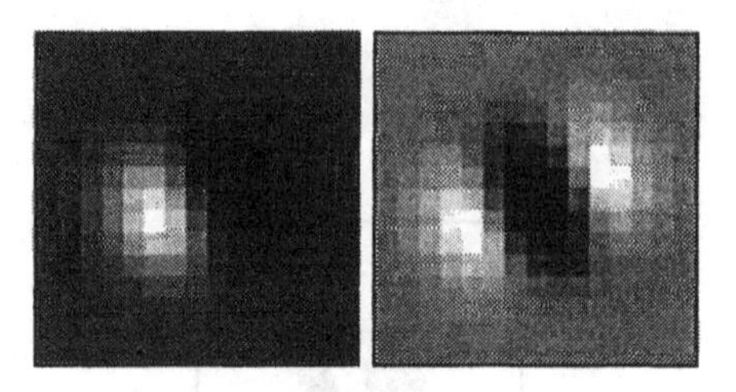

Figure 5.6. First- and second-order directional derivative approximation kernels in the 22.5 degree direction computed from (5.54). (Scale level $t = 4.0$, image size 15×15 pixels.)

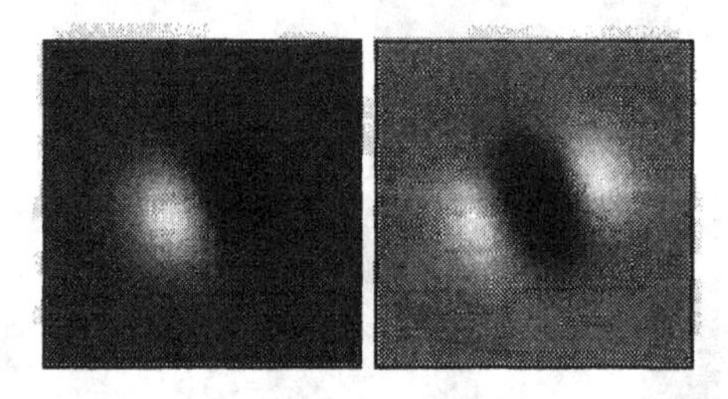

Figure 5.7. First- and second-order directional derivative approximation kernels in the 22.5 degree direction computed from (5.54). (Scale level $t = 64.0$, image size 127×127 pixels.)

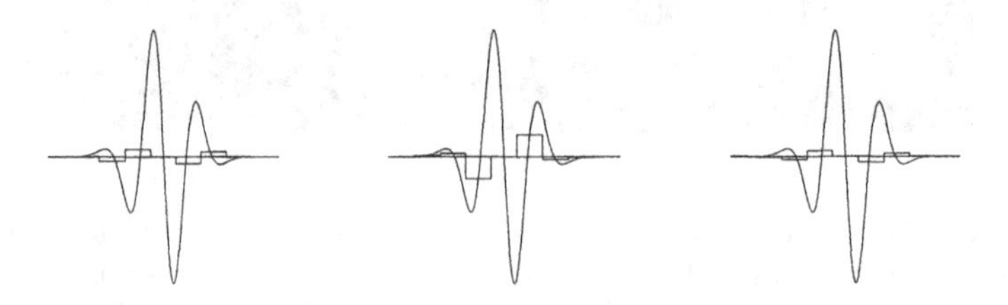

Figure 5.8. Different discrete approximations of the fifth-order derivative of the Gaussian at a fine scale ($t = 0.46$); (a) fifth-order difference of the discrete analogue of the Gaussian kernel, (b) sampled fifth-order derivative of the Gaussian kernel, and (c) fifth-order difference of the sampled Gaussian kernel. Observe that the slope at the origin of the kernel in (b) differs from the slopes of the other ones. This means that if this filter is used for derivative approximations, then, in general, even the sign of the derivative estimate may be wrong.

5.6. Summary and discussion

The object of chapters 3–5[13] has been to describe a canonical way to discretize the primary components in scale-space theory: the convolution smoothing, the diffusion equation, and the smoothed derivatives, such that the scale-space properties hold exactly in the discrete domain.

A theoretical advantage of the proposed discrete theory is that several algebraic scale-space properties in the continuous case transfer directly to the discrete domain, and operators that commute in the continuous case, commute (exactly) also after the discretization. Examples of this are the non-enhancement property of local extrema and the semi-group/cascade smoothing property of the smoothing operator.

A computational advantage of the proposed discrete derivative approximation kernels is that there is no need for re-doing the smoothing part of the transformation when computing several derivative approximations. Exactly the same results are obtained by smoothing the signal with the (large support) discrete analogue of the Gaussian kernel (once), and then computing the derivative approximations by applying different (small support) difference operators to the output from the smoothing operation. (For numerical issues of implementation see appendix A.1.)

The specific difference operators, $\delta_{x^i y^j}$, used here have been selected such that they in a numerical sense constitute consistent discrete approximations to the continuous derivative operators. This means that the discrete approximations will approach the continuous results when the grid effects get smaller, i.e., when the grid spacing becomes small compared to a characteristic length in the data. Hence, with increasing scale, the output from the proposed discrete operators can be expected to approach the corresponding continuous theory results.

It is difficult to say generally how large the numerical effects are in an actual implementation and how seriously they affect the output, since this is very much determined by the algorithms working on the scale-space representation as well as the goal of the analysis in which the scale-space part is just one of the modules. In addition, any specific convergence result depend upon what assumptions are posed on the continuous signal and the sampling method. However, in figures 5.9 and 5.10 some measures have been illustrated of how the difference between the sampled Gaussian kernel and the discrete analogue of the Gaussian kernel behaves as a function of the scale parameter. The graphs verify that the difference is largest for small values of t. Note also how the magnitude of the difference increases with the derivation order.

[13]Although the treatment in this chapter, because of simplicity of presentation, has been concerned with one-dimensional and two-dimensional signals, the methodology is general. Both the sufficiency result corollary 5.9 and the necessity result theorem 5.7 generalize to arbitrary dimensions.

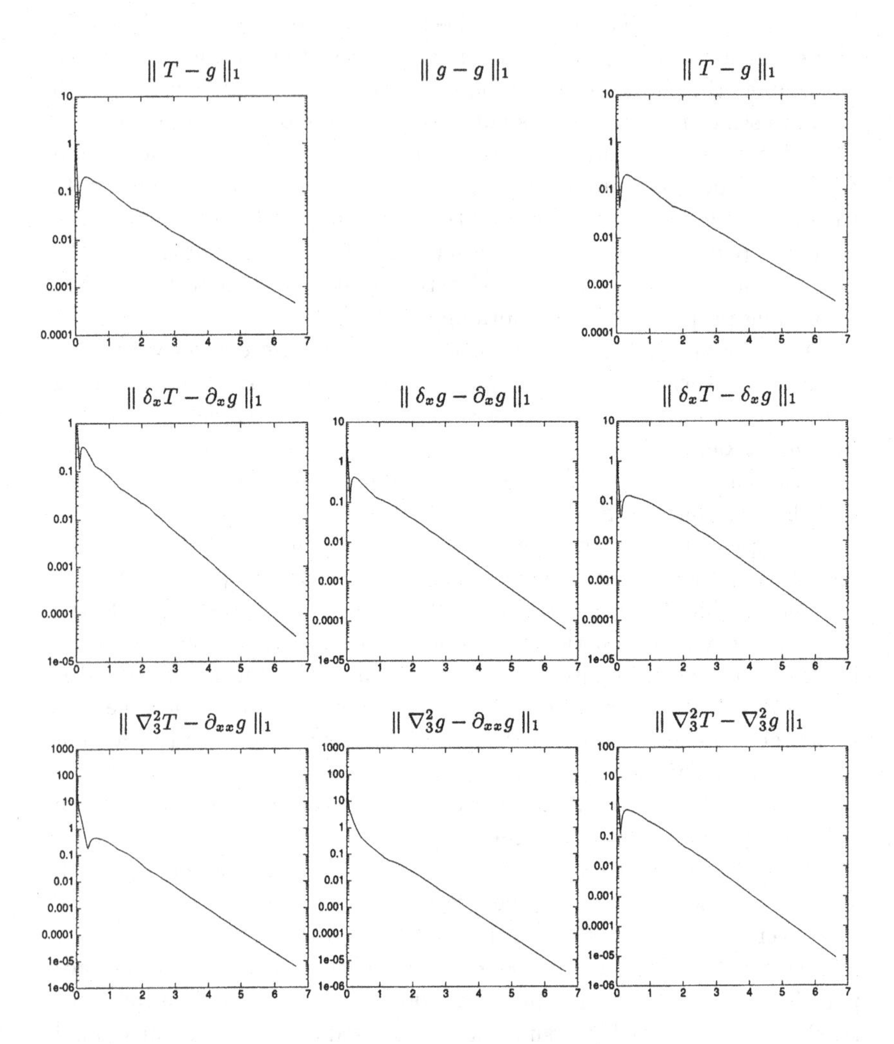

Figure 5.9. Comparisons between various discrete approximations to the continuous Gaussian kernel g and the discrete analogue of the Gaussian kernel T (measured in the l_1 norm in the one-dimensional case); (left column) central differences of T and sampled derivatives of g, (middle column) sampled derivatives and central differences of g, (right column) central differences of T and g. The order of the differences and derivatives increases from top to bottom. The scaling on the horizontal scale axis ($t \in [0, 256]$) is essentially logarithmic (it represents effective scale; see section 7.7.1.1).

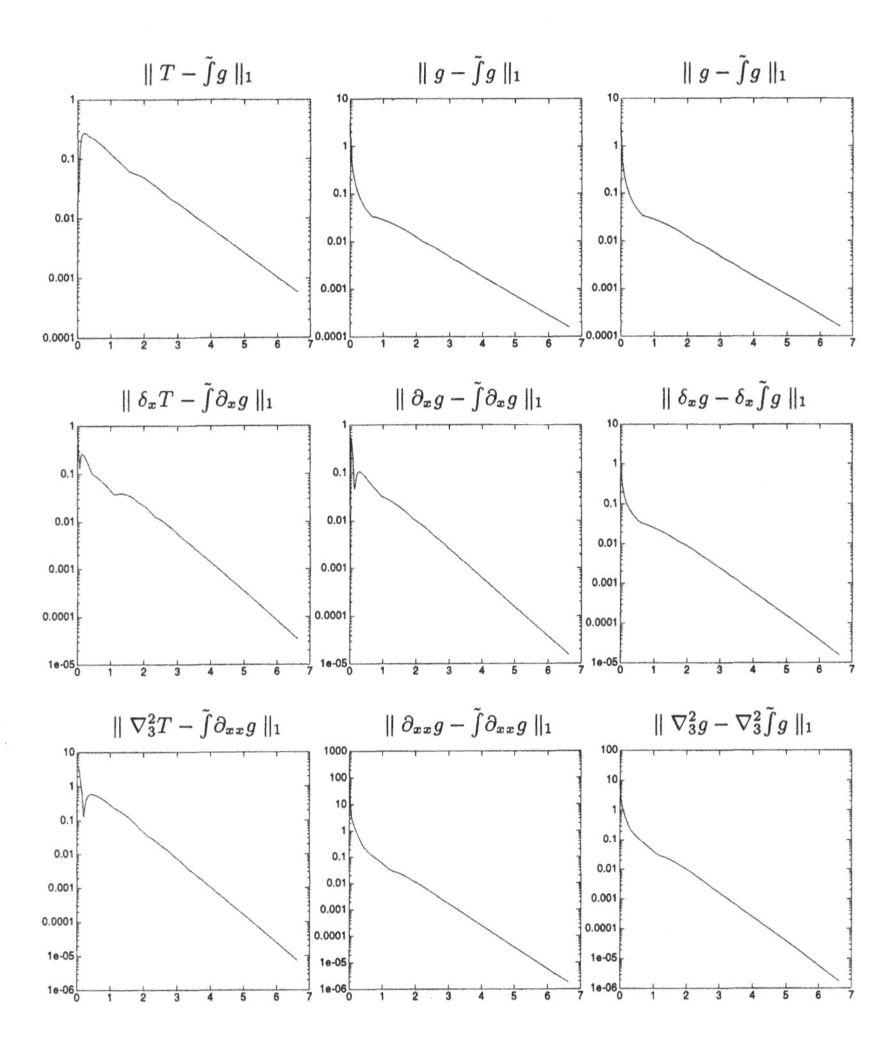

Figure 5.10. Comparisons between various discrete approximations to the continuous Gaussian kernel g and the discrete analogue of the Gaussian kernel T (measured in the l_1 norm in the one-dimensional case). Here, the modified integration sign $\tilde{\int}$ represents integration over each pixel support region. (left column) central differences of T and integrated derivatives g, (middle column) sampled and integrated derivatives of g, (right column) central differences and integrated derivatives of g. The order of the differences and derivatives increases from top to bottom. The scaling on the horizontal scale axis ($t \in [0, 256]$) is essentially logarithmic (it represents effective scale; see section 7.7.1.1).

The proposed framework has been derived from a set of theoretical properties postulated on the first stages of visual processing. In practice, it leads to a conceptually very simple scheme for computing low level differential geometric descriptors from raw (discrete) image data; which lends itself to direct implementation in terms of natural operations in a visual front-end; (i) *large support linear* smoothing, (ii) *small support linear* derivative approximations, and (iii) *pointwise non-linear* computations of differential geometric descriptors. In next chapter, it will be demonstrated how this scheme can be used for feature detection, by adding one more processing step, (iv) *nearest neighbour comparisons*.

6

Feature detection in scale-space

The treatment in previous chapters gives a formal justification for using linear filtering as an initial step in early processing of image data. More important, it provides a catalogue of what filter kernels are natural to use, as well as an extensive theoretical explanation of how smoothing kernels of different order and at different scales can be related. In particular, the discretization problem is extensively treated. This forms the basis for a theoretically well-founded modelling of the smoothing operation.

Of course, linear filtering cannot be used as the only component in a vision system aimed at deriving symbolic representations from images; some non-linear steps must be introduced into the analysis. More concretely, some mechanism is required for combining the output from the Gaussian derivative operators of different order and at different scales into some more explicit descriptors of the image geometry.

6.1. Differential geometry and differential invariants

An approach that has been advocated by Koenderink and his co-workers is to describe image properties in terms of differential geometric descriptors, i.e., different (possibly non-linear) combinations of derivatives. A basic motivation for this position is that differential equations and differential geometry constitute natural frameworks for expressing both physical processes and geometric properties. More technically, and as we have seen in chapter 2, it can also be shown that spatial derivatives are natural operators to derive from the scale-space representation.

When using such descriptors, it should be observed that a single partial derivative, e.g. L_{x_1}, does not represent any geometrically meaningful information, since its value is crucially dependent on the arbitrary choice of coordinate system. In other words, it is essential to base the analysis on descriptors that do not depend on the actual coordinatization of the spatial and intensity domains. Therefore, it is natural to require the representation to be invariant with respect to primitive[1] transformations,

[1] In fact, it would be desirable to directly compute features that are invariant under perspective transformations. Since, however, such differential invariants contain

like translations, rotations, scale changes, and certain intensity transformations. In fact, as we shall also see indications of below, quite a few types of low-level operations can expressed in terms of such multi-scale differential invariants defined from (non-linear) combinations of Gaussian derivatives at multiple scales. Examples of these are feature detectors, feature classification methods, and primitive shape descriptors. In this sense, the scale-space representation can serve as a useful basis for expressing a large number of early visual operations.

Kanatani (1990) and Florack *et al.* (1992, 1993) have pursued this approach of deriving *differential invariants* in an axiomatic manner, and considered image properties defined in terms of directional derivatives along certain preferred coordinate directions. If the direction, along which a directional derivative is computed, can be uniquely defined from the intensity pattern, then rotational invariance is obtained automatically, since the preferred direction follows any rotation of the coordinate system. Similarly, any derivative is translationally invariant. These properties hold both concerning transformations of the original signal f and the scale-space representation L of f generated by smoothing with the rotationally symmetric Gaussian.

Detailed studies of differential geometric properties of two-dimensional and three-dimensional scalar images are presented by Saldens *et al.* (1992), who make use of classical techniques from differential geometry (Spivak 1975; Koenderink 1990), algebraic geometry, and invariant theory (Grace and Young 1965; Weyl 1946) for classifying geometric properties of the N-jet of a signal at a given scale in scale-space.

Here, a short description will be given concerning some elementary results. Although the treatment will be restricted to the two-dimensional case, the ideas behind it are general and can be easily extended to higher dimensions.

6.1.1. *Local directional derivatives*

One way of choosing preferred directions is to introduce a local orthonormal coordinate system (u, v) at any point P_0, with the v-axis parallel to the gradient direction of the brightness L at P_0, and the u-axis perpendicular, i.e., $e_v = (\cos\beta, \sin\beta)^T$ and $e_u = (\sin\beta, -\cos\beta)^T$, where

$$e_v|_{P_0} = \begin{pmatrix} \cos\beta \\ \sin\beta \end{pmatrix} = \frac{1}{\sqrt{L_x^2 + L_y^2}} \begin{pmatrix} L_x \\ L_y \end{pmatrix}\Bigg|_{P_0}. \tag{6.1}$$

derivatives of much higher order, most work has so far been restricted to invariants of two-dimensional Euclidean operations and natural linear extensions thereof, like uniform rescaling and affine transformations of the spatial coordinates.

In terms of Cartesian coordinates, which arise frequently in standard digital images, these local directional derivative operators can be written

$$\partial_{\bar{u}} = \sin\beta\,\partial_x - \cos\beta\,\partial_y, \qquad \partial_{\bar{v}} = \cos\beta\,\partial_x + \sin\beta\,\partial_y. \tag{6.2}$$

This coordinate system is characterized by the fact that one of the first-order directional derivatives, $L_{\bar{u}}$, is zero.

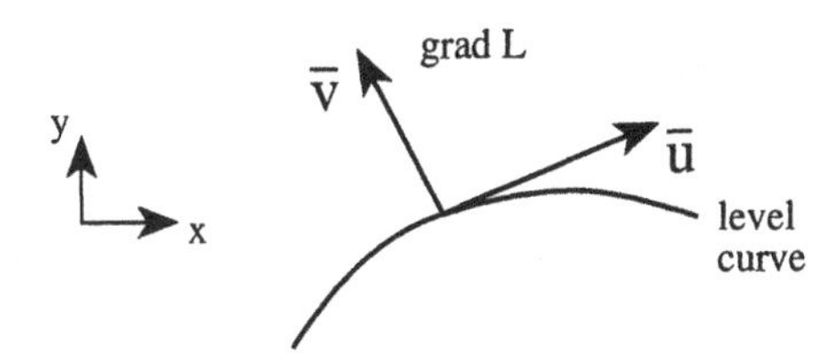

Figure 6.1. In the (u, v) coordinate system, the v-direction is at every point parallel to the gradient direction of L, and the u-direction is parallel to the tangent to the level curve.

Another natural choice is a coordinate system in which the mixed second-order derivative is zero. Such coordinates are named (p, q) by Florack *et al.* (1992). In these coordinates, in which $L_{\bar{p}\bar{q}} = 0$, the explicit expressions for the directional derivatives become slightly more complicated.

6.1.2. Monotonic intensity transformations

One approach to deriving differential invariants is by requiring the differential entities to be invariant with respect to arbitrary *monotonic intensity transformations.* Then, any property that can be expressed in terms of the level curves of the signal is guaranteed to be invariant. A classification by Kanatani (1990) and Florack *et al.* (1992), which goes back to the classical classification of polynomial invariants by Hilbert (1893), shows that concerning derivatives up to order two, there are only two irreducible differential expressions that are invariant to these transformations

$$\kappa = \frac{L_{\bar{u}\bar{u}}}{L_{\bar{v}}} = \frac{L_x^2 L_{yy} + L_y^2 L_{xx} - 2 L_x L_y L_{xy}}{(L_x^2 + L_y^2)^{3/2}}, \tag{6.3}$$

$$\mu = \frac{L_{\bar{u}\bar{v}}}{L_{\bar{v}}} = \frac{(L_x^2 - L_y^2) L_{xy} - L_x L_y (L_{yy} - L_{xx})}{(L_x^2 + L_y^2)^{3/2}}. \tag{6.4}$$

Here, κ is the curvature of level curves in the smoothed signal, and μ the curvature of the flow lines (integral paths) of the gradient vector field. A general scheme for extending this technique to higher-order derivatives and arbitrary dimensions has been proposed by Florack *et al.* (1993).

6.1.3. *Affine intensity transformations*

Another approach is restrict the invariance to *affine intensity transformations*. Then, the class of invariants becomes larger. A natural condition to impose is that a differential expression $\mathcal{D}L$ should (at least) be a *relative invariant* with respect to scale changes, i.e., under rescalings of the spatial coordinates $L'(x) \hat{=} L(sx)$ the differential entity should transform as $\mathcal{D}L' = s^k \mathcal{D}L$ for some k. Trivially, this relation holds for any product of mixed directional derivatives, and extends to sums (and rational functions) of such expressions provided that the differential expression is homogeneous.

In order to give a formal description of this, let $L_{\bar{u}^m \bar{v}^n} = L_{\bar{u}^\alpha}$ denote a mixed directional derivative of order $|\alpha| = m + n$, and let $\mathcal{D}L$ be a (possibly non-linear) homogeneous differential expression of the form

$$\mathcal{D}L = \sum_{i=1}^{I} c_i \prod_{j=1}^{J} L_{\bar{u}^{\alpha_{ij}}}, \tag{6.5}$$

where $c \in \mathbb{R}$, $|\alpha_{ij}| > 0$ for all $i = 1, \ldots, I$ and $j = 1, \ldots, J$, and

$$\sum_{j=1}^{J} |\alpha_{ij}| = N \tag{6.6}$$

for all $i \in [1..I]$. Then, $\mathcal{D}L$ is invariant with respect to translations and rotations, and $\mathcal{D}L$ is relative invariant to uniform rescalings of the spatial coordinates and affine intensity transformations. (Expressions of this form can be written more compactly using tensor notation.)

6.1.4. *Feature detection from differential singularities*

The *singularities* (zero-crossings) of such expressions play an important role. This is a special case of a more general principle of using zero-crossings of differential geometric expressions for describing geometric features; see e.g., Bruce and Giblin (1984) for an excellent tutorial. If a feature detector can be expressed as a zero-crossing of such a differential expression (or a combination), then the feature will also be absolute invariant to uniform rescalings of the spatial coordinates, i.e., size changes. Formally, this invariance property can be expressed as follows:

Let $\mathcal{S}_{\mathcal{D}}L$ denote the *singularity set* of a differential operator of the form (6.5), i.e.,

$$\mathcal{S}_{\mathcal{D}}L = \{(x;\ t) \in \mathbb{R}^2 \times \mathbb{R}_+ : \mathcal{D}L(x;\ t) = 0\}, \tag{6.7}$$

and let $\mathcal{G}$ be the Gaussian smoothing operator, i.e., $L = \mathcal{G}f$. Then, under a number of natural transformations of the spatial domain (represented by $x \in \mathbb{R}^2$) and the intensity domain (represented by either the unsmoothed f or the smoothed L) the singularity sets transform in the following way:

- Under a *translation* $(\mathcal{T}L)(x;\ t) = L(x + \Delta x;\ t)$, where $\Delta x \in \mathbb{R}^2$ is the translation vector, the singularity set is translated to $\mathcal{T}\mathcal{S}_{\mathcal{D}}L = \{(x;\ t)\colon \mathcal{D}L(x + \Delta x;\ t) = 0\}$,
- under a *rotation* $(\mathcal{R}L)(x;\ t) = L(Rx;\ t)$, where R is a 2×2 rotation matrix, the singularity set follows the rotation $\mathcal{R}\mathcal{S}_{\mathcal{D}}L = \{(x;\ t)\colon \mathcal{D}L(Rx;\ t) = 0\}$,
- under a *uniform rescaling* $(\mathcal{U}L)(x;\ t) = L(sx;\ t)$, where $s \in \mathbb{R}_+$ is the scaling constant, also the scale levels are affected $\mathcal{U}\mathcal{S}_{\mathcal{D}}L = \{(x;\ t)\colon \mathcal{D}L(sx;\ \underline{s^2t}) = 0\}$, while
- under *affine intensity transformations* $(\mathcal{A}L)(x;\ t) = aL(x;\ t) + b$, where $a, b \in \mathbb{R}$, the singularity set remains the same.

In more compact operator notation, these properties can be summarized by the following commutative relations:

Transformation		Invariance	
translation	$\mathcal{S}_{\mathcal{D}}\,\mathcal{G}\,\mathcal{T}\,f =$	$\mathcal{S}_{\mathcal{D}}\,\mathcal{T}\,\mathcal{G}\,f$	$= \mathcal{T}\,\mathcal{S}_{\mathcal{D}}\,\mathcal{G}\,f$
rotation	$\mathcal{S}_{\mathcal{D}}\,\mathcal{G}\,\mathcal{R}\,f =$	$\mathcal{S}_{\mathcal{D}}\,\mathcal{R}\,\mathcal{G}\,f$	$= \mathcal{R}\,\mathcal{S}_{\mathcal{D}}\,\mathcal{G}\,f$
uniform scaling	$\mathcal{S}_{\mathcal{D}}\,\mathcal{G}\,\mathcal{U}\,f =$	$\mathcal{S}_{\mathcal{D}}\,\mathcal{U}\,\mathcal{G}\,f$	$= \mathcal{U}\,\mathcal{S}_{\mathcal{D}}\,\mathcal{G}\,f$
affine intensity	$\mathcal{S}_{\mathcal{D}}\,\mathcal{G}\,\mathcal{A}\,f =$	$\mathcal{S}_{\mathcal{D}}\,\mathcal{A}\,\mathcal{G}\,f$	$= \mathcal{S}_{\mathcal{D}}\,\mathcal{G}\,f$

In other words, feature detectors formulated in terms of differential singularities by definition commute with a number of elementary transformations of the spatial and intensity domains, and it does not matter whether the transformation is performed before or after the smoothing step. Some simple examples of feature detectors that can be expressed in this way will be presented in the next section.

6.2. Experimental results: Low-level feature extraction

6.2.1. Edge detection based on local directional derivatives

A natural way to define edges from a continuous grey-level image $L\colon \mathbb{R}^2 \to \mathbb{R}$ is as the set of points for which the gradient magnitude assumes a maximum in the gradient direction. This method is usually referred to as *non-maximum suppression* (Canny 1986; Korn 1988).

6.2.1.1. Differential geometric definition of edges

To give a differential definition of this concept, introduce a curvilinear coordinate system (u, v), such that at every point the v-direction is parallel to the gradient direction of L, and at every point the u-direction is perpendicular. Moreover, at any point $P = (x, y) \in \mathbb{R}^2$, let $\partial_{\bar{v}}$ denote the directional derivative operator in the gradient direction of L at P and $\partial_{\bar{u}}$ the directional derivative in the perpendicular direction. Then, at any $P \in \mathbb{R}^2$ the gradient magnitude is equal to $\partial_{\bar{v}} L$, denoted $L_{\bar{v}}$, at that point. Assuming that the second- and third-order directional derivatives of L in the v-direction are not simultaneously zero, the condition for P_0 to be a gradient maximum in the gradient direction may be stated as;

$$\begin{cases} L_{\bar{v}\bar{v}} = 0, \\ L_{\bar{v}\bar{v}\bar{v}} < 0. \end{cases} \tag{6.8}$$

By expressing the directional derivatives in terms of derivatives with respect to the Cartesian coordinates (x, y), $\partial_{\bar{u}} = \sin\beta\,\partial_x - \cos\beta\,\partial_y$, $\partial_{\bar{v}} = \cos\beta\,\partial_x + \sin\beta\,\partial_y$, where $(\cos\beta, \sin\beta)$ is the normalized gradient direction of L at P_0 this condition can be expressed as

$$\begin{aligned} L_{\bar{v}\bar{v}} &= \frac{L_x^2 L_{xx} + 2L_x L_y L_{xy} + L_y^2 L_{yy}}{L_x^2 + L_y^2} = 0, \\ L_{\bar{v}\bar{v}\bar{v}} &= \frac{L_x^3 L_{xxx} + 3L_x^2 L_y L_{xxy} + 3L_x L_y^2 L_{xyy} + L_y^3 L_{yyy}}{L_x^2 + L_y^2} < 0. \end{aligned} \tag{6.9}$$

Since only the sign information is important, it is sufficient to study the numerators when implementing this operation

$$\begin{aligned} \tilde{L}_{\bar{v}\bar{v}} &= L_{\bar{v}}^2 L_{\bar{v}\bar{v}} = L_x^2 L_{xx} + 2L_x L_y L_{xy} + L_y^2 L_{yy}, \\ \tilde{L}_{\bar{v}\bar{v}\bar{v}} &= L_{\bar{v}}^3 L_{\bar{v}\bar{v}\bar{v}} = L_x^3 L_{xxx} + 3L_x^2 L_y L_{xxy} + 3L_x L_y^2 L_{xyy} + L_y^3 L_{yyy} \end{aligned} \tag{6.10}$$

where $L_{\bar{v}} = (L_x^2 + L_y^2)^{1/2}$. By reinterpreting L as the scale-space representation of a signal f, it follows that the edges in f at any scale t can be defined as the points on the zero-crossing curves of the numerator of $\tilde{L}_{\bar{v}\bar{v}}$ for which $\tilde{L}_{\bar{v}\bar{v}\bar{v}}$ is strictly negative.

Note that with this formulation there is no need for any explicit estimate of the gradient direction. Moreover, there are no specific assumptions about the shape of the intensity profile perpendicular to the edge.

6.2.1.2. *Discrete approximation and interpolation scheme*

Given discrete data, we shall now investigate the effect of approximating derivatives $L_{x^i y^j}$ by using the above discrete derivative approximations. From these in turn, discrete approximations to $\tilde{L}_{\bar{v}\bar{v}}$ and $\tilde{L}_{\bar{v}\bar{v}\bar{v}}$ will be computed as *pointwise* polynomials.

Of course, there do not exist any exact discrete correspondences to zero-crossing curves in the discrete case. Nevertheless, a *sub-pixel* edge detector can be defined by interpolating for zero-crossings in the discrete data. A natural way of implementing this, is by considering each cell with four neighbouring points in the image,

$$\{(x, y), (x + 1, y), (x, y + 1), (x + 1, y + 1)\}, \tag{6.11}$$

and performing a two-step linear interpolation. The idea is basically as follows. For any pair of (four-)adjacent points having opposite sign of $\tilde{L}_{\bar{v}\bar{v}}$, introduce a zero-crossing point on the line between, with the location set by linear interpolation. Then, connect any pair of such points within the same four-cell by a line segment, see figure 6.2(a) for an example.

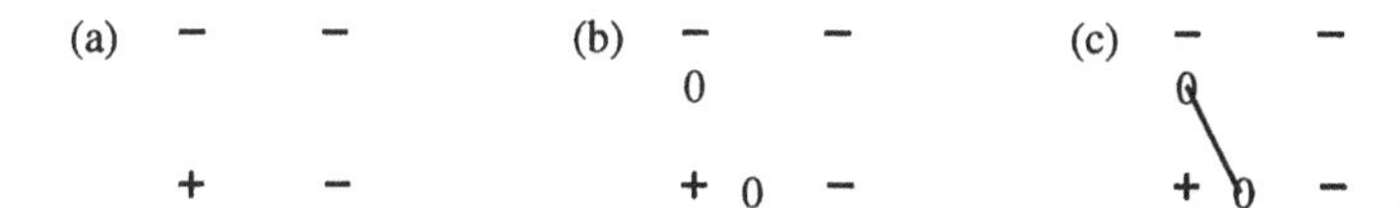

Figure 6.2. The sub-pixel edge interpolation procedure in a simple situation; a four-point cell in which $\tilde{L}_{\bar{v}\bar{v}}$ is positive in one point and negative in all the other ones, while $\tilde{L}_{\bar{v}\bar{v}\bar{v}}$ is negative in all points. (a) Sign pattern for $\tilde{L}_{\bar{v}\bar{v}}$. (b) Estimated locations of zero-crossings from linear interpolation. (c) Zero-crossings connected by a straight line segment. ("+" denotes a pixel for which $\tilde{L}_{\bar{v}\bar{v}} > 0$ and "−" a pixel for which $\tilde{L}_{\bar{v}\bar{v}} < 0$).

The extension to other types of situations is straightforward (see figure 6.3 for examples).[2] If this operation is performed on all four-cells in an image, then edge segments are obtained that can be easily linked into polygons by an edge tracker, and will, by definition, be continuous in space.

Figure 6.3. Topologically different sign patterns for $\tilde{L}_{\bar{v}\bar{v}}$ under the assumption that $\tilde{L}_{\bar{v}\bar{v}\bar{v}}$ is strictly negative at all points in a four-cell. (a) If all points have the same sign (in $\tilde{L}_{\bar{v}\bar{v}}$), then no zero-crossing should be generated. (b) If one point has a sign different from the three other points, then generate one edge segment between the two zero-crossings. (c) If there are two negative and two positive points, and if diagonally opposite points have the same sign, then connect the two zero-crossings. (d) On the other hand, if there are two negative and two positive points, and if diagonally opposite points have the same sign, then the situation is ambiguous. Here, such cases are suppressed.

[2]Cases in which $\tilde{L}_{\bar{v}\bar{v}\bar{v}}$ is positive at some point(s) are somewhat more complicated (there are 240 possible cases). The methodology that has been adopted in this implementation is to exclude all points for which $\tilde{L}_{\bar{v}\bar{v}\bar{v}} > 0$, and then for each pair of (remaining) neighbouring points having opposite sign in $\tilde{L}_{\bar{v}\bar{v}}$ introduce a (preliminary) zero-crossing point along the cell boundary. If exactly two such zero-crossing lines are found in a cell, then an edge segment is generated between them. All other cases are suppressed.

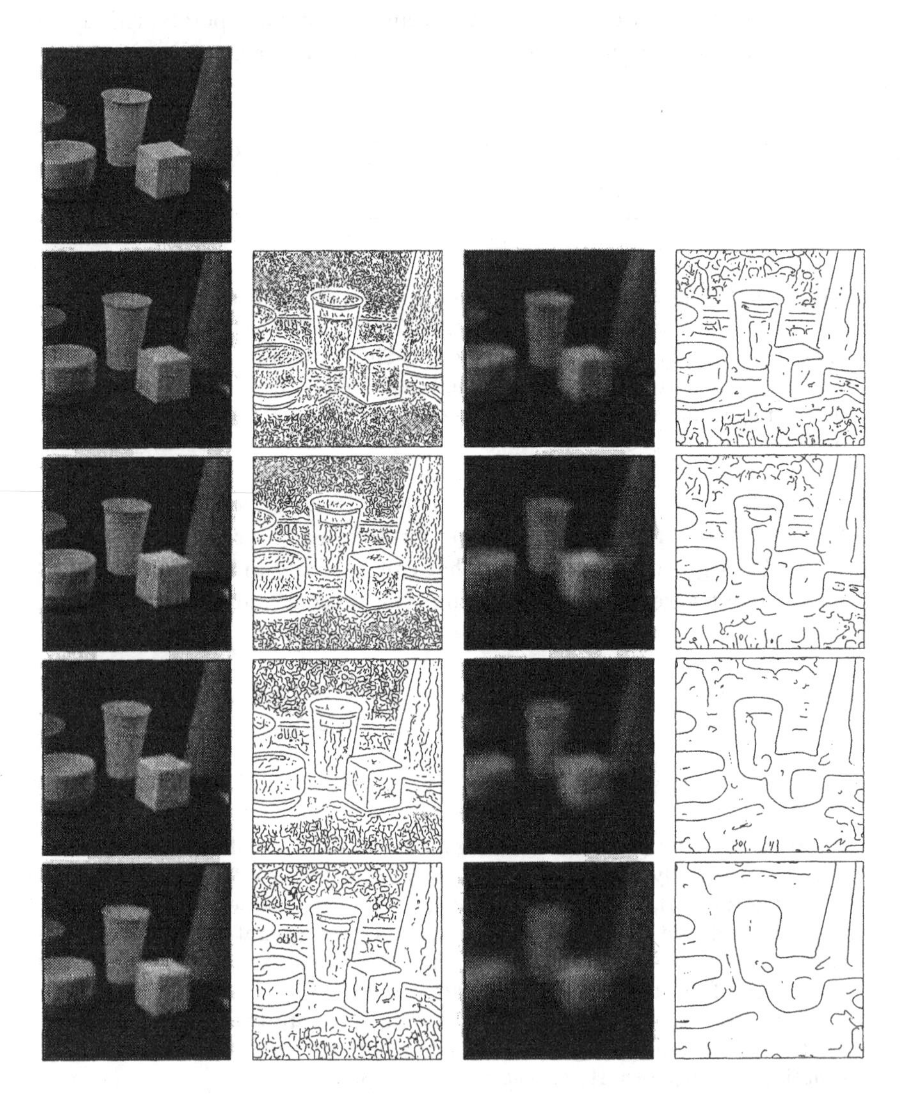

Figure 6.4. Edges detected by applying the presented sub-pixel edge detection scheme at scales $t = 1, 2, 4, 8, 16, 32, 64$, and 128 respectively (from top to bottom). (No thresholding has been performed.)

Figure 6.5. A comparison between the sub-pixel edge detection scheme based on discrete derivative approximations (middle row) and Canny–Deriche edge detection (bottom row). The scale values used for smoothing in the left column were t=4.0 (middle left) and α=0.7 (bottom left), while the corresponding values in the right column were t=1.4 (middle right) and α=1.2 (bottom right).

Figure 6.5. (From previous page.) In the left column no thresholding has been performed on the gradient magnitude or the length of the edge segments, while in the right column hysteresis thresholds on gradient magnitude (low = 3.0, high = 5.0) and a threshold on the length of the edge segments (5.0) were selected manually. Image size: left 256 × 256 pixels, right 512 × 512 pixels.

6.2.1.3. Experimental results

Figure 6.4 displays an example of applying this edge detection scheme to an image of a table scene at a number of different scales, while figure 6.5 shows a simple comparison with a traditional implementation of the Canny–Deriche edge detector (Deriche 1987). Of course, it is not easy to make a fair comparison between the two methods, since the Canny–Deriche method is pixel oriented and uses a different smoothing filter. Moreover, the usefulness of the output is ultimately determined by the algorithms that use the results from this operation as input, and can hardly be measured in isolation.

Finally, it should be remarked that the main intention with experiment is not to argue that this edge detection method is entirely new. It is rather to demonstrate that useful results can be obtained by using the proposed derivative approximations[3] up to order three, and that by using the scheme in figure 5.1, results qualitatively comparable to a state-of-the-art detector can be obtained by very simple means.

6.2.2. Junction detection

A commonly used entity for junction detection is the curvature of level curves in intensity data multiplied by the gradient magnitude. The motivation behind this approach is that corners can basically be characterized by two properties; (i) high curvature in the grey-level landscape, and (ii) high intensity gradient. Using just the level curve curvature is not sufficient, since then a large number of false alarms would be obtained in regions with smoothly varying intensity. Different version of this operator, usually the level curve curvature multiplied by the gradient magnitude raised to the power of one, have been used by several authors (Kitchen and Rosenfeld 1982; Dreschler and Nagel 1982; Koenderink and Richards 1988; Noble 1988; Deriche and Giraudon 1990; Florack *et al.* 1992). In terms of derivatives of the intensity function with respect to the (x, y)-, and (u, v)-coordinates respectively, the level curve curvature can be expressed as (6.3):

$$\kappa = \frac{L_y^2 L_{xx} - 2 L_x L_y L_{xy} + L_x^2 L_{yy}}{(L_x^2 + L_y^2)^{3/2}} = \frac{L_{\bar{u}\bar{u}}}{L_{\bar{v}}}. \tag{6.12}$$

[3]Which were derived solely from theoretical scale-space considerations.

To give a stronger response near edges, the curvature entity is usually multiplied by the gradient magnitude raised to some power, k. A natural choice is $k = 3$. This is the smallest value of k that leads to a polynomial expression

$$\tilde{\kappa} = L_{\bar{v}}^3 \, \kappa = L_{\bar{v}}^2 \, L_{\bar{u}\bar{u}} = L_y^2 L_{xx} - 2 L_x L_y L_{xy} + L_x^2 L_{yy}. \tag{6.13}$$

Moreover, the resulting operator then possesses a certain *skew invariance* property (Blom 1992): If the x- and y-coordinate axes are rescaled by different scaling factors s_1 and s_2, then (because of the homogeneity in terms of derivatives with respect to x and y) $\tilde{\kappa}$ will be multiplied by $s_1^2 s_2^2$. This means that any maximum remains a maximum under non-uniform rescalings. Since (6.13) is valid in an arbitrarily oriented coordinate system, this invariance holds in any direction. Hence, if a junction candidate is defined as a maximum in $\tilde{\kappa}$, a qualitatively similar type of response can be expected for junctions with different angles between the edges.

Assuming that the first- and second-order differentials of $\tilde{\kappa}$ are not simultaneously degenerate, a necessary and sufficient condition for a point P_0 to be a maximum in this rescaled level curve curvature is that:

$$\begin{cases} \partial_{\bar{u}}(\tilde{\kappa}) = 0, \\ \partial_{\bar{v}}(\tilde{\kappa}) = 0, \\ \det \mathcal{H}(\tilde{\kappa}) = \tilde{\kappa}_{\det \mathcal{H}} = \tilde{\kappa}_{\bar{u}\bar{u}} \tilde{\kappa}_{\bar{v}\bar{v}} - \tilde{\kappa}_{\bar{u}\bar{v}}^2 > 0, \\ \operatorname{sign}(\tilde{\kappa}) \tilde{\kappa}_{\bar{u}\bar{u}} < 0. \end{cases} \tag{6.14}$$

Interpolating for simultaneous zero-crossings in $\partial_{\bar{u}}(\tilde{\kappa})$ and $\partial_{\bar{u}}(\tilde{\kappa})$ gives a potential method for sub-pixel junction detection.

Figure 6.6 displays the result of computing the absolute value of this *rescaled level curve curvature* at a number of different scales. As a means of enhancing the maxima in the output, a certain type of blob detection, *grey-level blob* detection (see chapter 7), has been applied to the curvature data. Basically, each grey-level blob corresponds to one local extremum and vice versa. Observe that at fine scales mainly blobs due to noise are detected, while at coarser scales the operator gives a stronger response in regions that correspond to meaningful junctions in the scene.

6.2.3. *Blob detection*

Zero-crossings of the Laplacian

$$\nabla^2 L = L_{\bar{u}\bar{u}} + L_{\bar{v}\bar{v}} = L_{xx} + L_{yy} = 0 \tag{6.15}$$

have been used for *stereo matching* (Marr 1982), and *blob detection* (Blostein and Ahuja 1987). Blob detection methods can also be formulated in terms of local extrema in the grey-level landscape (see chapter 7 and chapter 10), as well as extrema in the Laplacian operator (see chapter 13).

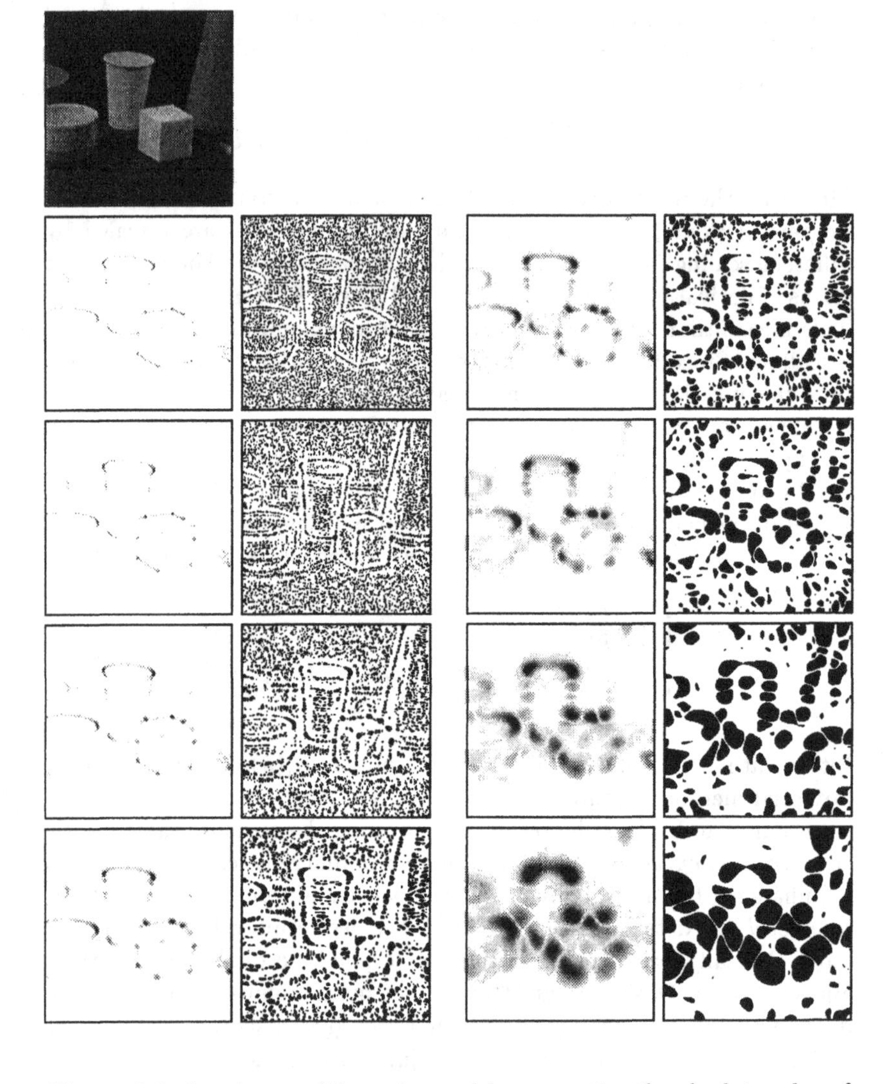

Figure 6.6. Junction candidates detected by computing the absolute value of the rescaled level curve curvature $\tilde{\kappa}^2$ at scales $t = 1, 2, 4, 8, 16, 32, 64$, and 128 respectively (from top to bottom). (No thresholding has been performed.) To enhance the maxima in the output, a certain type of blob detection, *grey-level blob* detection (see chapter 7), has been applied to the curvature data. Basically, each dark region is the support region of a grey-level blob and corresponds to a unique local extremum and vice versa. Observe that the junctions in the image appear as single maxima at certain scales. Methods for selecting significant junction candidates and scale levels from this data are described in section 11.3.7 and chapter 13.

Zero-crossings of the Laplacian have been used also for edge detection, although the localization is poor at curved edges. This property can be understood from the relation between the Laplacian operator and the second derivative in the gradient direction

$$\nabla^2 L = L_{\bar{u}\bar{u}} + L_{\bar{v}\bar{v}} = L_{\bar{v}\bar{v}} + \kappa L_{\bar{v}}, \tag{6.16}$$

which follows from the rotational invariance of the Laplacian operator and (6.12). This example constitutes a simple indication of how theoretical analysis of feature detectors becomes tractable when expressed in terms of the differential geometric framework.

6.3. Feature detection from differential singularities

The applications above exemplify how low-level feature detectors can be formulated in terms of singularities of differential entities. By this, one more step has been added to the flow of computations illustrated in figure 5.1, namely *singularity detection*, which in these cases is equivalent to the detection of zero-crossings and/or local extrema; operations that correspond to *nearest-neighbour processing*; see figure 6.7.

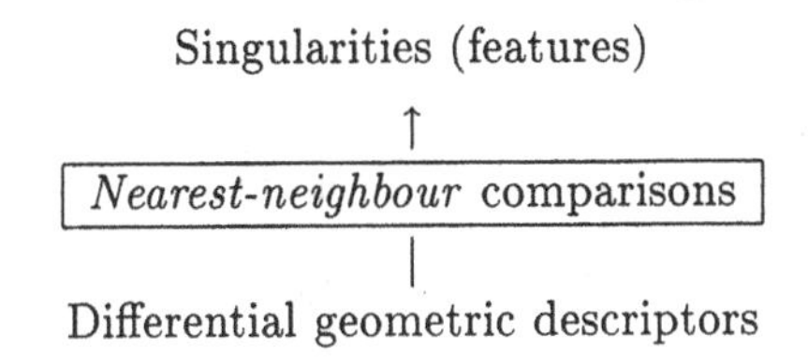

Figure 6.7. The low-level feature extractors have been expressed in terms of singularities of different differential geometric entities. This corresponds to the addition of one more processing step to the flow of computations illustrated in figure 5.1, namely singularity detection. This operation can (at least in these cases) be equivalently expressed in terms of detection of zero-crossings (and/or local extrema) corresponding to comparisons between nearest-neighbour pixels.

One of the main reasons why the formulation in terms of singularities is important is because these singularities do not depend on the actual numerical values of the differential geometric entities, but only on their *relative* relations. By this, they will be less affected by scale-space smoothing, which is well-known to decrease the amplitude of the variations in a signal and its derivatives. As we have seen in section 6.1.4, differential singularities are also invariant to certain primitive transformations.

6.4. Selective mechanisms

In the treatment so far, the major aim has been to demonstrate what information can be obtained from the computed data without introducing any commitments in the processing. Therefore, no attempts have been made to suppress "irrelevant" edges or junction candidates by thresholding or by other means. Nevertheless, when to use the output from these processes modules as input to other ones, there is an obvious need for some *selective mechanisms* for deciding what structures in the image should be regarded as significant, and what scales should be used for treating those. Figure 6.8 shows the result of applying a standard hysteresis thresholding step on gradient magnitude to the data in figure 6.4.

Figure 6.8. Edges selected from the data in figure 6.4 by applying hysteresis thresholding on gradient magnitude with manually selected thresholds. (left) Grey-level image. (middle) Edges detected at scale t=2 (low = 1.0, high = 3.0). (right) Edges detected at scale t=4 (low = 1.0, high = 2.0).

More generally, some method is necessary for selecting interesting scale levels for further analysis. The differential singularities play an important role in this context. As we shall see in chapter 8, such features can be easily related and linked across scales in a well-defined manner. For example, given a pointwise entity, like a local extremum or a junction candidate given by a maximum in $\tilde{\kappa}^2$, the implicit function theorem allows for the definition of a trajectory across scales along which the topology is locally the same. A methodology that will be explored in the following chapters is to use the *lifetime* of such structures in scale-space as an important ingredient when formulating a significance measure, and select scale levels along such trajectories where the magnitude of the response is maximal.

Part II

The scale-space primal sketch: Theory

7

The scale-space primal sketch

Scale-space theory provides a well-founded framework for dealing with image structures, which naturally occur at different scales. According to this theory, from a given signal one can generate a family of derived signals by successively removing structures when moving from finer to coarser scales. At any scale in this representation, features can then be defined either by geometric constructions or by combining Gaussian derivatives into differential invariants. In contrast to other multi-scale or multi-resolution representations, scale-space is based on a precise mathematical definition of causality, or scale invariance, and the behaviour of structure as scale changes can be analytically described. However, the information in the scale-space embedding is only *implicit* in the grey-level values. The smoothed images in the raw scale-space representation contain no *explicit* information about the features in them or the relations between features at different levels of scale.

The goal of the second part of this book is to present a theory for constructing such an explicit representation, the *scale-space primal sketch,* based on formal scale-space theory. This material constitutes the framework for the third part, where it will be demonstrated that the suggested representation allows for extraction of significant image structures in such a way that the output can serve as a guide to later stage visual processes.

We shall treat grey-level images, and the chosen features will be blobs, i.e., bright regions on dark backgrounds or vice versa. In this respect, the analysis is based on what can be termed the zero-order image structure, i.e., the raw smoothed grey-levels before differentiation. However, the methodology applies to any bounded function and is therefore useful in many tasks occurring in computer vision, such as the study of level curves, spatial derivatives in general, depth maps, and histograms, point clustering and grouping, in one or several dimensions.

From experiments one can (visually and subjectively) observe that the main features that arise in this scale-space representation seem to be blob-like, i.e., they are regions either brighter or darker than the background (see figure 7.10). Especially, such regions that stand out from the surroundings in the original image seem to be further enhanced by scale-space smoothing. The suggested scale-space primal sketch focuses on this aspect of image structure. The purpose is to build a representation for making such information in scale-space explicit. Therefore, there is a need to formalize what should be meant by a "blob."

7.1. Grey-level blob

What properties should be required from a blob definition? Intuitively, one would generally like a blob to be a connected region that is either significantly brighter or significantly darker than its neighbourhood. It should have a sufficiently large area and be stable over some sufficiently large interval in scale-space. Also, a blob should have some kind of natural significance measure associated with it.

It is clear that a blob should be a region associated with (at least) one local extremum. However, it is also essential to define the spatial extent of the region around the blob, and to associate a significance measure with it. Ehrich and Lai (1978) considered the extent problem. They allowed peaks to extend to valleys, a definition that will give non-intuitive results, e.g., for small peaks on large slopes. Koenderink and van Doorn (1984) briefly touched upon the problem with reference to work by Maxwell (1870) concerning level curves and critical points. The definition proposed here is related to these arguments.

7.1.1. Definition of grey-level blob

The blob definition this work is based on should be evident from figure 7.1. The basic idea is to let the blob extend "until it would merge with another blob."

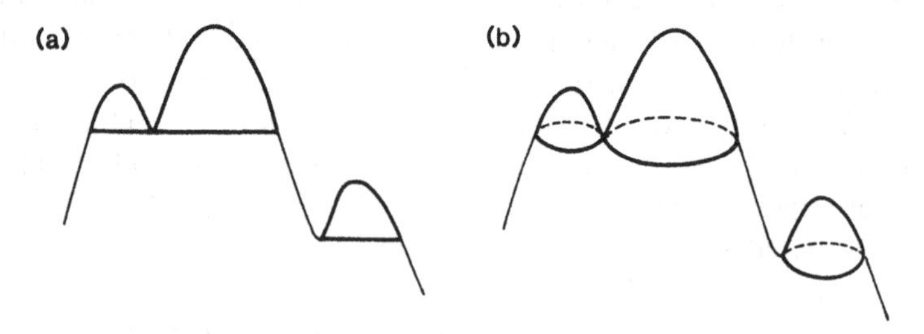

Figure 7.1. Illustration of the grey-level blob definition for (a) a one-dimensional signal, and (b) a two-dimensional signal. This figure shows bright blobs on a dark background. In one dimension a bright grey-level blob is generically given by a pair consisting of one local maximum and one local minimum, in two dimensions by a pair consisting of a maximum and a saddle.

To illustrate this notion, consider a grey-level image at a fixed level of scale, and study the case with bright blobs on a dark background. Imagine the image function as a flooded grey-level landscape. If the water level sinks gradually, peaks will appear. At some instances two different peaks become connected. The corresponding elevation levels or grey-levels are called the *base-levels* of the blobs, and are used for delimiting the spatial extent of the blobs. The *support region* of the blob is defined to consist of those points that have a grey-level exceeding the base-level and can be reached from the local maximum point without descending below the base-level of the blob.

Hence, a bright blob will grow and include points having lower grey-levels until it would meet with another blob. In this sense the blob definition can be regarded as conservative, since no attempt is made to include points in other directions. From this construction, the *grey-level blob* is defined as the three-dimensional volume delimited by the grey-level surface and the base-level. The three-dimensional grey-level blob volume constitutes a combined measure of the contrast and the spatial extent (area) of the blob.

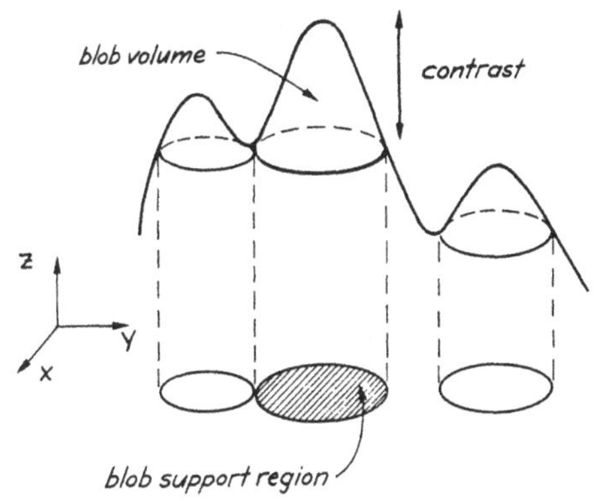

Figure 7.2. Illustration of the grey-level blob definition for a two-dimensional signal, with some descriptive quantities of a grey-level blob; volume, area, and contrast. This figure shows bright blobs on a dark background. Generically, a grey-level blob is given by a pair consisting of one extremum and one saddle point, denoted delimiting saddle point.

7.1.2. Mathematical definition

A precise mathematical definition of the grey-level blob concept can be stated as follows: Consider again the case with bright blobs on dark background, and assume a continuous generic (non-degenerate)[1] grey-level function $f: \mathbb{R}^2 \to \mathbb{R}$ at a fixed level of scale. Consider a local maximum

[1]Unless otherwise stated, the signals are throughout assumed to be *Morse*, i.e., all critical points are assumed to be non-degenerate, and all critical values are assumed to be distinct.

$A \in \mathbb{R}^2$. For any grey-level $z < f(A)$ let

$$X_z^{(A)} = \{\text{the connected component of } \{(x;\ \zeta) \in \mathbb{R}^2 \times \mathbb{R}: z \leq \zeta \leq f(x)\} \text{ that contains } (A, f(A))\}, \tag{7.1}$$

and define the sets $G_z^{(A)}$ and $H_z^{(A)}$ as follows: A point $(B, \zeta_0) \in X_z^{(A)}$ belongs to $G_z^{(A)}$ ($H_z^{(A)}$) if and only if there exists a path $p_{(A,f(A)),(B,\zeta_0)}$ from $(A, f(A))$ to (B, ζ_0) such that (i) every point on the path belongs to $X_z^{(A)}$, and (ii) the derivative of ζ along this path $\zeta|'_{p(A,f(A)),(B,\zeta_0)} < 0$ ($\zeta|'_{p(A,f(A)),(B,\zeta_0)} \leq 0$). The base-level of the blob $z_{base}(A)$ is then defined as the maximum value of z such that

$$z_{base}(A) = \max_{z<f(A)} z: \ \overline{G_z^{(A)}} \neq H_z^{(A)}, \tag{7.2}$$

where the notation $\overline{C}$ stands for the closure of a set C. $z_{base}(A)$ is the grey-level value of the *delimiting saddle point* $S = S_{delimit}(A)$ associated with A. The grey-level blob associated with the local maximum A is the set of points

$$G_{blob}(A) = \overline{G_{z_{base}(A)}^{(A)}}, \tag{7.3}$$

with the (three-dimensional) *grey-level blob volume*

$$G_{vol}(A) = \int_{(x;\ z) \in G_{blob}(A)} dx\, dz. \tag{7.4}$$

The projection of this region onto the spatial plane is called the *support region,*

$$D_{supp}(A) = \{x \in \mathbb{R}^2 : (x;\ \zeta) \in G_{blob}(A) \text{ for some } \zeta\}, \tag{7.5}$$

and the difference in grey-level between the extremum point and the base-level gives the *blob contrast*

$$C_{blob}(A) = f(A) - z_{base}(A). \tag{7.6}$$

It is worth stressing that the grey-level blob is treated as an object with extent both in space and grey-level. The definition is expressed in terms of two-dimensional continuous signals, but applies in arbitrary dimensions. Similarly, it can be extended to discrete signals by replacement of $\mathbb{R}^2$ by $\mathbb{Z}^2$, and by letting the paths be given by a suitable connectivity concept (e.g., eight-connectivity for a two-dimensional square grid). In the discrete case, the derivative condition $f|'_{p_{A,B}} < 0$ is replaced by a difference condition $f(x^{(k+1)}) - f(x^{(k)}) < 0$ along the path $\{x^{(k)}\}$.

Local minima can be treated analogously, and every local minimum point gives rise to a dark blob on bright background.

7.1.3. Properties

It can be easily verified that a blob will be connected. Moreover, the base level of a bright blob is in one dimension attained at a minimum point, in two dimensions at a saddle point. Consequently, the blobs are directly determined from topological properties of the grey-level landscape, namely the first-order singularities.

These blobs are not purely local features, as are extrema, but regional. An inherent property of the stated definition is that it leads to a competition between parts; the presence of another nearby blob might neutralize a blob or reduce its size. In other words, features manifest themselves only relative to the background. These aspects reflect important principles of the approach.

Note that this definition leads to separate systems for bright and dark blobs. This implies that some points may be left unclassified. Consequently, the given definition will in contrast to, e.g., the sign of the Laplacian of the Gaussian only attempt to make a partial (and hopefully safer) classification of the grey-level landscape.

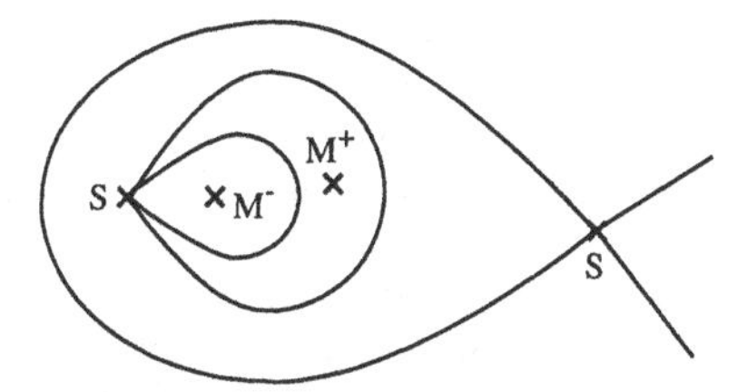

Figure 7.3. Example with a dark blob contained in a bright blob. This phenomenon can be avoided if the blob definition is modified such that a blob is allowed to delimit its own extent in such situations. Then, it will be guaranteed that no point belongs to both a dark and a bright blob. (M^+ = maximum point, M^- = minimum point, S = saddle point).

In one dimension the bright and dark blobs of a signal will be closely related, since a minimum point which delimits the extent of a bright blob will also constitute the seed of a dark blob. In two dimensions the situation is slightly different, since a saddle point that delimits the extent of a bright blob will not delimit the extent of any dark blob, unless the signal is degenerate. Therefore, in two dimensions, a point in a blob will in general belong to either a dark blob or a bright blob. In certain types of situations, however, it may indeed happen that some points are classified as belonging to both a dark blob and a bright blob (see figure 7.3 for an example). If for some reason this type of phenomenon is not desired, then it can be easily prevented from happening if the blob definition is modified slightly so that a blob is allowed to "delimit its own extent."

7.2. Grey-level blob tree

If the imaginary water level used for constructing grey-level blobs in figure 7.1 is allowed to decrease below the base level of a blob, then the grey-level blob will merge with the adjacent region sharing the same saddle point. By considering all such events under variations of the water level, a tree-like structure can be defined with successive inclusion relations. Every arc corresponds to a range in grey-level where the topology is locally the same, and the grey-level blobs constitute the leaves.

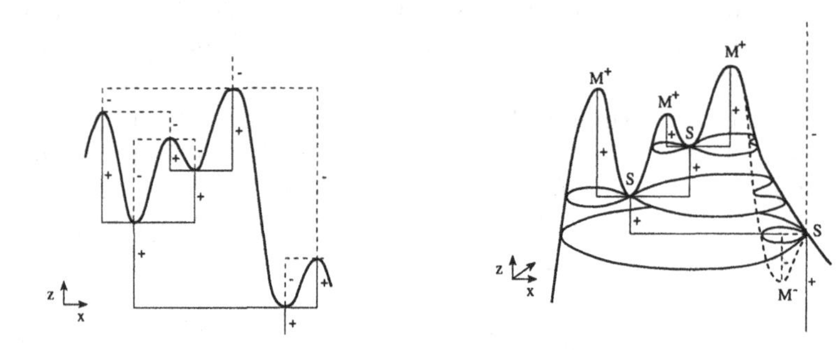

Figure 7.4. Examples of grey-level blob trees: (left) for a one-dimensional signal, and (right) for a two-dimensional signal. The right figure shows a mountain-like grey-level landscape with three main peaks (marked by "M^+"), and one hole (marked by "M^-"). Arcs originating from bright blobs are drawn with filled lines (marked by "+"), and arcs corresponding to dark blobs are drawn by dashed lines (marked by "−").

This representation, termed *grey-level blob tree*, has a qualitative similarity with the relational tree studied by Ehrich and Lai (1978). Simple self-explanatory examples demonstrating its construction are given in figure 7.4. Similarly to a leaf, every arc in the tree can be associated with a three-dimensional volume, as well as a support region and an area measure. The formal procedure for defining such relations is by treating every encountered delimiting saddle point as the seed of a new region, and then proceeding with the successive construction of grey-level blobs and arcs with decreasing water level. In this approach, every delimiting saddle point representing two merging regions is treated in the same way as an ordinary maximum, once a merge between the two regions at the saddle point has been registered.

By simultaneously considering the bright and dark grey-level blob trees of a signal, it is possible to express formal relations between blobs of reverse polarity. The saddle points and the level curves through those constitute the links, since these are the only entities occurring in both systems.

Finally, it should be pointed out that what has been defined here is a grey-level blob (and a corresponding tree) at one level of scale. When such objects are linked across scales, they result in scale-space blobs (and corresponding trees), which will be described after the next section.

7.3. Motivation for introducing a multi-scale hierarchy

It is easy to realize that the concept of a grey-level blob at a single level of scale is not powerful enough for stable extraction of image structures. It leads to an extreme degree of noise sensitivity, since two closely situated local extrema will neutralize each other. This means that a large peak distorted by a few superimposed local extrema of low amplitude will not be detected as one unit; only the fine scale blobs will be found (see figure 7.5).

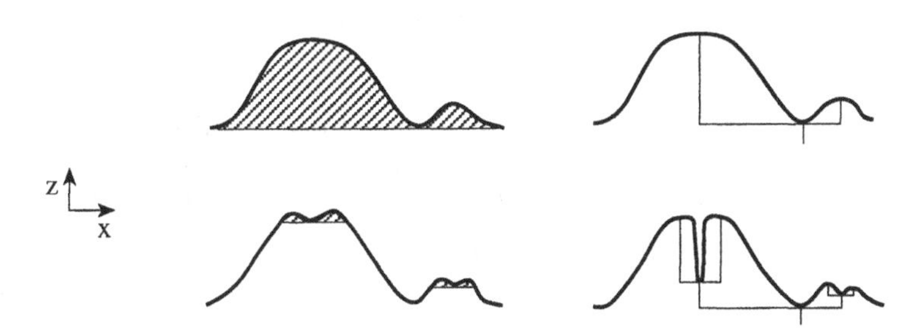

Figure 7.5. (left) A high-contrast large peak with two superimposed low-contrast fine scale peaks will not be detected as a grey-level blob if the signal is considered at one scale only. (right) A single noise spike can also substantially affect the relational tree.

Also the grey-level blob tree (and the relational tree) will be noise sensitive when considered at a single level of scale, since the hierarchical relations between different blobs are determined directly by the grey-levels in the valleys of the original signal. For example, a thin elongated structure superimposed onto the data may completely change the topological relations (see figure 7.5). A naive observer might say that the problems can be easily solved by thresholding, but how to select a proper threshold automatically?

To obtain more stable descriptors, it is natural to consider the behaviour of the grey-level blobs and the grey-level blob tree in scale-space. Since no *a priori* information can be expected about what scales are relevant, the only reasonable approach is to consider *all* scales simultaneously.

7.4. Scale-space blob

Given a grey-level blob existing at some level of scale, there will in general be a corresponding blob both at a slightly finer scale and a slightly coarser scale. Linking such grey-level blobs across scales, gives four-dimensional objects, called *scale-space blobs.* (A formal definition of how the linking is performed is given in section 8.2.1).

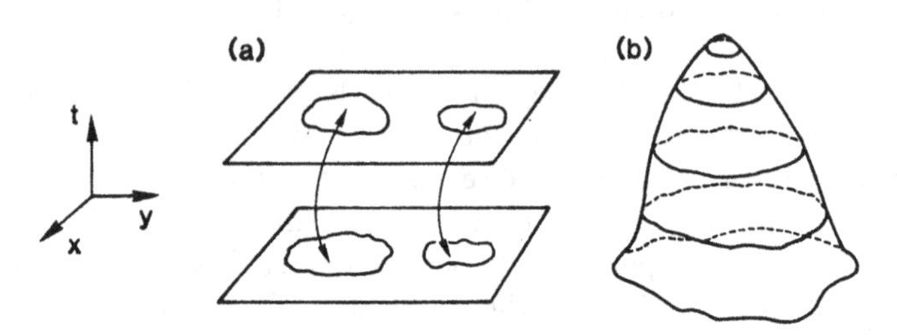

Figure 7.6. (a) Linking similar grey-level at adjacent levels of scale gives (b) scale-space blobs, which are objects with extent both in space, grey-level and scale. (In this figure the grey-level coordinate has been omitted. The slices illustrate the support regions of the grey-level blobs.)

At some level of scale it might be impossible to accomplish a plain link between a grey-level blob at that scale and a corresponding blob at a slightly coarser or finer scale. A *blob event* has occurred affecting the connectivity of the blobs. According to a classification in section 8.3, there are four possible types of blob events with increasing scale;

- *annihilation*—a blob disappears,
- *merge*—two blobs merge into one,
- *split*—one blob splits into two,
- *creation*—a new blob appears.

In summary, the classification of blob events means that each blob event corresponds to an annihilation or a creation of a pair consisting of one saddle point and one extremum point. For example, the difference between a blob annihilation and a blob merge, is that in the first case, the delimiting saddle point of the blob is contained in only one grey-level blob, while in the second case, it is part of two different grey-level blobs (see figure 7.7).

The scale levels where these singularities take place delimit the extent of the scale-space blobs in the scale direction. Consequently, every scale-space blob will be associated with a minimum scale, denoted appearance scale t_A, and a maximum scale, denoted disappearance scale t_D.

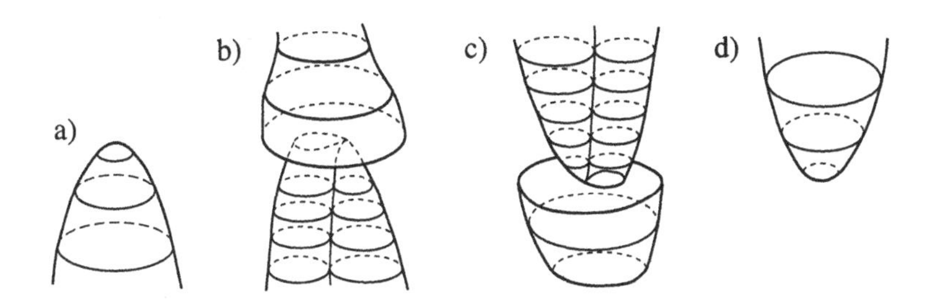

Figure 7.7. Generic blob events in scale-space: (a) annihilation, (b) merge, (c) split, (d) creation.

The difference[2] between the disappearance scale and the appearance scale gives the *scale-space lifetime* of the blob.

In merge and split situations the grey-level blobs existing before the bifurcation are regarded as belonging to different scale-space blobs than the grey-level blobs existing after the bifurcation.

In special configurations it may happen that a blob without a hole forms a torus, or that a torus fills in its hole. These events are also stable in the sense that a small disturbance of the original signal will not affect the qualitative behaviour. Here, such events will be considered not to affect the scale-space blobs; the grey-level blobs existing before such an event will be regarded as belonging to the same scale-space blob as the grey-level blobs existing after.

7.5. Scale-space blob tree

Similar considerations can be applied to the evolution properties over scales of the grey-level blob tree. The corresponding representation is called a scale-space blob tree. Interestingly, the only bifurcations that can occur in scale-space are those who affect the leaves, i.e., the grey-level blobs and the scale-space blobs. This is a direct consequence of the fact that both an extremum point and a saddle point must be involved in a bifurcation (section 8.3), and every extremum point corresponds to a unique grey-level blob, and hence a unique scale-space blob.

The additional complexity that arises when considering grey-level blob trees over scales compared to grey-level blobs is that a structural event called *reordering* may occur. It is the result of a relative change in grey-level between two different saddle points that directly determine the ordering relations in the grey-level blob tree, see figure 7.8.

[2]It turns out that some transformation of the scale parameter is necessary to capture the concept of scale-space lifetime properly (see section 7.7).

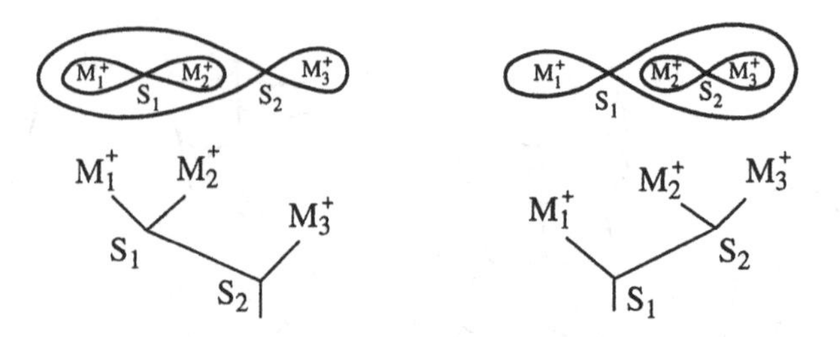

Figure 7.8. The additional complexity that occurs when considering grey-level blob trees is the introduction of reorderings, which are changes in the ordering relations in the grey-level blob tree resulting from relative changes in grey-level between saddle points. This figure shows a simple example with bright blobs; in the left case $z(S_1) > z(S_2)$, while in the right case $z(S_1) < z(S_2)$.

7.6. Grey-level blob extraction: Experimental results

Figure 7.9 displays an example of extracting (dark) grey-level blobs at different[3] scales in scale-space. It can be seen that at fine scales mainly small blobs due to noise and surface texture are detected. When the scale parameter increases, the noise blobs disappear gradually although much faster in regions near steep gradients. Notable in this context is that blobs due to noise can survive for a long time in scale-space if located in regions with slowly varying grey-level intensity. This observation shows that *scale-space lifetime alone cannot be used as the basis for a significance measure*, since then the significance of blobs due to noise[4] would be substantially overestimated. At coarser scales, the blocks appear as single blob objects. Finally, at very coarse scales, adjacent blocks become grouped into larger entities.

Figure 7.10 shows a similar scale-space sequence for a telephone and calculator image. Also in this case, mainly small blobs due to noise and surface texture are detected at fine scales. Moreover, a hierarchical behaviour between blobs at different scales appears again. The buttons on the keyboard manifest themselves as blobs after a small amount of smoothing. At coarser levels of scale, they merge into one unit (the keyboard). One can also observe that some other dark details in the image, the calculator, the cord, and the receiver, appear as single blobs at coarser levels of scale.

[3]This behaviour of the grey-level blobs over scales may be regarded as somewhat complex by a reader unfamiliar with these concepts. A detailed theoretical analysis is given in chapter 8.

[4]Of course, the contrast of such blobs decreases, but it is far from clear that it is possible to set a threshold on objective grounds.

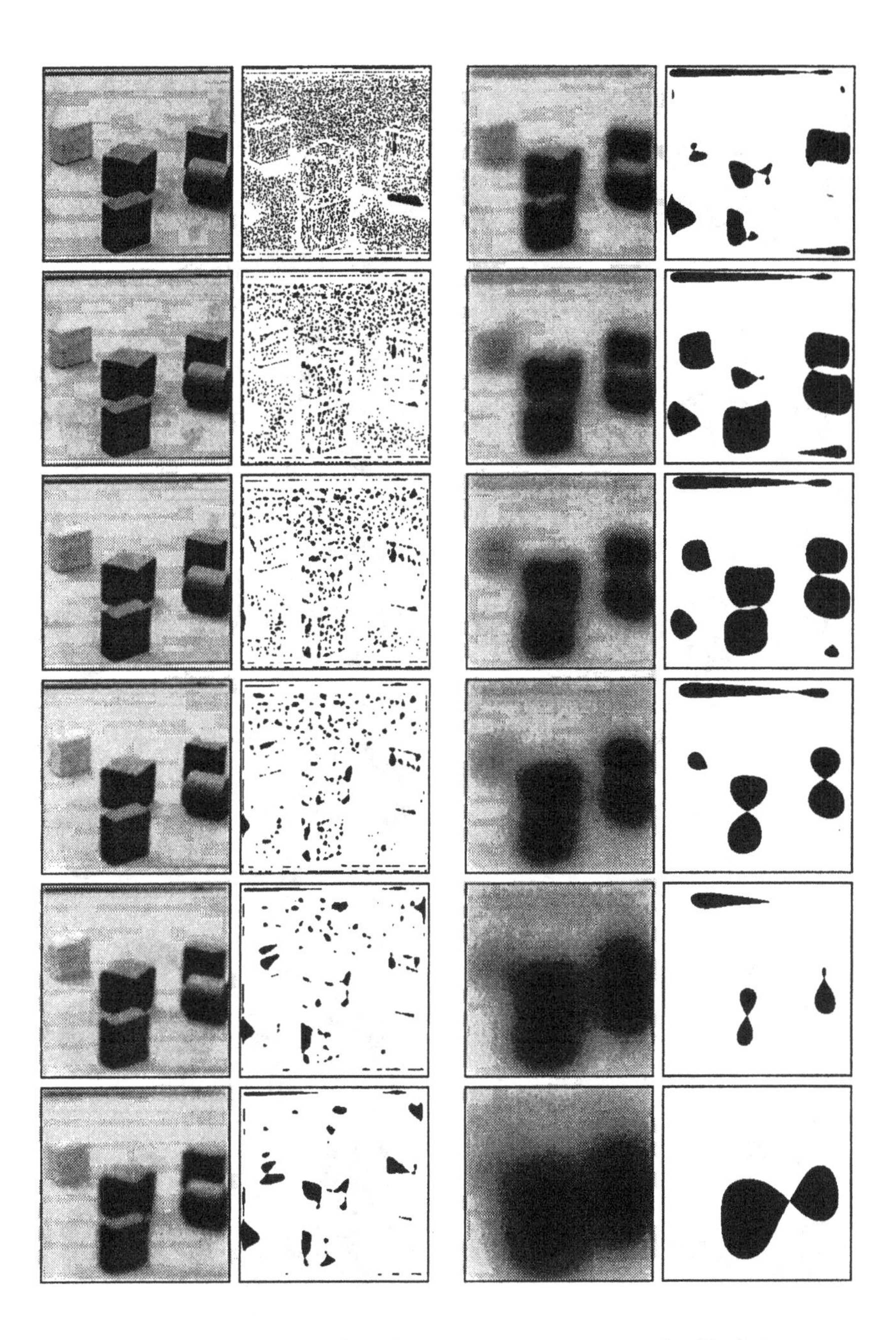

Figure 7.9. Grey-level and (dark) grey-level blob images of a block image at scale levels $t = 0, 1, 2, 4, 8, 16, 32, 64, 128, 256, 512$, and 1024 (from top left to bottom right). Each dark region in the right columns corresponds to the support region of a grey-level blob.

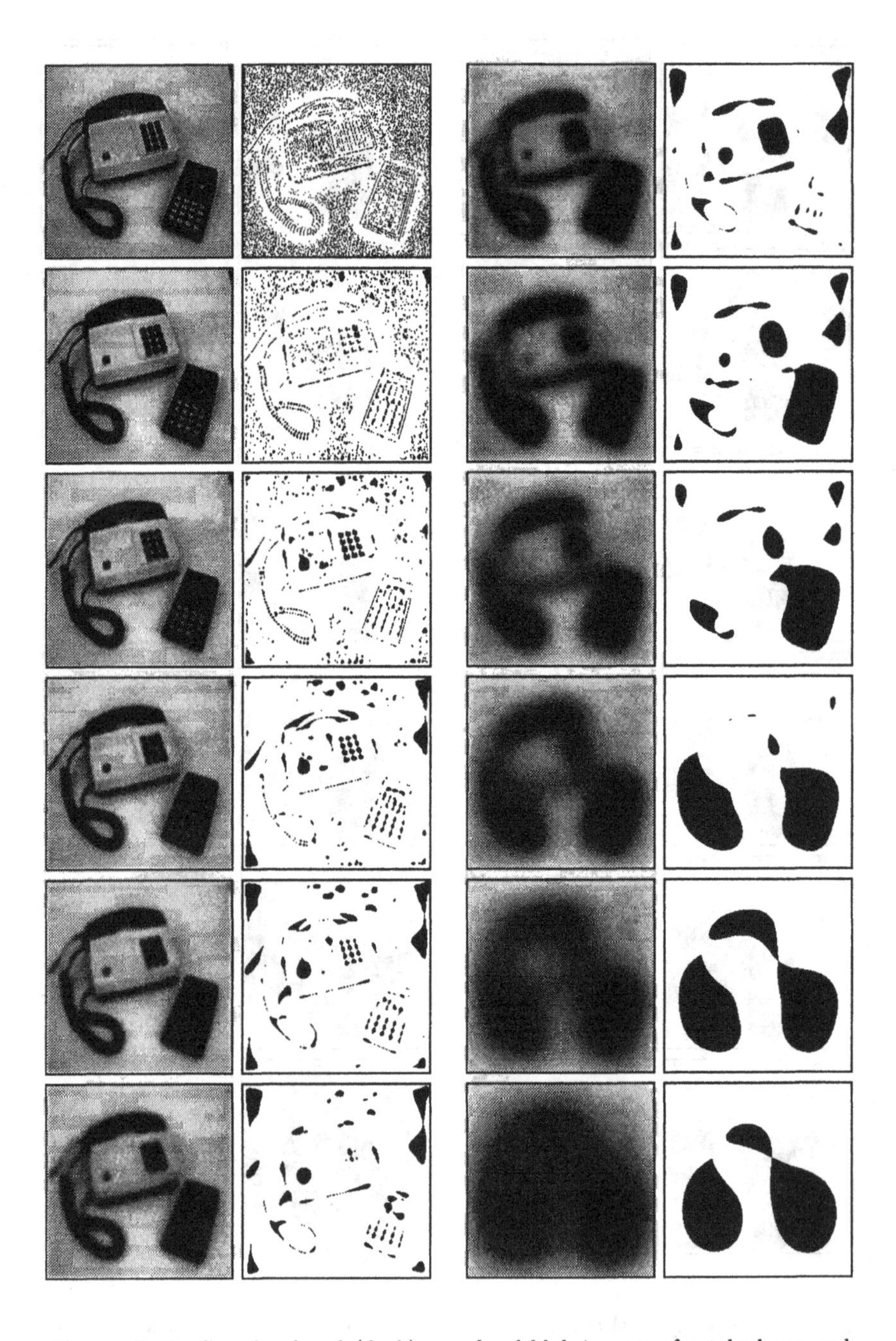

Figure 7.10. Grey-level and (dark) grey-level blob images of a telephone and calculator image at scale levels $t = 0, 1, 2, 4, 8, 16, 32, 64, 128, 256, 512$, and 1024 (from top left to bottom right).

These examples demonstrate that, as anticipated, the grey-level blob concept shows an extreme degree of noise sensitivity, which can be circumvented by the scale-space smoothing. But it is certainly far from a trivial problem to determine a proper amount of smoothing *automatically*, based on previous conventional methods.

The aim with the suggested blob linking across scales is to determine which blobs in the scale-space representation can be regarded as significant, without any *a priori* information about neither scale, spatial location nor the shape of the primitives.[5] As we shall see later, the output from the linking procedure also enables determination of a suitable scale level for handling each *individual*[6] blob.

7.7. Measuring blob significance

Since the ultimate goal of this analysis is to extract important structures in the image based on the appearance and significance of scale-space blobs in the scale-space representation, there is an absolute need for some methodology for comparing significance *between* different levels of scale. In other words, what is desired is a mechanism for judging whether a blob existing only at coarse levels of scale can be regarded as more significant or less significant than a blob with extent primarily at fine levels of scale.

The approach proposed here is to use the four-dimensional volumes of the scale-space blobs in scale-space (a precise definition is given in section 8.2.1). I suggest that it is a useful entity for constituting a significance measure, since it comprises both the spatial extent, the contrast, and the lifetime of the blob. Qualitative motivations for incorporating these entities into the significance measure can be summarized as follows:

spatial extent x **:** In the absence of further information, a blob having large spatial extent may be treated as more significant than a corresponding smaller blob.

contrast z **:** In the absence of further information, a high contrast blob may be treated as more significant than a similar blob with lower contrast.

lifetime t **:** In the absence of further information, a blob having a long lifetime in scale-space may be treated as more significant than a

[5]Except for the fact that scale-space smoothing favours blob-like bell-shaped objects.

[6]The word individual is emphasized here, since stable scales when they exist are, in general, local properties associated with object (or parts of objects) — not with entire images. However, the assumption of a globally stable scale is sometimes used implicitly in computer vision algorithms, for example, when edge detection is performed using uniform smoothing all over an image. Instead, it is argued that better performance can be obtained by adapting the scale levels to the local image structure, see chapter 11 for examples.

corresponding blob having a shorter lifetime. In general, a blob B_1 far away from another blob B_2 will survive longer in scale-space than a blob B_3, similar to B_1, but nearer to B_2. Moreover, the lifetime of a blob will in general be longer if there are no spatially coincident structures at other scales. Hence, two special cases implied by this heuristic principle are that:

- a blob B_1 *far away* from another blob B_2 will be treated as more significant than a blob B_3 similar to B_1, but nearer[7] to B_2,
- a blob, for which there are no spatially overlapping finer or coarser scale structures, will be treated as more significant than a similar blob, for which such *interfering structures* at nearby scales exist.

If the significance measure, however, is to be based on the scale-space blob volume, it is of crucial importance that the coordinates are measured in proper units, since in principle they could be transformed by arbitrary monotone functions.

7.7.1. Measuring scale-space lifetime

Consider first the measurement of scale-space lifetime. A natural choice of scale parameter for a continuous signal case is the logarithm of the ordinary scale parameter. For example, a scaling argument, requiring that structures at different scales should be treated in a similar manner, directly implies that a natural way to define scale-space lifetime is by letting the effective scale parameter be the logarithm of the ordinary scale parameter:

> Consider a structure existing at a certain scale and assume that the structure can be associated with a characteristic length (similar to a coarse characteristic length descriptor as used in dimensional analysis in physics) x. If a similar structure existing at a different level of scale is to be treated in a similar manner, then the *relative* change in characteristic length, Δx, of that structure caused by some amount of smoothing, $\Delta\tau$ (expressed in effective scale), should be independent of both the size of that structure and the current level of scale. In other words, the following relation must hold:
>
> $$\frac{\Delta x}{x} = C_1 \Delta\tau \qquad (7.7)$$

[7]Note in this context that if two blobs (B_2 and B_3 above) are closely located, there will in general be a large blob corresponding to the union of these two blobs at coarser scales in scale-space. Hence, although the smaller one of these blobs (B_3 above) may be assigned a small significance value, the union of these two blobs will be assigned a larger significance value, and hence attract the focus-of-attention to the union of the two adjacent structures.

for some arbitrary (non-zero) constant C_1. Assuming that the standard deviation of the Gaussian kernel, $\sigma = \sqrt{t}$, can be linearly related to a characteristic length in a grey-level image at that scale, it follows that

$$\frac{\Delta\sigma}{\sigma} = C_1 \Delta\tau. \tag{7.8}$$

By taking the limit of this expression as $\Delta\sigma$ and Δx simultaneously tend to zero, and then integrating we obtain

$$\tau = C_2 + \frac{1}{C_1} \log \sigma = C_2 + \frac{1}{2C_1} \log t \tag{7.9}$$

for some arbitrary integration constant C_2. This shows that for continuous signals the natural scale parameter is essentially the logarithm of the ordinary scale parameter.

This relation is well-known and has been used for example in pyramid representations, which usually comprise a logarithmic sampling along the scale direction. More compactly, the result can be derived by noting that the only way to introduce a dimensionless scale parameter is by introducing a logarithmic measure (Florack *et al.* 1992)

$$d\tau = C\frac{dt}{t}. \tag{7.10}$$

Based on this idea, one could be inspired to define scale-space lifetime as

$$t_{life} = \log t_D - \log t_A, \tag{7.11}$$

where t_D and t_A denote the disappearance and appearance scales of the scale-space blob respectively. It seems reasonable that this would give a good description at coarse scales, since it is well-known that changes in scale-space occur logarithmically with scale. For example, the scale parameter is usually sampled such that the ratio between successive scale values is constant.

Such an approach would, however, lead to unreasonable results for discrete signals at fine levels of scale, since then a blob existing in the original signal (at $t = 0$) would be assigned an infinite lifetime. Similarly, it can be observed that

$$t_{life} = t_D - t_A \tag{7.12}$$

does not work either, since then the lifetime of blobs at coarse scales in scale-space would be substantially overestimated.

7.7.1.1. *Effective scale*

Consequently, there is a need for introducing a transformed scale parameter $\tau = \tau_{eff}(t)$ such that scale-space lifetime measured by

$$\tau_{life} = \tau_D - \tau_A = \tau_{eff}(t_D) - \tau_{eff}(t_A) \tag{7.13}$$

gives a proper description of the behaviour in scale-space also for discrete signals. This scale parameter should neither favour fine scales to coarse scales nor the opposite.

In this section, a formal treatment will be given showing how the notion of "effective scale" can be defined in a precise way, so that the definition is valid both for continuous and discrete signals. Experimental results will also be presented illustrating how the major blob descriptors (volume, area, contrast) can be expected to behave with scale, and it will be explained how these results can be used for rescaling the relevant descriptors. Some other facts that will be illuminated are that the inner scale and the outer scale of an image really must be taken into account in an actual implementation.

At first glance the problem of transforming the scale parameter may seem somewhat ad hoc. What properties are required from an effective scale parameter? Intuitively, structures at different scales should be treated in a way as uniform as possible, so that neither the lifetime of fine scale structures is overestimated compared to the lifetime of coarse scale structures, nor the opposite.

Assuming that there would be a way to measure the amount of structure in an image, then a natural requirement would be that the amount of structure that is destroyed when the effective scale increases one unit should be independent of the current scale. In other words, if the amount of structure is plotted as a function of effective scale, then the graph would be a straight line. However, what does one mean with the amount of structure in an image? Moreover, even if there were such a definition, it would be possible to transform this measure by an arbitrary monotonical transformation, and the effective scale accordingly, while the graph would remain a straight line. Hence, there would still be an arbitrary transformation function to determine.

Another natural requirement is that the expected lifetime of a scale-space blob in scale-space should not vary over scales. As we shall return to in section 7.7.2, this leads to the problem of what should be meant by expected lifetime. Such an entity can be expected to vary strongly from one image to another. The approach that will be adopted here is to consider a reference signal and make a probabilistic approximation of behaviour over scales. It will be assumed that the expected remaining lifetime of a local extremum should not vary with scale. Intuitively this means that the probability that a certain local extremum disappears[8] after a small amount of smoothing $\Delta\tau$, expressed in effective scale, should be constant over scales. More precisely, we state the requirement that the *relative decay rate* in a *reference signal* should be *independent of scale.*

Assume now that it is known how the expected number of local extrema per unit area varies with scale (e.g., in a reference signal). In other words, assume that

$$p(t) = \{\text{expected density of extrema at scale } t\} \tag{7.14}$$

is known. What we want to define is a transformation function h such that the effective scale can be written $\tau = h(t)$. The relative decay rate requirement can be stated as

$$\frac{\partial_\tau p}{p} = \partial_\tau(\log p) = C = \text{constant}. \tag{7.15}$$

Integration and introduction of new arbitrary constants C_1 and C_2 gives

$$\tau = \tau_{eff}(t) = C_1 + C_2 \log p(t). \tag{7.16}$$

Without loss of generality, C_1 can be set to zero. It is just an offset coordinate, and cancels in the scale-space lifetime. Similarly, C_2 just corresponds to an arbitrary but unessential linear rescaling of the effective scale parameter.

So far no assumptions have been made about the dimensionality of the signal, or whether it is continuous or discrete. What is left to determine is how the density of local extrema can be expected to vary with scale.

For a large class of continuous signals, the number of local extrema decreases with scale approximately as t^α (see section 8.7). This result can be derived from a one-dimensional stationary normal white noise process as well as a corresponding process with a spectral density of the form $\omega^{-\beta}$ with $0 \le \beta < 3$. It can also be shown from dimensional analysis that in arbitrary dimensions N, the density of local extrema can be expected decrease with scale as $t^{-N/2}$ (see section 8.7). Under these conditions, the effective scale is given by a logarithmic transformation

$$\tau_{eff}(t) = -C_2 \alpha \log t. \tag{7.17}$$

For discrete signals, the density of local extrema can be expected to show the same qualitative behaviour at coarse scales, where the grid effects

[8]For one-dimensional signals, the number of local extrema in a signal is guaranteed to decrease monotonically with scales. In two and higher dimensions the situation is more complicated, since the number of local extrema can in fact increase locally with scale-space smoothing due to creations of saddle-extremum pairs. However, the expected number of local extrema, treated as an average over many signals can always be expected to decrease.

are negligible. At fine scales, however, the $t^{-\alpha}$ behaviour cannot hold, since it is based on the assumption that the original signal contains equal amounts of structure at all scales. The discrete signal is limited by its finite sampling density.

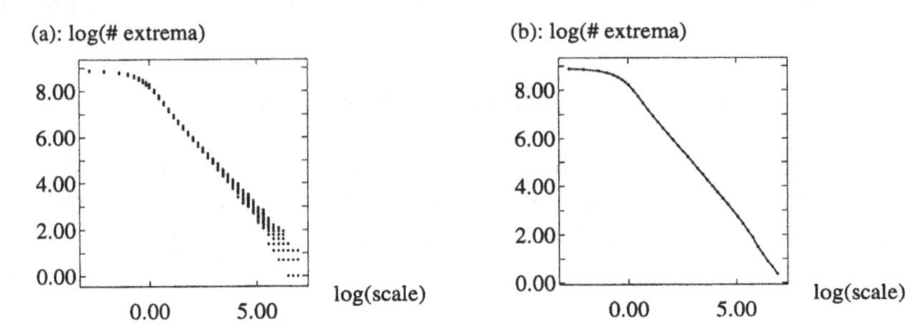

Figure 7.11. Experimental results showing the number of local extrema as function of the scale parameter t in log-log scale (a) measured values (b) accumulated mean values. Note that a straight-line approximation is valid only over a limited range of scales.

These ideas are illustrated in figure 7.11, which shows the logarithm of the number of local extrema in a finite image as function of the logarithm of the ordinary scale parameter t. The left diagram shows simulated results for a large number of white noise images generated from three different distributions, normal, rectangle, and exponential distribution. The right diagram shows the average of these results. Note that a straight-line approximation is valid only in an interior scale interval. At fine scales there is interference with the *inner scale* of the image given by its sampling density, and at coarse scales there is interference with the *outer scale* of the image given by its finite size.

The notion of effective scale takes the inner scale of the image into account, and enables a precise definition of scale-space lifetime also at fine levels of scale. Combined with the concept of scale-space for discrete signals it provides the necessary tool for investigating fine scale structures. For implementational purpose, $p(t)$ is estimated from synthetic simulation results for a set of reference data. Then, the transformation function is determined by

$$\tau_{eff}(t) = \log \frac{p_{ref}(0)}{p_{ref}(t)}, \tag{7.18}$$

where $p_{ref}(t)$ denotes the average density of local extrema in the simulations on the reference data. For the sake of implementation, this reference data has been selected as a large set ($\approx 10^2$) of white noise images. A motivation for this choice is given in the next section.

7.7.2. Transformed grey-level blob volumes

Similarly, the grey-level blob volumes need to be transformed, since the average volume can be expected to vary substantially with scale. When the scale parameter increases, the average contrast can be expected to decrease, and the average area to increase. What about the grey-level blob volume? Experimental results demonstrate that it actually decreases at fine scales and increases at coarser scales, see figure 7.12.

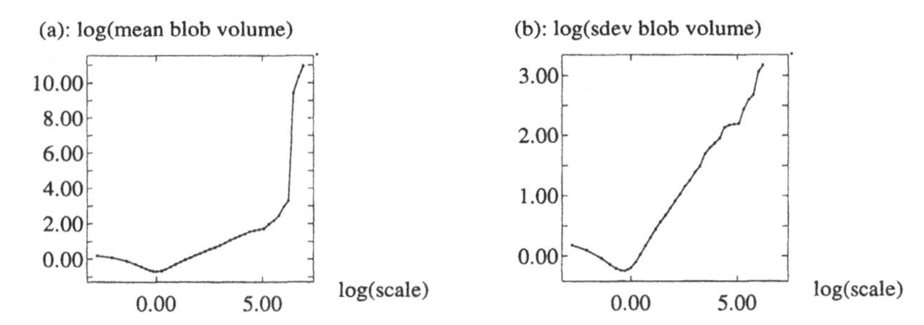

Figure 7.12. Experimental results showing (a) the mean value, and (b) the standard deviation of the grey-level blob volumes as function of scale for white noise images of different distribution.

Within the parts of the graphs where a linear approximation is valid, the mean value, $V_m(t)$, and the standard deviation of the grey-level blob volume, $V_\sigma(t)$, vary with scale approximately as

$$V_m(t) \sim \sqrt{t}, \qquad V_\sigma(t) \sim \sqrt{t}, \tag{7.19}$$

while corresponding experiments demonstrate that the variation of the area and contrast is of the form

$$A(t) \sim t, \qquad C(t) \sim \frac{1}{\sqrt{t}}, \tag{7.20}$$

as can be expected from dimensional analysis or a study of a single Gaussian blob.

If these effects are not taken into account, then the significance of coarse scale blobs will be substantially overestimated. It is clear that the blob behaviour depends strongly upon the image (since we actually want to use it for segmentation). Is it then possible to talk about expected behaviour? A natural approach would be to accumulated statistics for a large number of (representative) images, and to use these as reference data. Here, a simple approach will be used.

A conservative approach is to consider white noise data, i.e., images without any structured relations between adjacent pixels. If statistics is accumulated of how blobs can be expected to behave in such images, then the result will be an estimate of to how large extent accidental groupings can be expected to occur in scale-space.

If a grey-level blob at some level of scale has a volume smaller than the expected volume for white noise data, then it can hardly be regarded as significant. On the other hand, if at some level of scale the blob volume is much larger than the expected blob volume, and in addition, the difference in blob volume is much greater than the expected variation around the average value, then it may be reasonable to treat the blob as significant.

A natural normalization to perform is to subtract a measured grey-level blob volume G_{vol} by the mean value, $V_m(t)$, and divide by the standard deviation, $V_\sigma(t)$. This gives a transformed grey-level blob volume

$$V_{prel} = \frac{G_{vol} - V_m(t)}{V_\sigma(t)}. \tag{7.21}$$

Since, however, such a quantity may take negative values, it is not suitable for integration (which is a necessary step in the computation of the scale-space blob volume). Therefore, in the current implementation, the *effective grey-level blob volume* is defined in the following way, which empirically turns out to give reasonable results

$$V_{eff} = V_{trans}(G_{vol};\ t) = \begin{cases} 1 + V_{prel} & \text{if } V_{prel} \geq 0, \\ e^{V_{prel}} & \text{otherwise.} \end{cases} \tag{7.22}$$

With this definition, the effective volume of the mean value is one. For larger values it grows linearly with V_{prel}. Thus, V_{eff} and V_{prel} show the same qualitative behaviour for the significant grey-level blobs. For smaller values of V_{prel}, V_{eff} decreases to zero, and the qualitative difference between V_{prel} and V_{eff}, increases as the significance decreases.

Hence, a qualitatively correct behaviour is obtained for the important blobs, and it can be expected that this solution should not affect the result too seriously. It should also be mentioned, that in order to adapt the amplitude of the signal to the reference data, V_m and V_σ are rescaled linearly from a least-squares fit between the actual and the expected behaviour of these entities.

Finally, these transformed grey-level blob volumes are integrated over the scale-space blob according to (8.7). A discussion about other possible approaches to normalizing the scale-space blob volume is given in section 12.5.1.

7.8. Resulting representation

To summarize, the data structure proposed is a tree-like multi-scale representation of blobs at all levels of scale in scale-space *including* the relations between blobs at different scales. Grey-level blobs[9] should be extracted at all levels of scale, the bifurcations in scale-space be registered, grey-level blobs stable over scales be linked across scales into scale-space blobs, and the normalized scale-space blob volumes be computed.

Since the representation tries to capture significant features and events in scale-space using a small set of primitives, it is called a *scale-space primal sketch*. In the resulting data structure constructed according to this description, every scale-space blob contains explicit information about which grey-level blobs it consists of. The grey-level blobs are detected at (sampled) scale levels obtained from an adaptive scale linking and refinement algorithm outlined in chapter 9. Further, the normalized scale-space blob volumes have been computed, and the scale-space blobs "know" about the type of bifurcations that have taken place at the appearance and disappearance scales. There are also links to the other scale-space blobs involved in the bifurcations. Hence, the representation[10] explicitly describes the hierarchical relations between blobs at different scales, see figure 7.13 for a schematic illustration, and section 9.6 for a detailed description.

The intention with this representation is to capture *inherent geometric* properties of the underlying grey-level image. I suggest that the representation as such is useful in itself. Worth emphasizing is that the involved quantities (grey-level blobs and scale-space blobs) are defined solely in terms of *singularities*, namely local extrema, saddle points and bifurcations in scale-space. Moreover, the representation is completely free from tuning parameters.

In chapter 10 it will be shown how some directly available information from this representation can be used for extracting significant image structures. Then, applications will be given of how such output from the scale-space primal sketch can be used for tuning later stage processes and guiding the focus-of-attention.

Before that, however, we will in the next chapter investigate some of the theoretical properties of the representation, and then in chapter 9 describe an algorithm for actually computing it.

---

[9]As explained in previous sections, grey-level blob trees can be treated in a similar way. Since, however, the problem of normalizing the spatial and grey-level coordinates has so far been studied only concerning the grey-level blobs and the scale-space blobs, the remaining part of this presentation will be concerned with these objects.

[10]More detailed information about what type of information can be contained in a data structure representing the scale-space primal sketch is given in section 9.6.

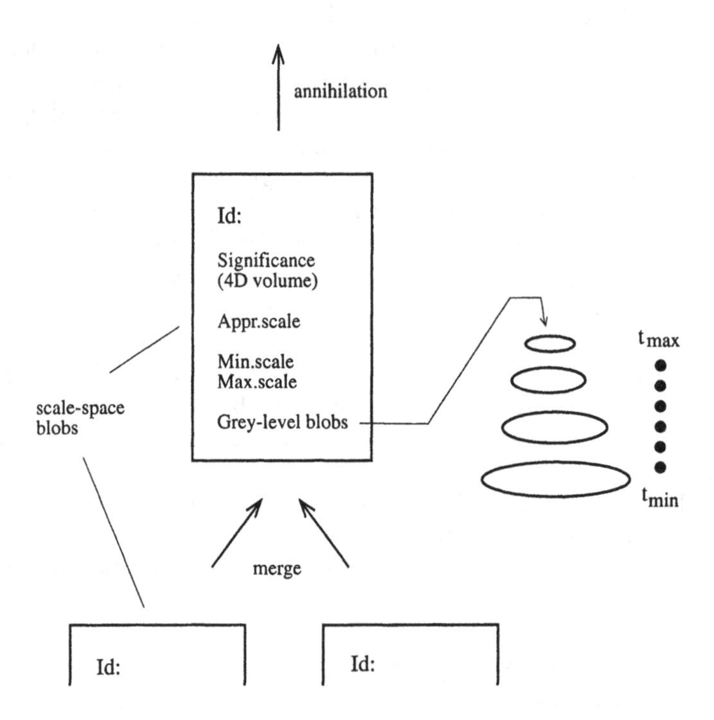

Figure 7.13. The scale-space primal sketch based on grey-level blobs can be seen as a tree-like multi-scale representation of blobs with the scale-space blobs as basic primitives (vertices) and the relations (bifurcations) between scale-space blobs at different levels of scale as branches. (This figure only shows an example of what type of information is made explicit by the representation; the example illustrates two scale-space blobs merging into one. A more detailed description of what information is available in the representation is given in section 9.6. Note that this illustration is only concerned with grey-level blobs. When grey-level blob trees are considered in the analysis, the scale-space primal sketch becomes a three-dimensional graph with hierarchical relations along the grey-level dimension as well.)

8

Behaviour of image structures in scale-space: Deep structure

The treatment so far has been mainly concerned with the formal definition of the scale-space representation and the definition of image descriptors at any single scale. A complementary problem concerns how to relate structures at different scales. This subject has been termed *deep structure* by Koenderink (1984). When a pattern is subjected to scale-space smoothing, its shape changes and may be distorted. For example, features like local extrema, edges, blobs, etc. can be expected to drift when the underlying grey-level image is subject to blurring. More generally, transitions between objects of qualitatively different appearance may also take place. This gives rise to the notion of *dynamic shape,* which as argued by Koenderink and van Doorn (1986) is an essential component of any shape description of natural objects.

Aspects of these phenomena have been studied by several authors from different viewpoints. Canny (1986) discussed the general trade-off problem between detection and localization occurring in edge detection. Bergholm (1987) estimated the drift velocity of edges for a set of plausible configurations with the aim of estimating a step size for scale changes in the edge focusing algorithm. Berzins (1984) analyzed the localization error for zero-crossings of the Laplacian of the Gaussian.

Other kinds of phenomena affecting the topology may also occur. As developed by Koenderink and van Doorn (1986), blobs can disappear, merge, and split. Similar transitions apply to edges, zero-crossings of the Laplacian, corners, etc. Such events are usually called *bifurcations.*

In this chapter we shall study critical points, that is, local extrema and saddle points, and investigate in detail what happens to those features when an image undergoes scale-space smoothing. We shall

- develop how these feature points can be expected to behave when the scale parameter in scale-space changes,
- derive an expression for their drift velocity,
- classify their behaviour at bifurcation situations into a discrete set of generic situations, and

- give a coarse estimate to the global problem of how the number of local extrema in a signal can be expected to vary with scale.

The results that will be derived are not based on any specific models for the intensity variations in the image but are generally valid under rather weak *a priori* assumptions. Although the results will be expressed in a general form, the primary intention with the study is to provide a further theoretical basis of the scale-space primal sketch concept developed in chapter 7. In this context, the results to be presented will find their main application in

- the formal construction and definition of the primitives (scale-space blobs) in the scale-space primal sketch. The scale-space blobs are defined as families of grey-level blobs, which in turn are directly determined by pairs of critical points. This treatment allows for precise mathematical definitions of those concepts.
- providing a theoretical basis for the linking algorithm necessary when computing the representation.
- giving further motivations for the normalization process with respect to "expected scale-space behaviour," which is necessary when defining the significance measures of the scale-space blobs.

In other words, we shall try to explain what happens when scale changes in scale-space, especially with application to the scale-space primal sketch. Therefore, special attention will be given to the primitive objects of that representation, i.e. the grey-level blobs and scale-space blobs. The methodology of analysis is, however, general and can be applied to, for example, any features that can be expressed as zero-crossings of differential expressions (differential singularities).

Before starting, let us point out that some of the results to be presented are (at least partly) known or touched upon before, see e.g., Koenderink (1984, 1990), Koenderink and van Doorn (1986), and Clark (1988). Bifurcations in scale-space have also been studied by Johansen *et al.* (1986), who have shown that a band-limited one-dimensional signal up to a multiplicative constant is determined by its "top points," that is the points in scale-space where bifurcations take place. A work by Johansen (1993) extends the analysis in an earlier version of this presentation (Lindeberg 1991, 1992) with a differential geometric study of trajectories of critical points in scale-space.

The purpose of this treatment is to develop systematically and comprehensively what can be said about the behaviour in scale-space of critical points using elementary mathematical techniques and to convey an intuitive feeling for the qualitative behaviour in the different generic cases. Detailed calculations will also be given showing the behaviour of blobs in a set of "characteristic examples."

8.1. Trajectories of critical points in scale-space

In many situations it is of interest to estimate the drift velocity of critical points when the scale parameter varies. Such information is useful, for instance, when estimating the localization error of feature points due to scale-space smoothing, or when tracking local extrema or related entities across scales. In non-degenerate situations, that is when the second differential is a non-degenerate quadratic form, such an analysis can be based on the implicit function theorem.

DEFINITION 8.1. (CRITICAL POINT) *A point $x_0 \in \mathbb{R}^N$ is a critical point of a mapping $f\colon \mathbb{R}^N \to \mathbb{R}$ if the gradient at this point*

$$(\nabla f)(x_0) = \left. \begin{pmatrix} \partial_{x_1} f \\ \vdots \\ \partial_{x_N} f \end{pmatrix} \right|_{x_0} \tag{8.1}$$

is zero. The critical point is said to be non-degenerate if the Hessian matrix in this point

$$(\mathcal{H}f)(x_0) = \left. \begin{pmatrix} \partial_{x_1 x_1} f & \dots & \partial_{x_N x_1} f \\ \vdots & \ddots & \vdots \\ \partial_{x_1 x_N} f & \dots & \partial_{x_N x_N} f \end{pmatrix} \right|_{x_0} \tag{8.2}$$

is non-singular. Otherwise it is said to be degenerate.

LEMMA 8.2. (BEHAVIOUR OF CRITICAL POINTS IN SCALE-SPACE)
Let $L\colon \mathbb{R}^N \times \mathbb{R}_+ \to \mathbb{R}$ be the scale-space representation of an N-dimensional continuous signal given by the diffusion equation (2.27), and assume that at some scale level $t_0 > 0$ a point $x_0 \in \mathbb{R}$ is a non-degenerate critical point for the mapping $x \mapsto L(x;\ t_0)$.

Then, there exist an open set $S_{(x_0;\ t_0)} \subset \mathbb{R}^N \times \mathbb{R}_+$ and an open interval $I_{t_0} \subset \mathbb{R}_+$ with $(x_0;\ t_0) \in S_{(x_0;\ t_0)}$ and $t_0 \in I_{t_0}$ having the following property: To every $t_1 \in I_{t_0}$ there corresponds a unique $x_1 \in \mathbb{R}^N$ such that $(x_1;\ t_1) \in S_{(x_0;\ t_0)}$ and x_1 is a non-degenerate critical point for the mapping $x \mapsto L(x;\ t_1)$.

If this x_1 is defined to be $r(t_1)$, then r is a continuously differentiable mapping $I_{t_0} \to \mathbb{R}^N$ such that

- *$r(t_0) = x_0$,*
- *$r(t_1)$ is for every $t_1 \in I_{t_0}$ a non-degenerate critical point for the mapping $x \mapsto L(x;\ t_1)$.*
- *the derivative of r with respect to t in the point x_0 is given by*

$$\partial_t r(t_0) = -\tfrac{1}{2}\, (\mathcal{H}L)(x_0)^{-1} (\nabla^2 (\nabla L))(x_0). \tag{8.3}$$

Proof. The result follows easily by applying the implicit function theorem to the gradient function. The standard version of the implicit function theorem (Rudin 1976) basically gives that there exists a path $r\colon I_{t_0} \to \mathbb{R}^N$ of critical points, and that the derivative of r is

$$\partial_t r(t_0) = -(\mathcal{H}L)(x_0)^{-1}(\partial_t(\nabla L))(x_0). \tag{8.4}$$

Then, the fact that L satisfies the diffusion equation can be used for replacing derivatives of L with respect to t by derivatives of L with respect to the spatial coordinates in order to arrive at (8.3). Since the Hessian $(\mathcal{H}L)(r(t))$ along this path is a continuous function of t, it follows that $r(t)$ will remain non-degenerate provided that the initial point x_0 is non-degenerate, and I_{t_0} is selected as a sufficiently short interval. □

8.1.1. *Interpretation: Drift velocity estimates*

This lemma expresses how critical points in general can be expected to behave in scale-space. One of the most immediate interpretations is that (8.3) gives a straightforward estimate of the *drift velocity of critical points* under scale-space smoothing.

This estimate can also be extended to comprise edges. For simplicity, assume that the edge under study is sufficiently long and sufficiently close to a straight line such that a one-dimensional analysis is a valid approximation. Further, without loss of generality, assume that the coordinate system is oriented such that the edge is perpendicular to the x_1-axis. Then, use non-maximum suppression to define the location of the edge as those points where the gradient magnitude (here the x_1-derivative) assumes a maximum along the gradient direction (here the x_1-direction). In other words, edge points are defined as those points where the second derivative along the gradient direction is zero. Now, since under these conditions, critical points are given by zeros in the first derivative and edge points by zeros in the second derivative, the one-dimensional version of (8.3) can be applied just by replacing L by L_{x_1}. Hence, the drift velocity in the direction perpendicular to a *straight edge* is

$$\partial_t r(t_0) = -\frac{1}{2}\frac{L_{x_1^4}(x_0;\ t_0)}{L_{x_1^3}(x_0;\ t_0)}. \tag{8.5}$$

A similar idea, although based on an approximate derivation, has been expressed by Zhuang and Huang (1986).

This analysis is applicable also to edges given by zero-crossings of the Laplacian, provided that the second derivative along the edge direction (here the x_2-direction) is sufficiently small to be neglected. Trivially, an identical result holds for edges of one-dimensional signals. Observe that there are no specific assumptions about the shape of the intensity profile perpendicular to the edge. Hence, the result is valid for any configuration that can be described by a one-dimensional analysis.

In particular, it means that the drift velocity may tend to infinity when two adjacent parallel edges are just about to merge into one. This result can, for example, be used for explaining an observation by Zhang and Bergholm (1991), who noted that configurations consisting of two adjacent edges (so-called "staircase edges"; see figure 8.1) can lead to a rapid edge drift when the scale parameter changes, which in turn violates the assumptions behind the estimate of the scale step used in the edge focusing algorithm (Bergholm 1987). In such situations the third derivative is in fact very close to zero. A more general analysis of curved edges is given in section 8.6.

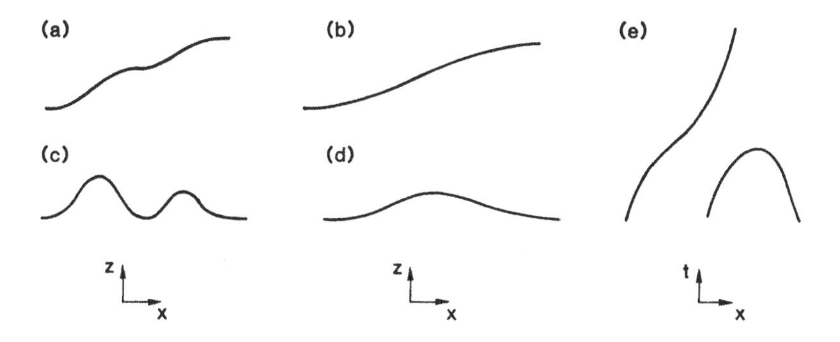

Figure 8.1. (a) A "staircase edge" can lead to a rapid edge drift. This behaviour can be explained by noting that (b) after sufficient amount of blurring the configuration will tend to a "diffuse step edge" as well as by studying the derivatives of (c) the original signal, and (d) the signal after strong smoothing. (e) By considering the paths the zero-crossings of the Laplacian describe as scale changes it is easy to realize that when the edge points tend to each other the drift velocity will tend to infinity. See also section 8.5 for a more detailed description of the behaviour at bifurcation situations, in particular section 8.5.4 concerning this configuration.

Finally, regarding the drift velocity estimates for local extrema and edges, it should be pointed out that although the drift velocity *momentarily* may tend to infinity, the total drift (integrated over some scale interval of finite length) will always be finite. What the results mean, is that it is not possible to derive any *uniform* upper bound for the drift velocity of these features. Given any scale interval of length Δt and any distance $|\Delta x|$ it is always possible to find a signal such that the total drift of a feature during the time Δt exceeds $|\Delta x|$. This property emphasizes the need for algorithms based on *adaptive* sampling along the scale direction.

8.1.2. Interpretation: Extremum paths

Another consequence of lemma 8.2 is that a non-degenerate critical point existing at a certain level of scale can, in general, be traced to a similar critical point both at a slightly coarser and a slightly finer scale. By continuation, such local paths obtained from the implicit function theorem can be extended to curves as long as the Hessian remains non-zero. It can be easily shown that the type of critical point remains the same as long the Hessian matrix is non-singular.

> It is obvious that a local maximum (minimum) cannot be transformed into a saddle point or vice versa. If the Hessian would change sign, then it would first become zero (since it is a continuous function of the scale parameter). Then, however, the trajectory would by definition be cut off by a degenerate critical point into two separate segments.
>
> Moreover, a maximum point cannot be transformed into a minimum point or opposite, since then (at least) the partial derivative $L_{x_1x_1}$ would need to change sign. Such a sign change implies that this derivative would first become zero (because of continuity), which in turn means that the quadratic form would become indefinite, i.e., the point would get transformed into a saddle point. This transition has to go through a degenerate critical point, which means that the trajectory would be cut off into at least two parts.

In other words, if $(x_0;\ t_0)$ is a local maximum (minimum/saddle), then there exists a curve through this point, such that every point on the curve is a local maximum (minimum/saddle) at that scale. The curve is delimited by two scale levels t_{min} and t_{max}, at which the Hessian matrix degenerates (except for the boundary cases $t_{min} = 0$ or $t_{max} = \infty$). At all interior points the extremum point is non-degenerate. Such a curve $r_0 \colon [t_{min}, t_{max}] \to \mathbb{R}^2$ is is called an *extremum path* (*saddle path*).

The situation in other dimensions is similar, although there are no stable saddle points in the one-dimensional case.

8.2. Scale-space blobs

The notion of extremum path in previous section allows for a formal definition of scale-space blob—the basic primitive in the scale-space primal sketch. In chapter 7, a grey-level blob was defined as a local extremum with extent and a scale-space blob in turn as a family of those. More precisely, a grey-level blob of a two-dimensional signal was given by a pair consisting of a local extremum and a saddle point, and in one dimension by a maximum and minimum point, implying a one-to-one correspondence between local extrema and grey-level blobs. The previous definition of scale-space blob was, however, intuitive: "similar blobs at adjacent levels of scale were linked into scale-space blobs." The linking process proceeded until no such linking could be performed, i.e., until a bifurcation was encountered. The idea behind this construction was to identify and group similar features at different scales into higher order and unified objects.

8.2.1. *Definition of scale-space blob*

We can now express the linking criterion in a more formal way. Consider the two-dimensional case, and study a local extremum x_0 with associated grey-level blob $G_{blob}(x_0)$ in a non-degenerate (Morse) signal at some scale t_0 in scale-space. Then, there is a unique extremum path $r_0 \colon [t_{min}, t_{max}] \to \mathbb{R}^2$ associated with the extremum $x_0 = r_0(t_0)$ of the grey-level blob, and for each extremum $r_0(t)$ along this path there is a corresponding grey-level blob $G_{blob}(r_0(t))$.

For every scale level $t \in [t_{min}, t_{max}]$ where the scale-space representation is non-degenerate, there is a unique delimiting saddle point[1] $S_{delimit}(r(t))$ associated with the local extremum $r_0(t)$. All such saddle points associated with an extremum path need, however, not be on the same saddle path. At certain scales transitions between different saddle paths can be expected to take place. Generically, this occurs at a discrete set of scales, at which the local extremum point is non-degenerate, and the extent of the grey-level blob is delimited by two non-degenerate saddle points having the same grey-level. Such transitions will not be regarded as affecting the scale-space blobs.

On the other hand, if the delimiting saddle (or the extremum) is involved in a bifurcation, then the local topology will be changed—a *blob event* has occurred. It is therefore natural to proceed with the linking as long as the extrema and their delimiting saddle points are non-degenerate, and to stop it when either of the critical points degenerates. Hence, consider a (maximal) scale interval $[t'_{min}, t'_{max}] \subset [t_{min}, t_{max}]$ such that

- for all interior scales, the (possible multiple) delimiting saddle points $S_{delimit}(r_0(t))$ are non-degenerate, and

- at the end points either of $r_0(t'_{min})$ and $S_{delimit}(r_0(t'_{min}))$ and also either of $r_0(t'_{max})$ and $S_{delimit}(r_0(t'_{max}))$ are degenerate critical points.

Then, the *scale-space blob* associated with the segment $r'_0 \colon [t'_{min}, t'_{max}] \to \mathbb{R}^2$ of the extremum path is the (four-dimensional) set

$$S_{blob}(r'_0) = \overline{\{(x;\ z;\ t) \in \mathbb{R}^2 \times \mathbb{R} \times \mathbb{R}_+ :} \\ \overline{(t'_{min} < t < t'_{max}) \wedge ((x;\ z) \in G_{blob}(r'_0(t)))\}}, \tag{8.6}$$

[1] If an extremum point E and a saddle point S together define the extent of a grey-level blob, then S is said to be the delimiting saddle point of E (section 7.1).

and the *support region* of the scale-space blob is in turn the (three-dimensional) region

$$S_{support}(r'_0) = \{(x;\ t) \in \mathbb{R}^2 \times \mathbb{R}_+ : (x;\ z;\ t) \in S_{blob}(r'_0) \text{ for some } z\}.$$

In most figures with scale-space blobs (e.g., in chapter 7) it is the latter descriptor that has been illustrated.

8.2.2. *Definition of scale-space blob volume*

Strictly, in this coordinate system the scale-space blob volume is

$$S_{vol}(r'_0) = \int_{(x;\ z;\ t) \in S_{blob}(r'_0)} dx\, dz\, dt = \int_{t \in [t'_{min}, t'_{max}]} G_{volume}(r'_0(t))\, dt,$$

where $G_{vol}(r'_0(t))$ is the grey-level blob volume of the grey-level blob associated with the extremum point $r'_0(t)$. However, when the scale-space blob volume is to be used as a significance measure in the scale-space primal sketch, it turns out that some transformations of the coordinate axes must be performed (see section 7.7). One would like structures at different scales to be treated uniformly, such that the significance measure neither favours fine scales to coarse scales nor the opposite. Therefore, the *normalized scale-space blob volume* is defined by

$$S_{vol,norm}(r'_0) = \int_{t \in [t'_{min}, t'_{max}]} V_{trans}(G_{vol}(r'_0(t));\ t)\, d\tau_{eff}(t), \qquad (8.7)$$

where τ_{eff}: $\mathbb{R}_+ \rightarrow \mathbb{R}$ is a transformation function mapping the ordinary scale parameter t to a transformed scale parameter τ called *effective scale* (see section 7.7.1), and V_{trans}: $\mathbb{R} \times \mathbb{R}_+ \rightarrow \mathbb{R}$ is a corresponding transformation function normalizing the variation of the grey-level blob volumes into a more uniform behaviour over scales (see section 7.7.2).

8.3. Bifurcation events for critical points: Classification

The implicit function theorem used in previous sections guarantees that linking of non-degenerate critical points is a well-defined operation. When the Hessian matrix becomes singular, the implicit function theorem is no longer applicable, and *bifurcations* may occur.

Useful tools for analysing the qualitative behaviour of functions around bifurcation points can be obtained from a branch of mathematics known as *singularity theory*; see Poston and Stewart (1978) or Gibson (1979) for application-oriented introductions, and Arnold (1981), Arnold *et al.* (1985, 1988), Golubitsky and Schaeffer (1985), or Lu (1976) for more rigorous treatments of the subject.

8.3.1. Thom's classification theorem

One of the fundamental results in singularity theory is that the typical qualitative behaviour of families given by a small number of parameters can be expressed completely by the qualitative behaviour of a finite set of families. A famous theorem by Thom classifies the generic behaviour of families of functions with the number of parameters $r \leq 4$ into seven elementary catastrophes. A summarizing result expressed by Poston and Stewart (1978) states that:

> THOM'S CLASSIFICATION THEOREM: *Typically an r-parameter family $\mathbb{R}^N \times \mathbb{R}^r \to \mathbb{R}$ of smooth functions $\mathbb{R}^N \times \mathbb{R}^r \to \mathbb{R}$, for any N and $r \leq 4$, is structurally stable and is in every point (locally) equivalent to one of the following forms:*
>
> - *non-critical:* x_1,
> - *non-degenerate critical, or Morse:* $x_1^2 + \cdots + x_i^2 - x_{i+1}^2 - \cdots - x_N^2 \; (0 \leq i \leq N)$,
> - *degenerate critical, catastrophe;*
>   - *fold:* $x_1^3 + u_1 x_1 + (M)$,
>   - *cusp:* $\pm(x_1^4 + u_2 x_1^2 + u_1 x_1) + (M)$,
>   - *swallowtail:* $x_1^5 + u_3 x_1^3 + u_2 x_1^2 + u_1 x_1 + (M)$,
>   - *butterfly:* $\pm(x_1^6 + u_4 x_1^4 + u_3 x_1^3 + u_2 x_1^2 + u_1 x_1) + (M)$,
>   - *elliptic umbilic:* $x_1^2 x_2 - x_2^3 + u_3 x_1^2 + u_2 x_2 + u_1 x_1 + (N)$,
>   - *hyperbolic umbilic:* $x_1^2 x_2 + x_2^3 + u_3 x_1^2 + u_2 x_2 + u_1 x_1 + (N)$,
>   - *parabolic umbilic:* $\pm(x_1^2 x_2 + x_2^4 + u_4 x_2^2 + u_3 x_1^2 + u_2 x_2 + u_1 x_2) + (N)$,
>
> *where (M) and (N) indicate Morse functions on the forms*
>
> $$(M) = x_2^2 + \cdots + x_i^2 - x_{i+1}^2 - \cdots - x_N^2 \quad (2 \leq i \leq N),$$
>
> $$(N) = x_3^2 + \cdots + x_i^2 - x_{i+1}^2 - \cdots - x_N^2 \quad (2 < i \leq N),$$
>
> *which must be added on to the previously mentioned expressions in order to match up the dimensions.*

The intuitive explanation of "structurally stable" is that a sufficiently small perturbation does not change the qualitative behaviour at the singularity. The term "locally equivalent" essentially means that the function can be (locally) transformed to the listed polynomial representative of the singularity by a change of variables (a diffeomorphism). For precise definitions of these concepts, the reader is referred to the sources cited above. A brief review can also be found in (Lindeberg 1992).

8.3.2. *Generic singularities of one-parameter families*

Applied to one-parameter families, like the scale-space representation of a signal, this result means that the natural type of singularity to expect is the *fold singularity*. It can be represented by the polynomial

$$G_N(x;\ u) = x_1^3 + 3x_1u + \sum_{i=2}^{N} \pm x_i^2, \tag{8.8}$$

where x_i should be interpreted as offset coordinates around the bifurcation point, here translated to the origin, and u as the parameter. This singularity means that at the bifurcation point the first and second order directional derivatives are zero along a certain direction, while the third derivative in that direction is non-zero.

Hence, if one is interested in the behaviour of the critical points of a signal during the evolution of the diffusion equation, it should in principle be sufficient to study this situation. For simplicity, consider from now on the two-dimensional case, and replace the notation (x_1, x_2) for coordinates by (x, y). Then, the *singularity set* is given by the solutions of

$$\partial_x G_2(x, y;\ u) = 3x^2 + u = 0, \tag{8.9}$$
$$\partial_y G_2(x, y;\ u) = \pm 2y = 0, \tag{8.10}$$

and the *bifurcation set* by the solution of

$$\partial_x G_2(x, y;\ u) = 3x^2 + u = 0, \tag{8.11}$$
$$\partial_y G_2(x, y;\ u) = \pm 2y = 0, \tag{8.12}$$
$$\partial_{xx} G_2(x, y;\ u) = 6x = 0. \tag{8.13}$$

Thus, the singularity set is given by

$$-(x_1(u), y_1(u)) = (x_2(u), y_2(u)) = (\sqrt{-u/3}, 0) \quad (u \leq 0), \tag{8.14}$$

and the bifurcation occurs at an isolated point $(x, y;\ u) = (0, 0;\ 0)$. From the sign of the Hessian

$$\det(\mathcal{H}G_2)(x, y;\ u) = \pm 12x, \tag{8.15}$$

it follows that $(x_1(u), y_1(u))$ are saddle/maximum points and $(x_2(u), y_2(u))$ are minimum/saddle points for every $u < 0$. At $u = 0$ the points merge along a parabola and then disappear; see figure 8.2(a).

Similarly, for a one-dimensional signal the fold singularity is represented by the polynomial $G_1(x;\ u) = x^3 + xu$, and corresponds to a maximum and a minimum point merging with increasing u; see figure 8.2(b).

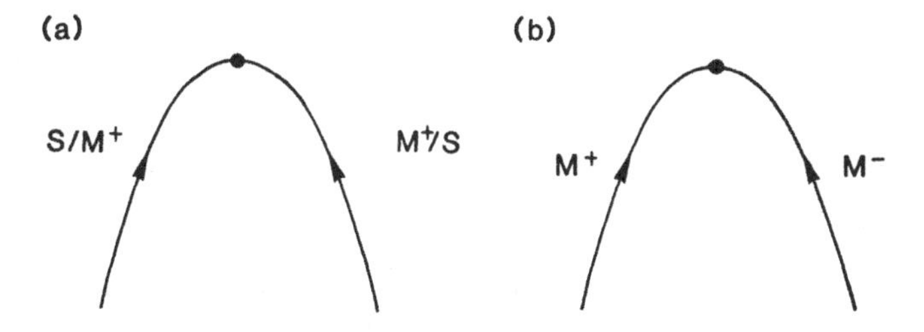

Figure 8.2. (a) The generic behaviour at a singularity of a one-parameter family of two-dimensional functions is described by the unfolding $G_2(x, y;\ u) = x^3 + ux \pm y^2$. The singularity set of this family, that is the set of critical points of the mapping $x \mapsto G_2(x, y;\ u)$, describes an extremum point and a saddle point that merge along a parabola and then disappear. (b) For a one-parameter family of one-dimensional functions the behaviour is instead given by $G_1(x;\ t) = x^3 + ux$. The singularity set in this case corresponds to a similar merge of a maximum point and a minimum point. (The notation S/M^+ means that the trajectory corresponds to either a saddle point or a maximum point etc.)

To summarize, the typical behaviour to be expected at singularities in a one-parameter family of continuous signals is:

- annihilations or creations of pairs of local extrema and saddle points in the two-dimensional case, and
- annihilations or creations of pairs of local maxima and local minima in the one-dimensional case.

8.3.3. Interpretations with respect to scale-space representation

By comparisons with earlier theoretical and experimental results we know that this describes the qualitative behaviour of critical points in scale-space. However, there is one apparent complication when to give a more detailed interpretation. Thom's classification theorem states that there exists a diffeomorphism such that the singularity set of a solution to the one-dimensional diffusion equation around a bifurcation point $(x_0;\ t_0)$ in scale-space can be transformed into the singularity set of G_2 around $(0;\ 0)$. However, there is obviously some directional information lost in the equivalence concept: In which direction should the u parameter be interpreted as running? If u and t are treated as increasing simultaneously, then the situation describes a minimum and a maximum merging with increasing t. On the other hand, if u runs in the reverse direction, then the interpretation would be that a pair with a minimum and a maximum would be created when t increases. The latter phenomenon is, however, impossible, since the number of local extrema cannot increase under scale-space smoothing in the one-dimensional case (see chapter 3).

The diffusion equation apparently introduces a directional preference to its solutions (due to the causality requirements), which makes such creations impossible. How should this information be incorporated into the analysis of the singularities in scale-space?

One way of avoiding the previous blindness of the equivalence concept to the structural property of the diffusion equation could be by trying to develop results similar to Thom's classification theorem, which instead of being expressed in terms of the ordinary standard basis of polynomials could be expressed in terms of polynomials satisfying the diffusion equation. Natural modified representatives of the fold singularities in the one-dimensional and two-dimensional cases would then be

$$G_1(x;\ t) = x^3 + 3xt, \tag{8.16}$$

$$G_2(x, y;\ t) = x^3 + 3xt \pm (y^2 + t). \tag{8.17}$$

Another approach is to use the previous classification to state what configurations are possible in general one-parameter families of functions. Then, the special structure of the diffusion equation can be taken into account for judging which cases apply to the scale-space representation when the directional constraint of the diffusion equation has been added.

Yet a third approach will be considered in the next section, where it will be demonstrated that the results from the second approach agrees with what can be obtained from a differential geometric analysis of the paths that critical points form in scale-space.

8.3.4. Algebraic classification of singularities

The bifurcation events between critical points in scale-space can also be classified by using elementary techniques. Following Johansen (1993), consider critical paths defined by the solutions to

$$\begin{cases} L_x(x, y;\ t) = 0, \\ L_y(x, y;\ t) = 0. \end{cases} \tag{8.18}$$

Bifurcation points are points $r_0 = (x_0, y_0;\ t_0)^T$ where

$$\det(\mathcal{H}L)(r_0) = L_{xx}(r_0)L_{yy}(r_0) - L_{xy}^2(r_0) = 0. \tag{8.19}$$

Without loss of generality assume that the coordinate system is rotated such that

$$L_{xy}(r_0) = 0. \tag{8.20}$$

Then, either L_{xx} or L_{yy} must be zero at the bifurcation point. Without loss of generality assume that

$$L_{xx}(r_0) = 0, \tag{8.21}$$

and let a critical path r through this point be parameterized by an arc length parameter s such that $r(s) = (x(s), y(s);\ t(s))^T$ with $r_0 = r(s_0)$

$$|r'(s)|^2 = x'(s)^2 + y'(s)^2 + t'(s)^2 = 1. \tag{8.22}$$

Implicit differentiation of $L_x(x(s), y(s);\ t(s)) = 0$ and $L_y(x(s), y(s);\ t(s)) = 0$ with respect to s then gives

$$\begin{cases} L_{xx}(r(s))\,x'(s) + L_{xy}(r(s))\,y'(s) + L_{xt}(r(s))\,t'(s) = 0, \\ L_{xy}(r(s))\,x'(s) + L_{yy}(r(s))\,y'(s) + L_{yt}(r(s))\,t'(s) = 0. \end{cases} \tag{8.23}$$

Using $L_{xx}(r_0) = 0$ and $L_{xy}(r_0) = 0$ these relations reduce to

$$\begin{cases} L_{xt}(r_0)\,t'(s_0) = 0, \\ L_{yy}(r_0)\,y'(s_0) + L_{yt}(r_0)\,t'(s_0) = 0. \end{cases} \tag{8.24}$$

In the generic case we can assume that $L_{xt}(r_0) \neq 0$ and $L_{yy}(r_0) \neq 0$. Then, (8.24) gives $y'(s_0) = 0$ and $t'(s_0) = 0$. By selecting the positive root from the normalization condition $x'(s_0)^2 = 1$ obtained from (8.22), we get $r'(s_0) = (x'(s_0), y'(s_0);\ t'(s_0))^T = (1, 0;\ 0)^T$. This shows that the curve intersects the bifurcation point with a horizontal tangent.

Concerning the second order structure at the bifurcation point, implicit differentiation of (8.23) combined with the expression for $r'(s_0)$ and the assumptions $L_{xx}(r_0) = 0$ and $L_{xy}(r_0) = 0$ gives

$$\begin{cases} L_{xxx}(r_0) + L_{xt}(r_0)\,t''(s_0) = 0, \\ L_{xxy}(r_0) + L_{yy}(r_0)\,y''(s_0) + L_{yt}(r_0)\,t''(s_0) = 0, \end{cases} \tag{8.25}$$

from which explicit expressions can be obtained for the curvature components of the critical path

$$x''(s_0) = 0, \tag{8.26}$$

$$\begin{aligned} y''(s_0) &= \frac{L_{xxx}(r_0)\,L_{yt}(r_0) - L_{xxy}(r_0)\,L_{xt}(r_0)}{L_{yy}(r_0)\,L_{xt}(r_0)} \\ &= \frac{L_{xxx}(r_0)\,L_{yyy}(r_0) - L_{xxy}(r_0)\,L_{xyy}(r_0)}{L_{yy}(r_0)\,(L_{xxx}(r_0) + L_{xyy}(r_0))}, \end{aligned} \tag{8.27}$$

$$t''(s_0) = \frac{L_{xxx}(r_0)}{L_{xt}(r_0)} = -\frac{2L_{xxx}(r_0)}{L_{xxx}(r_0) + L_{xyy}(r_0)}. \tag{8.28}$$

The result $x''(s_0) = 0$ is a direct consequence of the fact that the derivative of a constant length vector is perpendicular to the original vector.

In the two other expressions alternative forms have also been given, in which derivatives with respect to t have been replaced by derivatives with respect to y and x using the fact that the Gaussian derivatives satisfy the diffusion equation. One example is $L_{xt} = \partial_t L_x = \frac{1}{2}(\partial_{xx} + \partial_{yy})L_x$.

By studying the sign of $t''(s_0)$ it can be seen that a pair of critical points is *annihilated* with increasing scale if

$$t''(s_0) = -\frac{2L_{xxx}(r_0)}{L_{xxx}(r_0) + L_{xyy}(r_0)} < 0, \tag{8.29}$$

while a pair of critical points is *created* with increasing scale if

$$t''(s_0) = -\frac{2L_{xxx}(r_0)}{L_{xxx}(r_0) + L_{xyy}(r_0)} > 0. \tag{8.30}$$

In order to analyse the types of critical points, consider a first order Taylor expansion of the coordinate functions

$$r(s) = (x(s), y(s);\ t(s))^T = (x'(s_0)\, s + \mathcal{O}(s^2), \mathcal{O}(s^2);\ \mathcal{O}(s^2))^T,$$

and approximate the partial derivatives of L linearly

$$L_{xx}(r) = L_{xx}(r_0) + (\nabla_{(x;\,t)} L_{xx})(r_0)^T r + \mathcal{O}(r^2), \tag{8.31}$$

$$L_{xy}(r) = L_{xy}(r_0) + (\nabla_{(x;\,t)} L_{xy})(r_0)^T r + \mathcal{O}(r^2), \tag{8.32}$$

$$L_{yy}(r) = L_{yy}(r_0) + (\nabla_{(x;\,t)} L_{yy})(r_0)^T r + \mathcal{O}(r^2), \tag{8.33}$$

where the symbol $\nabla_{(x;\,t)} = (\partial_x, \partial_y;\ \partial_t)^T$ means that differentiation should be performed with respect to x, y, and t. Then, the first order Taylor expansion of the Hessian can be written

$$\begin{aligned} \det(\mathcal{H}L)(r(s)) &= L_{xx}(r(s))L_{yy}(r(s)) - L_{xy}^2(r(s)) \\ &= L_{xxx}(r_0)\, L_{yy}(r_0)\, s + \mathcal{O}(s^2). \end{aligned} \tag{8.34}$$

Hence, provided that $L_{xxx}(r_0) \neq 0$ and $L_{yy}(r_0) \neq 0$, the critical points on one side of the bifurcation point are local extrema, and the critical points on the other side are saddle points.

To summarize, this analysis gives an alternative verification that the generic singularities for critical points are annihilations and creations of saddle-extremum pairs. Moreover, it states an explicit condition for when creations can occur. In a coordinate system selected such that $L_{pp}(r_0) = L_{pq}(r_0) = 0$ at the bifurcation point r_0, this condition can be expressed:

$$\begin{aligned} &((L_{ppp}(r_0) > 0) \wedge (L_{pqq}(r_0) < -L_{ppp}(r_0)) \vee \\ &((L_{ppp}(r_0) < 0) \wedge (L_{pqq}(r_0) > -L_{ppp}(r_0)). \end{aligned} \tag{8.35}$$

In one dimension, (8.28) reduces to $t'' = -2 < 0$, showing that critical points are always annihilated and never created in this case.

8.4. Bifurcation events for grey-level blobs and scale-space blobs

A natural question that arises in connection with the scale-space primal sketch concerns what types of blob events are possible in bifurcation situations. Since scale-space blobs are defined in terms of paths of critical points, the behaviour of a scale-space blob at a singularity is solely determined by the behaviour of the extremum/saddle paths during a short scale interval around the bifurcation moment.

Compared to the previous treatment, where critical points were analysed, there is one additional factor that must be taken into account when dealing with scale-space blobs, namely the fact that a saddle point delimiting the extent of a grey-level blob involved a bifurcation can be associated with other grey-level blobs as well. This leads to natural coupling between scale-space blobs sharing the same saddle path (of delimiting saddle points) in a neighbourhood of a bifurcation.

8.4.1. Shared and non-shared saddle path

In view of this observation let us define the following: A saddle path involved in a structurally stable bifurcation is said to be *non-shared* before (after) the bifurcation if there exists some scale interval before (after) the bifurcation during which every saddle point of the saddle path is not contained in more than one grey-level blob. Otherwise, the saddle path is said to be *shared.*

More precisely, a saddle path is said to be non-shared before (after) a bifurcation at t_{bifurc} if there exists some $\epsilon > 0$ such that for all scales in the interval $t \in]t_{bifurc} - \epsilon, t_{bifurc}[$ ($t \in]t_{bifurc}, t_{bifurc} + \epsilon[$) the saddle point of the saddle path at that scale does not belong to more than one grey-level blob, see figure 8.3. Another way to express this property is that a shared saddle point is the delimiting saddle point of two (or more) grey-level

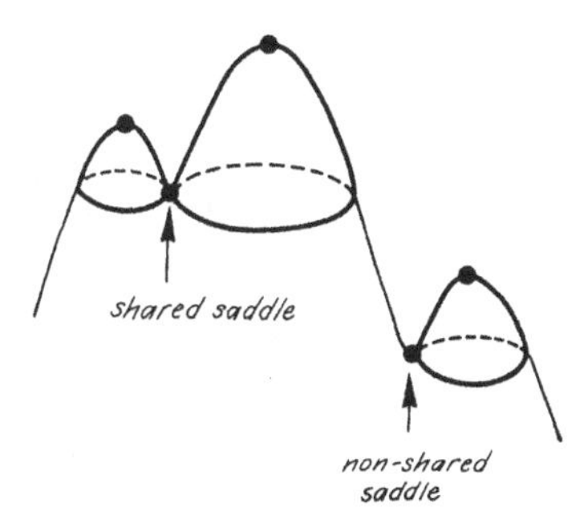

Figure 8.3. The definition of grey-level blob for a two-dimensional signal. Every local extremum gives rise to a blob and the extent of the blob is given by a saddle point. A saddle point is said to be shared if it is contained in more than one grey-level blob, i.e., if it is a delimiting saddle point of two (or more) grey-level blobs of the same polarity.

blobs of the same polarity, while a non-shared saddle point either is the delimiting saddle point of one or no grey-level blob.

8.4.2. Generic bifurcation events for scale-space blobs

Hence, depending on whether a extremum-saddle pair is annihilated or created, and depending on whether the saddle path is shared or non-shared, four generic cases can be distinguished:

- A *non-shared* saddle path participating in the *annihilation* of an extremum-saddle pair with increasing scale describes an isolated blob that disappears. Such a blob event can aptly be called a *blob annihilation.*

- On the other hand, a *shared* saddle path involved in a similar *annihilation* describes a blob disappearing under the influence of a neighbour blob—a *blob merge.*

- Similarly, a *shared* saddle point taking part in an extremum-saddle pair that is *created* with increasing scale describes a *blob split.*

- Finally, a *non-shared* saddle path participating in an extremum-saddle creation describes an isolated blob which appears— a *blob creation.*

To summarize, (Below, the term annihilation (creation) of an extremum-saddle pair means that a pair consisting of an extremum path and a saddle path disappears (appears) when the scale parameter increases.)

PROPOSITION 8.3. (CLASSIFICATION OF BLOB EVENTS; 2D)
In the scale-space representation of two-dimensional continuous signal, the following blob events are possible at a structurally stable bifurcation:

- *blob annihilation—annihilation of an extremum-saddle pair, where the saddle path is non-shared before the bifurcation,*

- *blob merge—annihilation of an extremum-saddle pair, where the saddle path is shared before the bifurcation,*

- *blob split—creation of an extremum-saddle pair, where the saddle path is shared with another scale-space blob after the bifurcation,*

- *blob creation—creation of an extremum-saddle pair, where the saddle path is non-shared after the bifurcation.*

These four cases constitute the definitions of the terms annihilation, merge, split and creation with respect to grey-level blobs and scale-space blobs in the two-dimensional case.

Proof. From section 8.3 it follows that the typical behaviour at singularities are pairwise annihilations and creations of extremum-saddle pairs. Combined with the definition of shared saddle path this means that the class of possible blob events is restricted to the given four types, provided that only structurally stable bifurcations are considered.

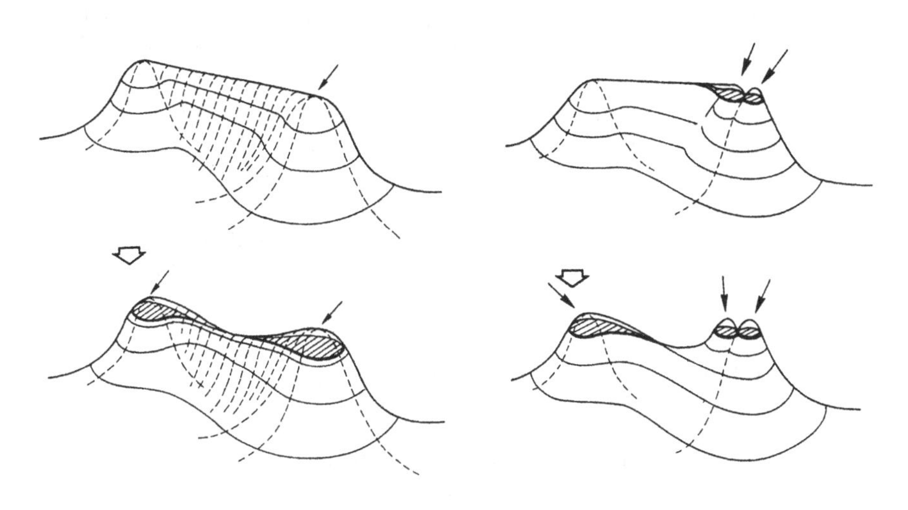

Figure 8.4. (a) New local extrema can be created with increasing scale in the scale-space representation of a two-dimensional signal. Interpreted in terms of blobs the configuration describes a blob split. (b) By modifying the example slightly (by replacing the higher one of the two peaks with a double peak) one realizes that blob creations can occur as well. The base levels of the different grey-level blobs have been indicated.

What remains to verify is that all these four types can be instantiated, and that they are structurally stable. It is well-known that blob annihilations and blob merges can take place in scale-space (see also section 8.4 for illustrative examples). The fact that splits can occur is known as well (see the example given by Lifshitz and Pizer (1987) illustrated in figure 8.4(a)). The latter configuration can be modified to describe a blob creation as well, if the higher of the two peaks is replaced by a double peak (see figure 8.4(b)). Then, the extent of the two smaller blobs at the higher peak will be delimited by the grey-level in the valley between them, which means that when the narrow ridge has eroded and given rise to the creation of a saddle-extremum pair in which the saddle path is not be shared by any other blob. □

The assumption of structural stability is important in this context, since otherwise, there is an infinite variety of possible events. For instance, three or more blobs could merge into one blob at the same moment. Such events will however be unstable,[2] since a small perturbation of the input signal would perturb such a simultaneous merge of three blobs into a sequence of two successive pairwise merges.

Algorithmically, this means that an encountered actual situation with, say, three blobs at a fine scale seeming to belong all to the same blob at a coarser scale, can in general be decomposed into transitions of the four primitive types. This principle forms the idea behind the automatic scale refinement algorithm described in chapter 9, which essentially refines the scale sampling until all relations between scale-space blobs in scale-space can be decomposed into events of the previously listed types.

For one-dimensional signals the possible events[3] are blob annihilations and blob merges. Splits and creations are impossible due to the earlier mentioned property that new local extrema cannot be created in the one-dimensional case.

8.5. Behaviour near singularities: Examples

In previous sections a general methodology has been described for analysing the evolution properties of critical points. Moreover, the qualitative behaviour at bifurcation points has been classified. Here, a number of examples will be given demonstrating how the blob descriptors vary with scale in characteristic bifurcation situations.

8.5.1. Third order Taylor expansion in one dimension

A special property in one dimension is that it is possible to arrive at the generic representative of the fold unfolding (8.16) by a simple qualitative study. Consider a third order Taylor expansion of the scale-space embedding around a given point x_0 at some scale t_0

$$f_{t_0}(x) = \alpha + \beta(x - x_0) + \gamma(x - x_0)^2 + \epsilon(x - x_0)^3, \qquad (8.36)$$

where

$$\alpha = L(x_0;\ t_0), \quad \beta = L_x(x_0;\ t_0), \quad \gamma = \tfrac{1}{2} L_{xx}(x_0;\ t_0), \quad \epsilon = \tfrac{1}{6} L_{xxx}(x_0;\ t_0).$$

[2]Note in this context that for Morse functions no pair of critical points will have the same values. In other words, for generic functions all critical points will be distinct. Although, by definition, the grey-level function will not be Morse at a bifurcation, we can, in general, assume this latter property to hold at bifurcations. This means that situations with three or more blobs simultaneously merging into one can be expected not to occur.

[3]A formal statement of this result, including the relevant definitions, can be found in (Lindeberg 1992).

The scale-space representation of this signal (with $L(x;\ t_0) = f_{t_0}(x)$) is

$$L(x;\ t) = \alpha + \beta(x - x_0) + \gamma(x - x_0)^2 + \epsilon(x - x_0)^3 + \delta_1(t - t_0) + \delta_2(x - x_0)(t - t_0),$$

where $\delta_1 = \gamma$ and $\delta_2 = 3\epsilon$. For simplicity, introduce new (offset) variables $u = x - x_0$ and $v = t - t_0$. Then,

$$\tilde{L}(u;\ v) = L(u + x_0;\ v + t_0) = \alpha + \beta u + \gamma(u^2 + v) + \epsilon(u^3 + 3uv).$$

The critical points of the function $u \mapsto \tilde{L}(u;\ t)$ are given by

$$\partial_u \tilde{L}(u;\ v) = \beta + 2\gamma u + 3\epsilon(u^2 + v) = 0. \tag{8.37}$$

If $\epsilon = 0$ there is one single root $x = -\frac{b}{2\gamma}$, whose location is independent of t. Obviously, this case is not interesting, since it implies a stationary solution. Therefore, assume that $\epsilon \neq 0$. Then, there are two trajectories of critical points

$$u_{1,2}(v) = -\frac{\gamma}{3\epsilon} \pm \sqrt{\frac{\gamma^2}{9\epsilon^2} - (v + \frac{\beta}{3\epsilon})}, \tag{8.38}$$

which only exist when $v \leq \frac{\gamma^2}{9\epsilon^2} - \frac{\beta}{3\epsilon}$. They meet at the bifurcation point

$$(u_{bifurc};\ v_{bifurc}) = \left(-\frac{\gamma}{3\epsilon};\ \frac{\gamma^2}{9\epsilon^2} - \frac{\beta}{3\epsilon}\right). \tag{8.39}$$

From the second derivative $\tilde{L}_{uu}(u_{1,2};\ v) = 2\gamma + 6\epsilon u_{1,2} = \pm 6\epsilon\sqrt{\frac{\gamma^2}{9\epsilon^2} - (v + \frac{\beta}{3\epsilon})}$ it can be seen that the second derivative has different sign in the two critical points. Thus, the bifurcation consists of one maximum point and one minimum point that meet and annihilate (see figure 8.5).

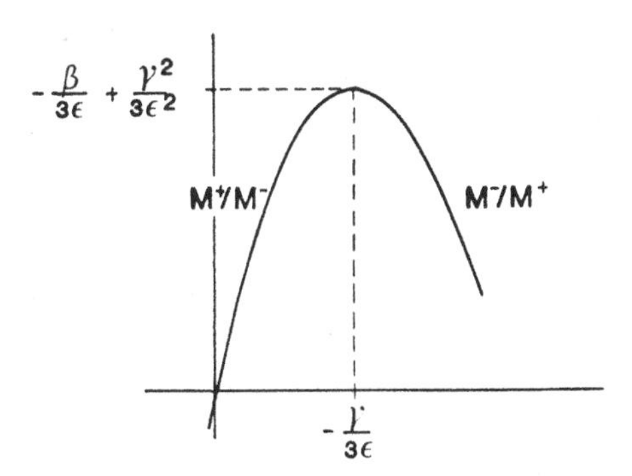

Figure 8.5. Third order Taylor expansion of the scale-space embedding: Schematic view over the loci of the critical points as scale changes. The bifurcation consists of a maximum point and a minimum point that meet and annihilate.

Drift velocity estimates. Assume for a moment that $x = x_0$ is a critical point for the mapping $x \mapsto L(x;\ t_0)$, i.e., that at $u = 0$ is a critical point for the mapping $u \mapsto \tilde{L}(u;\ 0)$. Then, $\beta = 0$ and (8.39) gives an estimate of the time Δt_{bifurc} and the distance Δx_{bifurc} to the bifurcation

$$(\Delta x_{bifurc};\ \Delta t_{bifurc}) = (\frac{\gamma}{3\epsilon};\ (\frac{\gamma}{3\epsilon})^2) = (\frac{L_{xx}}{L_{xxx}};\ (\frac{L_{xx}}{L_{xxx}})^2). \tag{8.40}$$

So far, no numerical experiments have been made testing the feasibility of using this estimate for scale step regulation in blob linking. Note, however, that despite the pessimistic upper bounds on the drift velocities discussed in section 8.1.1, the local extremum will hardly escape far outside the support region of the grey-level blob. This property turns out to be very useful in the blob linking algorithm to be described in section 9.2.

OBSERVATION 8.4. (COARSE BOUND ON THE DRIFT OF LOCAL EXTREMA) *Although the drift velocity of a local extremum point may momentarily be very large (tend to infinity near a bifurcation), when scale changes, the grey-level blob support region defines a natural spatial region to search for blobs in at the next level of scale.*

Reduction to the fold unfolding. To simplify further considerations, introduce again new coordinates by

$$\xi = u + \frac{\gamma}{3\epsilon}, \quad \tau = v + \frac{\beta}{3\gamma} - \frac{\gamma^2}{9\epsilon^2}, \quad \tilde{\tilde{L}}(\xi;\ \tau) = \tilde{L}(\xi - \frac{\gamma}{3\epsilon};\ \tau - \frac{\beta}{3\gamma} - \frac{\gamma^2}{9\epsilon^2}).$$

Then, the expression for the scale-space representation reduces to polynomial representative of the fold unfolding

$$\lambda(\xi;\ \tau) = \tilde{\tilde{L}}(\xi;\ \tau) - (\alpha - \frac{\gamma\beta}{3\epsilon} + \frac{2\gamma^3}{27\epsilon^2}) = \epsilon(\xi^3 + 3\xi\tau). \tag{8.41}$$

All these coordinate shifts only mean that the coordinate axes have been translated such that the bifurcation occurs at $(\xi;\ \tau) = (0;\ 0)$, and a constant has been subtracted to achieve $\lambda(0;\ 0) = 0$. Hence, $\lambda\colon \mathbb{R}\times\mathbb{R}_+ \to \mathbb{R}$ still satisfies the diffusion equation.

8.5.2. *The fold singularity in one dimension*

Consider again the generic unfolding of the scale-space embedding in the neighbourhood of a bifurcation

$$L(x;\ t) = x^3 + 3xt, \tag{8.42}$$

where x and t should be interpreted local coordinates in a coordinate system centered at the bifurcation point. As mentioned above, the critical points of this function are given by

$$\partial_x L(x;\ t) = 3(x^2 + t) = 0, \tag{8.43}$$

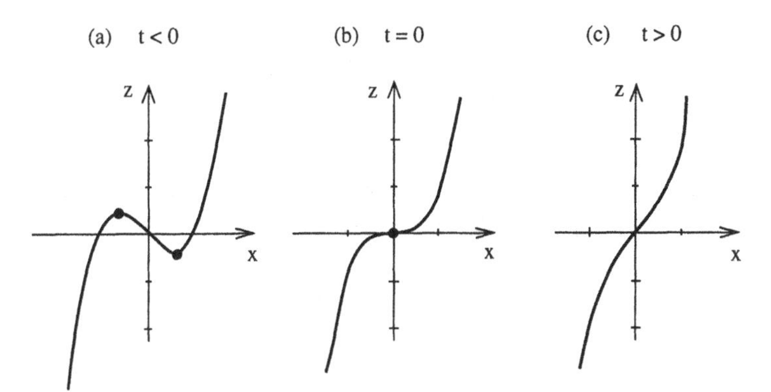

Figure 8.6. Fold unfolding in the one-dimensional case: Schematic view of the smoothed signal; (a) before the bifurcation, (b) at the bifurcation, and (c) after the bifurcation.

that is by

$$\xi_1(t) = -\sqrt{-t}, \quad \xi_2(t) = +\sqrt{-t}. \tag{8.44}$$

Moreover, the critical values are

$$\lambda_1(t) = -\lambda_2(t) = L(\xi_1(t);\ t) = -L(\xi_2(t);\ t) = +2(-t)^{\frac{3}{2}}. \tag{8.45}$$

In terms of grey-level blobs, this bifurcation describes the simultaneous annihilation of a bright and a dark blob. Obviously, the contrasts of the two one-dimensional blobs have equal magnitude

$$C_1(t) = C_2(t) = \mid \lambda_2(t) - \lambda_1(t) \mid = 4(-t)^{\frac{3}{2}}, \tag{8.46}$$

and the boundaries ρ_1 and ρ_2 of the support regions are determined by the equations $\lambda(\rho_1, t) = \lambda_2(t)$ and $\lambda(\rho_2, t) = \lambda_1(t)$, with solutions $\rho_1(t) = +2\sqrt{-t}$ and $\rho_2(t) = -2\sqrt{-t}$. Hence, the variation of the area of the support region follows

$$A_1(t) = |\xi_2(t) - \rho_2(t)| = A_2(t) = |\rho_1(t) - \xi_1(t)| = 3\sqrt{-t}, \tag{8.47}$$

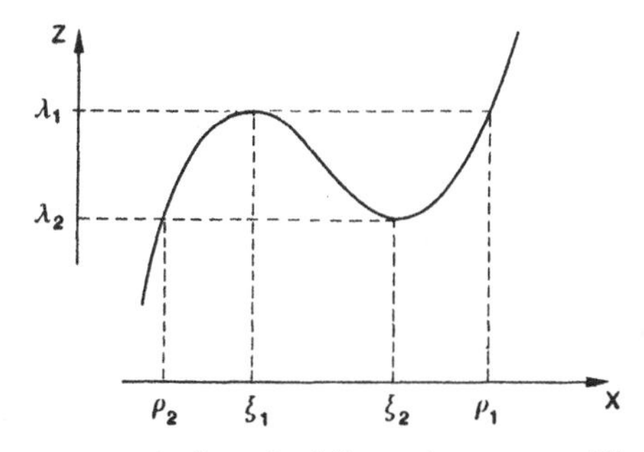

Figure 8.7. The situation before the bifurcation occurs. Illustration of the definitions of ξ_1, ξ_2, λ_1, λ_2, ρ_1 and ρ_2.

and the grey-level blob volume is

$$V_1(t) = V_2(t) = \int_{x=\rho_2(t)}^{\xi_2(t)} |\lambda(x;\ t) - \lambda_2(t)|\, dx = \frac{27(-t)^2}{4}. \tag{8.48}$$

Assuming that the scale-space blob is delimited by some minimum scale $t_{min} < 0$, its scale-space blob volume can be computed by

$$S_1 = \int_{t_{min}}^{0} V_1(t)\, dt = \frac{9(-t_{min})^3}{4}. \tag{8.49}$$

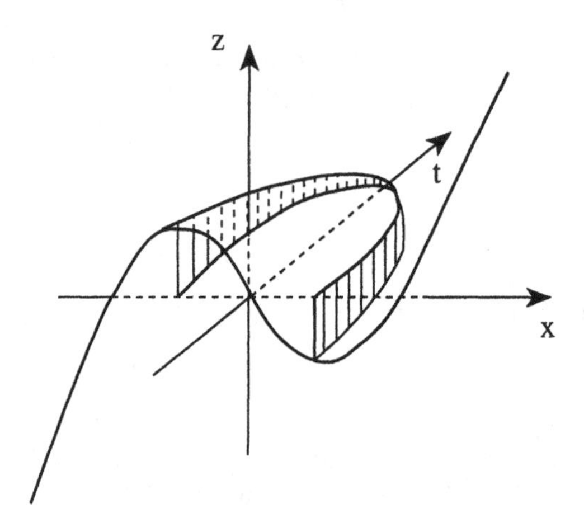

Figure 8.8. The fold singularity in the one-dimensional case.

8.5.3. The fold singularity in two dimensions

In the two-dimensional case the generic fold unfolding is

$$x^3 + ux \pm y^2. \tag{8.50}$$

By replacing each polynomial by a corresponding polynomial satisfying the diffusion equation (i.e., x^3 by $x^3 + 3xt$, (x by x), and y^2 by $y^2 + t$), we get

$$L(x, y;\ t) = x^3 + (u + 3t)x \pm (y^2 + t). \tag{8.51}$$

Obviously, u corresponds to an unessential translation of the scale parameter and can be omitted. Moreover, assume that the sign of the $\pm(y^2+t)$ term is positive. Then, the scale-space family to be studied is

$$L(x, y;\ t) = x^3 + 3xt + y^2 + t, \tag{8.52}$$

where x, y and t should again be interpreted as offset coordinates. The critical points of this mapping are given by

$$\begin{cases} \partial_x L = 3(x^2 + t) = 0, \\ \partial_y L = 2y = 0, \end{cases} \tag{8.53}$$

and their type by the sign of

$$\det(\mathcal{H}L) = L_{xx}L_{yy} - L_{xy}^2 = 12x. \tag{8.54}$$

If $t < 0$ there are two solutions:

$$r_1(t) = (x_1(t), y_1(t)) = (-\sqrt{-t}, 0), \quad r_2(t) = (x_2(t), y_2(t)) = (+\sqrt{-t}, 0),$$

where r_1 describes a saddle path, and r_2 a minimum path. The values at the critical points are

$$L_1(t) = L(r_1(t);\ t) = t - 2t\sqrt{-t}, \quad L_2(t) = L(r_2(t);\ t) = t + 2t\sqrt{-t}.$$

In terms of grey-level blobs, this singularity describes the annihilation of a dark grey-level blob. The boundary of the support region is obtained by solving the equation $L(x, y;\ t) = L_1(t)$, which can be reduced to

$$x^3 + 3tx + y^2 - 2(-t)^{\frac{3}{2}} = 0 \tag{8.55}$$

(see figure 8.9). Let y^- and y^+ denote the results of solving for y as a function of x and t in (8.55). Closed-form expressions for the blob descriptors can then be calculated as

$$C(t) = L_1(t) - L_2(t) = 4(-t)^{\frac{3}{2}}, \tag{8.56}$$

$$A(t) = \int_{x=-\sqrt{-t}}^{2\sqrt{-t}} \int_{y=y^-(x;\ t)}^{y=y^+(x;\ t)} dy\, dx = \frac{24\sqrt{3}(-t)^{\frac{5}{4}}}{5}, \tag{8.57}$$

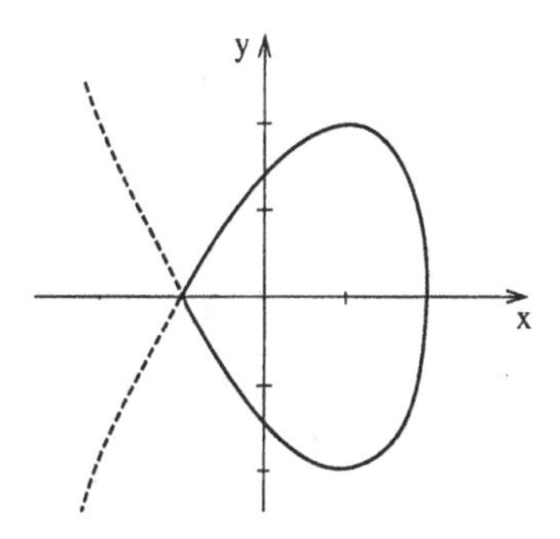

Figure 8.9. The support region of the grey-level blob. (The dashed line indicates the continuation of the level curve corresponding to the base level of the blob).

$$V(t) = \int_{x=-\sqrt{-t}}^{2\sqrt{-t}} \int_{y=y^-(x;\, t)}^{y^+(x;\, t)} (2(-t)^{\frac{3}{2}} - 3tx - x^3 - y^2)\, dy\, dx = \frac{3456\sqrt{3}(-t)^{\frac{11}{4}}}{385}.$$

If the sign of the $(y^2 + t)$ term in (8.51) instead is selected negative, the trajectories of critical points will be similar. The only difference is that the minimum point is replaced by a saddle point and the saddle point by a maximum point. In terms of blobs, that bifurcation corresponds to the annihilation of a bright blob.

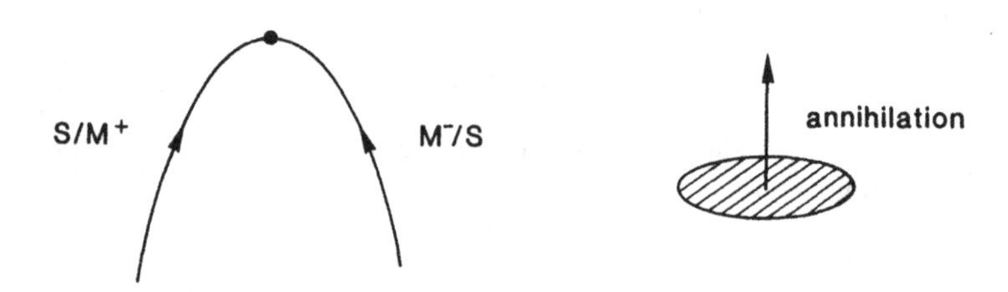

Figure 8.10. The fold unfolding in two dimensions $L(x, y;\ t) = x^3 + 3xt \pm (y^2 + t)$ describes (a) a saddle point and a minimum (maximum) point merging with increasing scale, i.e., (b) the annihilation of a dark (bright) grey-level blob.

8.5.4. The cusp singularity in two dimensions

Consider next the generic unfolding of the cusp singularity

$$x^4 + ux^2 + vx \pm y^2. \tag{8.58}$$

To make this function satisfy the diffusion equation, replace x^4 by $x^4 + 6tx^2 + 3t^3$, x^2 by $x^2 + t$, and y^2 by $y^2 + t$. This gives

$$L(x, y;\ t) = x^4 + (6t + u)x^2 + vx + ut + 3t^2 \pm (y^2 + t). \tag{8.59}$$

Here, the parameter u corresponds to a non-essential translation of the scale parameter and can be disregarded. Moreover, assume that the sign of the $\pm(y^2 + t)$ term is positive. Then, we get the unfolding

$$L(x, y;\ t) = x^4 + 6x^2t + vx + 3t^2 + y^2 + t, \tag{8.60}$$

where v is a free parameter. The critical points satisfy

$$\begin{cases} \partial_x L = 4x^3 + 12tx + v = h(x) = 0, \\ \partial_y L = 2y = 0, \end{cases} \tag{8.61}$$

and their type is determined by the sign of the Hessian

$$(\mathcal{H}L)(x, y;\ t) = 24(x^2 + t). \tag{8.62}$$

After some calculations it can be shown that the roots to $h(x) = 4x^3 + 12tx + v = 0$ obey the following qualitative behaviour:

- If $t > -(\frac{v}{8})^{\frac{2}{3}}$ then the equation $h(x) = 0$ has only one real root, and the unique stationary point is a local minimum.

- If $t < -(\frac{v}{8})^{\frac{2}{3}}$ then $h(x) = 0$ has three distinct roots x_i, satisfying $x_1(t) < -\sqrt{-t} < x_2(t) < +\sqrt{-t} < x_3(t)$. From (8.62) it follows that $x_1(t)$ and $x_3(t)$ are local minima and that $x_2(t)$ is a saddle.

- If $t = -(\frac{v}{8})^{\frac{2}{3}}$ then $h(x)$ has either one root of multiplicity three or two roots of multiplicity one and two. The bifurcation occurs at $x = (\frac{v}{8})^{\frac{1}{3}}$, and corresponds to the root with multiplicity greater than one. The behaviour around this point depends on the value of v, see figure 8.11.

Hence, the singularity describes a minimum-saddle pair annihilating under the influence of another minimum. In terms of blobs, this corresponds to two grey-level blobs merging into one.

Note that variation of the parameter v affect the bifurcation diagram of the critical points (figure 8.11), while the bifurcation diagram for the grey-level blobs remains the same (figure 8.12). This demonstrates that *bifurcation relations between grey-level blobs are more stable than bifurcation relations between critical points.*

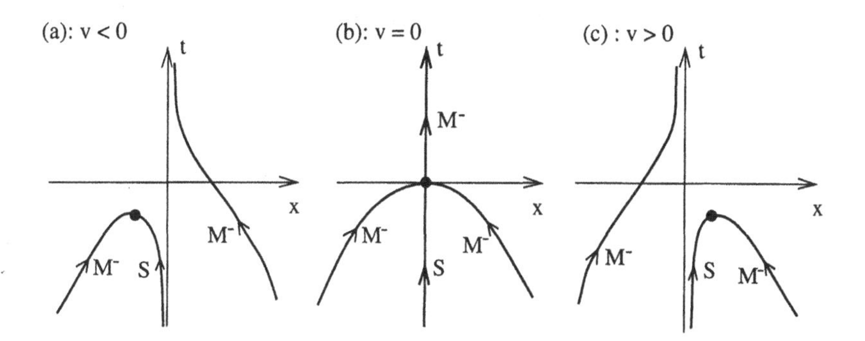

Figure 8.11. The cusp unfolding in two dimensions $L(x, y;\ t) = x^4 + 6x^2t + vx + 3t^2 + (y^2 + t)$ describes a minimum and a saddle merging under the influence of another minimum. Different events may occur depending on the value of v.

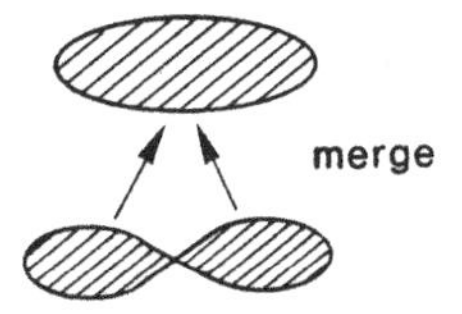

Figure 8.12. In terms of grey-level blobs, the three situations in figure 8.11 correspond to two dark grey-level blobs merging into one (independent of the value of v).

If the sign of the $\pm(y^2+t)$ is instead selected negative, then x_1 and x_2 will be saddle points, and x_2 a maximum, while if the sign of the entire unfolding is changed, all maxima are replace by minima and vice versa.

Zero-Crossings of the Laplacian. Introduce a parameter α such that

$$L_\alpha(x, y;\ t) = x^4 + 6x^2t + vx + 3t^2\alpha(y^2 + t), \tag{8.63}$$

and analyse the zero-crossings of the Laplacian, which are given by

$$\frac{\partial^2 L_\alpha}{\partial x^2} + \frac{\partial^2 L_\alpha}{\partial y^2} = 12x^2 + 12t + 2\alpha = 0. \tag{8.64}$$

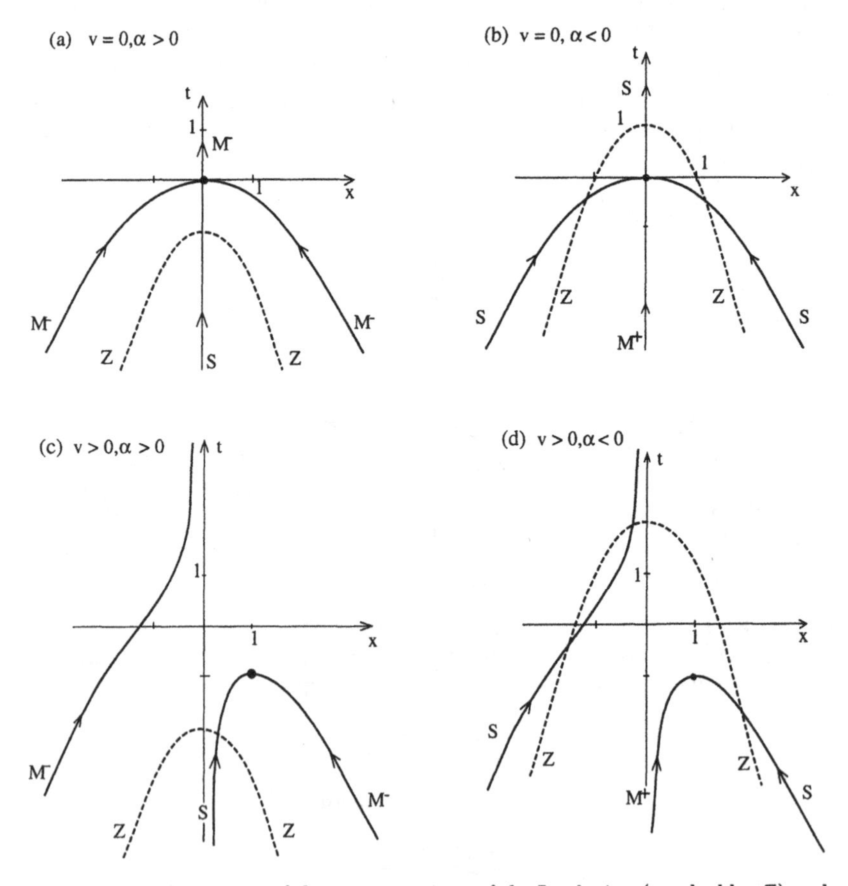

Figure 8.13. Locations of the zero-crossings of the Laplacian (marked by Z) and critical points (marked by M^+, M^-, and S) for the cusp unfolding (8.63) in the two-dimensional case. Note that during a certain scale interval the zero-crossings of the Laplacian fail to enclose isolated local extrema.

Obviously, there are two solutions $x = \pm\sqrt{-t - \alpha/6}$ if $t \leq -\alpha/6$, and we can observe that these zero-crossing curves do not give a correct subdivision around the local extrema for all t (see figure 8.13).

This example demonstrates that *in two (and higher) dimensions there is no absolute relationship between the locations of the Laplacian zero-crossing curves and the local extrema of a signal.* A Laplacian zero-crossing curve may enclose either no extremum, one extremum, or more than one local extremum. Only in the one-dimensional case it holds that there is exactly one local extremum point between two zero-crossings of the second derivative.

Drift velocity analysis. To analyse the drift velocity of the local extremum point *not* involved in the bifurcation, consider the one-dimensional version of (8.60) (delete the $(y^2 + t)$ term), and differentiate with respect to t:

$$\partial_t x = -\frac{x}{x^2 + t}. \tag{8.65}$$

To find the scale where the drift velocity is maximal, differentiate again and set the derivative to zero:

$$\partial_{tt} x = \frac{2xt}{(x^2 + t)^3} = 0. \tag{8.66}$$

Here, we are not interested in the case $x = 0$, since the behaviour at the bifurcation has already been analysed. Thus, as expected, the maximum drift velocity occurs when $t = 0$. Then, $x = (-\frac{v}{4})^{1/3}$ and

$$|\partial_t x|_{max} = -\frac{1}{x} = (\frac{4}{v})^{\frac{1}{3}}, \tag{8.67}$$

which shows that the maximum drift velocity tends to infinity as v tends to zero. This example demonstrates a further consequence of the results in section 8.1.1, namely that *even for critical points not directly involved in bifurcations there is no absolute upper bound one their drift velocity*, a conclusion which valid both in one and two dimensions.

Application to edge tracking. This analysis gives a further explanation to some of the problems that occur when edge focusing is applied to "staircase edges" (see figure 8.1 and the brief discussion in section 8.1.1). From experiments (Bergholm 1989) it is known that, in general, only one of the two edges in such configurations will be found by the focusing algorithm, and that in certain cases even that edge might get lost.

The fact that only one of the edges will be found is obvious from the bifurcation diagram in figure 8.11 provided that the focusing procedure is initiated from a sufficiently coarse scale and the bifurcation takes places sufficiently far away from the edge subject to tracking. The bifurcation diagram and the previous analysis for local extrema also indicate that the drift velocity of an edge point may increase rapidly even though the edge is not directly involved in any bifurcation, and hence exceed the finite drift velocity estimate used by the edge focusing algorithm.

8.6. Relating differential singularities across scales

Although the analysis so far has been concerned with critical points, the ideas behind it are general, and can be extended to other aspects of image structures. Differential singularities (features which can be defined as zero-crossings of (possibly non-linear) differential expressions) are conceptually easy to relate across scales.

For simplicity, let us restrict the analysis to the two-dimensional case, and consider features that at any scale t can be defined by

$$h(x, y;\ t) = 0 \tag{8.68}$$

for some function $h\colon \mathbb{R}^2 \times \mathbb{R}_+ \to \mathbb{R}^N$, where N is either 1 or 2. Using the implicit function theorem it is easy to analyze the dependence of (x, y) on t in the solution to (8.68). The results to be derived give estimates of the drift velocity of different features due to scale-space smoothing, and provides a theoretical basis for linking and identifying corresponding features at adjacent scales.

8.6.1. Zero-dimensional entities (points)

Assume first that $N = 2$, and write $h(x, y;\ t) = (h_1(x, y;\ t), h_2(x, y;\ t))$ for some functions $h_1, h_2\colon \mathbb{R}^2 \times \mathbb{R}_+ \to \mathbb{R}$. The derivative of the mapping h at a point $P_0 = (x_0, y_0;\ t_0)$ is

$$h'|_{P_0} = \begin{pmatrix} \partial_x h_1 & \partial_y h_1 & \partial_t h_1 \\ \partial_x h_2 & \partial_y h_2 & \partial_t h_2 \end{pmatrix}\Bigg|_{P_0} = \begin{pmatrix} \frac{\partial(h_1,h_2)}{\partial(x,y)} & \frac{\partial(h_1,h_2)}{\partial(t)} \end{pmatrix}\Bigg|_{P_0}. \tag{8.69}$$

If the matrix $\partial(h_1, h_2)/\partial(x, y)$ is non-singular at P_0, then the solution (x, y) to $h(x, y;\ t_0) = 0$ will be an isolated point. Moreover, the implicit function theorem guarantees that there exists some local neighbourhood around P_0 where (x, y) can be expressed as a function of t. The derivative of that mapping $t \mapsto (x, y)$ is:

$$\begin{pmatrix} \partial_t x \\ \partial_t y \end{pmatrix}\Bigg|_{P_0} = -\begin{pmatrix} \partial_x h_1 & \partial_y h_1 \\ \partial_x h_2 & \partial_y h_2 \end{pmatrix}\Bigg|_{P_0}^{-1} \begin{pmatrix} \partial_t h_1 \\ \partial_t h_2 \end{pmatrix}\Bigg|_{P_0}. \tag{8.70}$$

If h is a function of the spatial derivatives of L only, which is the case, for example, for the feature extractors treated in section 6.1.4, then the fact that spatial derivatives of L satisfy the diffusion equation

$$\partial_t L_{x^i y^j} = \tfrac{1}{2}(\partial_{xx} + \partial_{yy}) L_{x^i y^j}, \tag{8.71}$$

can be used for replacing derivatives with respect to t by derivatives with respect to x and y. In this way, closed form expression can be obtained containing only partial derivatives of L with respect to x and y.

For example, the *junction candidates* given by (6.14) satisfy $(\tilde{\kappa}_{\bar{u}}, \tilde{\kappa}_{\bar{v}}) = (0,0)$. In terms of directional derivatives, (8.70) can then be written

$$\left. \begin{pmatrix} \partial_t u \\ \partial_t v \end{pmatrix} \right|_{P_0} = - \left. \begin{pmatrix} \tilde{\kappa}_{\bar{u}\bar{u}} & \tilde{\kappa}_{\bar{u}\bar{v}} \\ \tilde{\kappa}_{\bar{u}\bar{v}} & \tilde{\kappa}_{\bar{v}\bar{v}} \end{pmatrix} \right|_{P_0}^{-1} \left. \begin{pmatrix} \partial_t \tilde{\kappa}_{\bar{u}} \\ \partial_t \tilde{\kappa}_{\bar{v}} \end{pmatrix} \right|_{P_0}. \tag{8.72}$$

By differentiating the expressions for $\tilde{\kappa}_{\bar{u}}$ and $\tilde{\kappa}_{\bar{v}}$ with respect to t, by using the fact that the spatial derivatives satisfy the diffusion equation, and by expressing the result in terms of directional derivatives along the preferred u- and v-directions (see section 6.1.1), the following expressions can be obtained (the calculations have been done using Mathematica[4])

$$\begin{aligned}
\tilde{\kappa}_{\bar{u}} &= L_{\bar{v}}^2 L_{\bar{u}\bar{u}\bar{u}}, \\
\tilde{\kappa}_{\bar{v}} &= L_{\bar{v}}^2 L_{\bar{u}\bar{u}\bar{v}} + 2L_{\bar{v}}(L_{\bar{u}\bar{u}}L_{\bar{v}\bar{v}} - L_{\bar{u}\bar{v}}^2), \\
\tilde{\kappa}_{\bar{u}\bar{u}} &= L_{\bar{v}}^2 L_{\bar{u}\bar{u}\bar{u}\bar{u}} + 2L_{\bar{u}\bar{u}}(L_{\bar{u}\bar{u}}L_{\bar{v}\bar{v}} - L_{\bar{u}\bar{v}}^2) + 2L_{\bar{v}}(L_{\bar{u}\bar{v}}L_{\bar{u}\bar{u}\bar{u}} - L_{\bar{u}\bar{u}}L_{\bar{u}\bar{u}\bar{v}}), \\
\tilde{\kappa}_{\bar{u}\bar{v}} &= L_{\bar{v}}^2 L_{\bar{u}\bar{u}\bar{u}\bar{v}} + 2L_{\bar{u}\bar{v}}(L_{\bar{u}\bar{u}}L_{\bar{v}\bar{v}} - L_{\bar{u}\bar{v}}^2) + 2L_{\bar{v}}(L_{\bar{v}\bar{v}}L_{\bar{u}\bar{u}\bar{u}} - L_{\bar{u}\bar{v}}L_{\bar{u}\bar{u}\bar{v}}), \\
\tilde{\kappa}_{\bar{v}\bar{v}} &= L_{\bar{v}}^2 L_{\bar{u}\bar{u}\bar{v}\bar{v}} + 2L_{\bar{v}\bar{v}}(L_{\bar{u}\bar{u}}L_{\bar{v}\bar{v}} - L_{\bar{u}\bar{v}}^2) \\
&\quad + 2L_{\bar{v}}(L_{\bar{u}\bar{u}}L_{\bar{v}\bar{v}\bar{v}} + 2L_{\bar{v}\bar{v}}L_{\bar{u}\bar{u}\bar{v}} - 3L_{\bar{u}\bar{v}}L_{\bar{u}\bar{v}\bar{v}}), \\
\partial_t \tilde{\kappa}_{\bar{u}} &= L_{\bar{v}}^2 (L_{\bar{u}\bar{u}\bar{u}\bar{u}\bar{u}} + L_{\bar{u}\bar{u}\bar{u}\bar{v}\bar{v}})/2 \\
&\quad + (L_{\bar{u}\bar{u}}L_{\bar{v}\bar{v}} - L_{\bar{u}\bar{v}}^2)(L_{\bar{u}\bar{u}\bar{u}} + L_{\bar{u}\bar{v}\bar{v}}) + L_{\bar{v}}(L_{\bar{u}\bar{u}\bar{u}}L_{\bar{v}\bar{v}\bar{v}} - L_{\bar{u}\bar{u}\bar{v}}L_{\bar{u}\bar{v}\bar{v}}), \\
\partial_t \tilde{\kappa}_{\bar{v}} &= L_{\bar{v}}^2 (L_{\bar{u}\bar{u}\bar{u}\bar{u}\bar{v}} + L_{\bar{u}\bar{u}\bar{v}\bar{v}\bar{v}})/2 + (L_{\bar{u}\bar{u}}L_{\bar{v}\bar{v}} - L_{\bar{u}\bar{v}}^2)(L_{\bar{u}\bar{u}\bar{v}} + L_{\bar{v}\bar{v}\bar{v}}) \\
&\quad + L_{\bar{v}}(L_{\bar{v}\bar{v}}(L_{\bar{u}\bar{u}\bar{u}\bar{u}} + L_{\bar{u}\bar{u}\bar{v}\bar{v}}) + L_{\bar{u}\bar{u}}(L_{\bar{v}\bar{v}\bar{v}\bar{v}} + L_{\bar{u}\bar{u}\bar{v}\bar{v}}) \\
&\quad - 2L_{\bar{u}\bar{v}}(L_{\bar{u}\bar{u}\bar{u}\bar{v}} + L_{\bar{u}\bar{v}\bar{v}\bar{v}})) \\
&\quad + L_{\bar{v}}(L_{\bar{u}\bar{u}\bar{v}}(L_{\bar{u}\bar{u}\bar{v}} + L_{\bar{v}\bar{v}\bar{v}}) - L_{\bar{u}\bar{v}\bar{v}}(L_{\bar{u}\bar{u}\bar{u}} + L_{\bar{u}\bar{v}\bar{v}})).
\end{aligned}$$

(These expressions simplify somewhat if we make use of $L_{\bar{u}\bar{u}\bar{u}}|_{P_0} = 0$, which follows from $\tilde{\kappa}_{\bar{u}} = 0$.) Note that as long as the Hessian matrix of $\tilde{\kappa}$ is non-degenerate, the sign of the $\tilde{\kappa}_{\mathcal{H}}$ and $\tilde{\kappa}_{\bar{u}\bar{u}}$ will be constant. This means that the type of extremum remains the same. For *local extrema* of the grey-level landscape, given by $(L_x, L_y) = (0,0)$, the expression for the drift velocity, of course, reduces to previously derived expression (8.3).

8.6.2. One-dimensional entities (curves)

If $N = 1$, then there will no longer be any unique correspondence between points at adjacent scales. An ambiguity arises, very similar to what is called the aperture problem in motion analysis. Nevertheless, we can determine the drift velocity in the normal direction of the curve.

[4]Mathematic is a registered trademark of Wolfram Research Research Inc., U.S.A.

Given a function $h\colon \mathbb{R}^2 \times \mathbb{R}_+ \to \mathbb{R}$ consider the solution to $h(x, y;\ t) = 0$. Assume that $P_0 = (x_0, y_0;\ t_0)$ is a solution to this equation and that the gradient of the mapping $(x, y) \mapsto h(x, y;\ t_0)$ is non-zero. Then, in some neighbourhood around (x_0, y_0) the solution (x, y) to $h(x, y;\ t_0) = 0$ defines a curve. Its normal at (x_0, y_0) is given by $(\cos\phi, \sin\phi) = (h_x, h_y)/(h_x^2 + h_y^2)^{1/2}$ at P_0. Consider next the function $\tilde{h}\colon \mathbb{R} \times \mathbb{R}_+ \to \mathbb{R}$ defined by $\tilde{h}(s;\ t) = h(x_0 + s\cos\phi, y_0 + s\sin\phi;\ t)$. It has the derivative

$$\tilde{h}_s(0;\ t_0) = h_x(x_0, y_0;\ t_0)\cos\phi + h_y(x_0, y_0;\ t_0)\sin\phi = \left.\sqrt{h_x^2 + h_y^2}\right|_{P_0}.$$

Since this derivative is non-zero, we can apply the implicit function theorem. It follows that there exists some neighbourhood around P_0 where $\tilde{h}(s;\ t) = 0$ defines s as a function of t. The derivative of this mapping is

$$\partial_t s|_{P_0} = -\left.\tilde{h}_s\right|_{P_0}^{-1} \left.\tilde{h}_t\right|_{P_0} = -\left.\frac{h_t}{\sqrt{h_x^2 + h_y^2}}\right|_{P_0}. \tag{8.73}$$

As an example of this, consider an *edge* given by non-maximum suppression

$$h = \alpha = L_x^2 L_{xx} + 2L_x L_y L_{xy} + L_y^2 L_{yy} = 0. \tag{8.74}$$

By differentiating (8.74), by using the fact that the derivatives of L satisfy the diffusion equation, and by expressing the result in terms of the directional derivatives we get

$$\alpha = L_{\bar{v}}^2 L_{\bar{v}\bar{v}} = 0, \tag{8.75}$$

$$\alpha_{\bar{u}} = L_{\bar{v}}^2 L_{\bar{u}\bar{v}\bar{v}} + 2L_{\bar{v}} L_{\bar{u}\bar{v}} L_{\bar{u}\bar{u}}, \tag{8.76}$$

$$\alpha_{\bar{v}} = L_{\bar{v}}^2 L_{\bar{v}\bar{v}\bar{v}} + 2L_{\bar{v}} L_{\bar{u}\bar{v}}^2, \tag{8.77}$$

$$\alpha_t = L_{\bar{v}}^2 (L_{\bar{u}\bar{u}\bar{v}\bar{v}} + L_{\bar{v}\bar{v}\bar{v}\bar{v}})/2 + L_{\bar{v}} L_{\bar{u}\bar{v}} (L_{\bar{u}\bar{u}\bar{u}} + L_{\bar{u}\bar{v}\bar{v}}). \tag{8.78}$$

To summarize, the drift velocity in the normal direction of a *curved edge* in scale-space is (with $\alpha_{\bar{u}}$ and $\alpha_{\bar{u}}$ according to (8.76)–(8.77))

$$(\partial_t u, \partial_t v) = -\frac{L_{\bar{v}}(L_{\bar{u}\bar{u}\bar{v}\bar{v}} + L_{\bar{v}\bar{v}\bar{v}\bar{v}}) + 2L_{\bar{u}\bar{v}}(L_{\bar{u}\bar{u}\bar{u}} + L_{\bar{u}\bar{v}\bar{v}})}{2((L_{\bar{v}} L_{\bar{u}\bar{v}\bar{v}} + 2L_{\bar{u}\bar{v}} L_{\bar{u}\bar{u}})^2 + (L_{\bar{v}} L_{\bar{v}\bar{v}\bar{v}} + 2L_{\bar{u}\bar{v}}^2)^2)} \left(\frac{\alpha_{\bar{u}}}{L_{\bar{v}}}, \frac{\alpha_{\bar{v}}}{L_{\bar{v}}}\right). \tag{8.79}$$

Unfortunately, this expression cannot be further simplified unless additional constraints are posed on L. For a *straight edge*, however, where all partial derivatives with respect to u are zero, it reduces to

$$(\partial_t u, \partial_t v) = -\frac{1}{2}\frac{L_{\bar{v}\bar{v}\bar{v}\bar{v}}}{L_{\bar{v}\bar{v}\bar{v}}}(0, 1), \tag{8.80}$$

which agrees with the result in (8.5). For a curve given by the *zero-crossings of the Laplacian* we have

$$(\partial_t u, \partial_t v) = -\frac{\nabla^2(\nabla^2 L)}{2((\nabla^2 L_{\bar{u}})^2 + (\nabla^2 L_{\bar{v}})^2)}(\nabla^2 L_{\bar{u}}, \nabla^2 L_{\bar{v}}), \tag{8.81}$$

which also simplifies to (8.80) if all directional derivatives in the u-direction are set to zero. Similarly, for a *parabolic curve*, given by $\det(\mathcal{H}L) = L_{xx}L_{yy} - L_{xy}^2 = 0$, the drift velocity in the normal direction is

$$(\partial_t p, \partial_t q) = -\frac{L_{qq}L_{pppp} + (L_{pp} + L_{qq})L_{ppqq} + L_{pp}L_{qqqq}}{2((L_{pp}L_{pqq} + L_{qq}L_{ppp})^2 + (L_{pp}L_{qqq} + L_{qq}L_{ppq})^2)} \times (L_{pp}L_{pqq} + L_{qq}L_{ppp}, L_{pp}L_{qqq} + L_{qq}L_{ppq}). \tag{8.82}$$

Here, the result has been expressed in a pq-coordinate system, with the p- and q-axes aligned to the principal axes of curvature such that the mixed second-order directional derivative $L_{\bar{p}\bar{q}}$ is zero.

8.7. Density of local extrema as function of scale

In some applications it is of interest to know how the density of local extrema can be expected to vary with scale. One example is the derivation of effective scale, a transformed scale parameter intended to capture the concept of "scale-space lifetime" in a proper manner (see section 7.7). Of course, this question seems to be very difficult or even impossible to answer to generally, since such a quantity can be expected to vary substantially from one image to another. How should one then be able to talk about "expected behaviour"? Should one consider all possible (realistic) images, study how this measure evolves with scale and then form some kind of average?

In this section a simple study will be performed. We will consider random noise data with normal distribution. Under these assumptions it turns out to be possible to derive a compact closed form expression for this quantity. The analysis will be based on a treatment by Rice (1945) about the expected density of zero-crossings and local maxima of stationary normal processes; see also (Papoulis 1972; Cramer and Leadbetter 1967).

8.7.1. Continuous analysis

The density of local maxima μ for a stationary normal process is given by the second and fourth derivatives of the autocorrelation function R

$$\mu = \frac{1}{2\pi}\sqrt{-\frac{R^{(4)}(0)}{R''(0)}}. \tag{8.83}$$

This expression can also be written as (Rice 1945; Papoulis 1972)

$$\mu = \frac{1}{2\pi}\sqrt{\frac{\int_{-\infty}^{\infty}\omega^4 S(\omega)d\omega}{\int_{-\infty}^{\infty}\omega^2 S(\omega)d\omega}}, \tag{8.84}$$

where S is the spectral density

$$S(\omega) = \int_{-\infty}^{\infty} e^{-i\omega\tau} R(\tau)\, d\tau. \tag{8.85}$$

Since the scale-space representation L is generated from the input signal f by a linear transformation, the spectral density of L, denoted S_L, is given by

$$S_L(\omega) = |H(\omega)|^2 S_f(\omega), \tag{8.86}$$

where S_f is the spectral density of f, and $H(\omega)$ is the Fourier transform of the impulse response h

$$H(\omega) = \int_{-\infty}^{\infty} h(t) e^{-i\omega t}\, dt \tag{8.87}$$

In our scale-space case, h is of course the Gaussian kernel

$$g(\xi;\; t) = \frac{1}{\sqrt{2\pi t}} e^{-\xi^2/2t}, \tag{8.88}$$

which has the Fourier transform

$$G(\omega;\; t) = e^{-\omega^2 t/2}. \tag{8.89}$$

Assuming that f is generated by white noise with $S_f(w) = 1$ this gives

$$S_L(\omega) = e^{-\omega^2 t}. \tag{8.90}$$

Using the formula (Spiegel 1968: 15.77)

$$\int_0^{\infty} x^m e^{-ax^2} dx = \frac{\Gamma((m+1)/2)}{2a^{(m+1)/2}}, \tag{8.91}$$

a closed form expression can be calculated for the density of local maxima of a continuous signal, $p_c(t)$:

$$p_c(t) = \frac{1}{2\pi}\sqrt{\frac{\int_{-\infty}^{\infty}\omega^4\, e^{-\omega^2 t} d\omega}{\int_{-\infty}^{\infty}\omega^2\, e^{-\omega^2 t} d\omega}} = \frac{1}{2\pi}\sqrt{\frac{2\frac{\Gamma(5/2)}{2t^{5/2}}}{2\frac{\Gamma(3/2)}{2t^{3/2}}}} = \frac{1}{2\pi}\sqrt{\frac{3}{2}}\frac{1}{\sqrt{t}}. \tag{8.92}$$

Of course, an identical result applies[5] to local minima. To summarize,

[5]Observe that the same type of qualitative behaviour ($p_c(t) \sim t^{-\frac{1}{2}}$) applies also to the local extrema in the *spatial derivatives* of the scale-space representation (just replace $H = G$ by $H = (i\omega)^n G$ in the previous analysis).

PROPOSITION 8.5. (DENSITY OF LOCAL EXTREMA; WHITE NOISE; 1D) *In the scale-space representation of a one-dimensional continuous signal generated by a white noise stationary normal process, the expected density of local maxima (minima) in a smoothed signal at a certain scale decreases with scale as* $t^{-1/2}$.

This scale dependence implies that a graph showing the density of local maxima (minima) as function of scale can be expected[6] to be a *straight line* in a log-log diagram

$$\log(p_c(t)) = \tfrac{1}{2}\log(\tfrac{3}{2}) - \log(2\pi) - \tfrac{1}{2}\log(t) = \text{constant} - \tfrac{1}{2}\log(t).$$

In section 7.7.1 it was shown that a natural way to convert the ordinary scale parameter t into a transformed scale parameter, effective scale τ, is by $\tau(t) = A + B\log(p(t))$, where $p(t)$ again denotes the expected density of local extrema at a certain scale t and A and B are arbitrary constants. This result gives:

COROLLARY 8.6. (EFFECTIVE SCALE FOR CONTINUOUS SIGNALS; 1D) *For continuous one-dimensional signals the effective scale parameter* τ_c *as function of the ordinary scale parameter* t *is (up to an arbitrary affine transformation, i.e., for some arbitrary constants* A' *and* $B' > 0$*) given by a logarithmic transformation*

$$\tau_c(t) = A' + B'\log(t). \tag{8.93}$$

An interesting question concerns what will happen if the uncorrelated white noise model for the input signal is changed. A self-similar spectral density that has been applied to fractals (Barnsley *et al.* 1988; Gårding 1988) is given by

$$S_f(w) = w^{-\beta}. \tag{8.94}$$

For one-dimensional signals, reasonable values of β are obtained between 1 and 3. Of course, such a distribution is somewhat non-physical, since $S_f(w)$ will tend to infinity as t tends to zero and neither one of the spectral moments is convergent. However, when multiplied by a Gaussian function the second and fourth order moments in (8.84) will converge provided that $\beta < 3$. We obtain,

$$p_{c,\beta}(t) = \frac{1}{2\pi}\sqrt{\frac{\int_{-\infty}^{\infty} \omega^4\, e^{-\omega^2 t}\omega^{-\beta} d\omega}{\int_{-\infty}^{\infty} \omega^2\, e^{-\omega^2 t}\omega^{-\beta} d\omega}} = \frac{1}{2\pi}\sqrt{\frac{3-\beta}{2}}\frac{1}{\sqrt{t}} \quad (\beta < 3). \tag{8.95}$$

[6] Of course, we cannot expect that a graph showing this curve for a particular signal to be a straight line, since this would require some type of ergodicity assumption that in general will not be satisfied. However, the average behaviour over many different types of imagery can be expected to be close to this situation.

PROPOSITION 8.7. (DENSITY OF LOCAL EXTREMA; FRACTAL NOISE) *In the scale-space representation of a one-dimensional continuous signal generated by a stationary normal process with spectral density* $\omega^{-\beta}$ ($\omega \in [0, 3[$)*, the expected density of local maxima (minima) in a smoothed signal at a certain scale decreases with scale as* $t^{-1/2}$.

Note that also this graph will be a straight line in a log-log diagram.

8.7.2. *Discrete analysis*

From the previous continuous analysis it is apparent that the density of local extrema may tend to infinity when the scale parameter tends to zero. As earlier indicated, this result is not applicable to discrete signals, since in this case the density of local extrema will have an upper bound because of the finite sampling. Hence, to capture what happens in the discrete case, a genuinely discrete treatment is necessary. The treatment will be based on the discrete scale-space concept developed in chapter 3. Given a discrete signal $f \colon \mathbb{Z} \to \mathbb{R}$ the scale-space representation $L \colon \mathbb{Z} \times \mathbb{R}_+ \to \mathbb{R}$ is defined by

$$L(x;\ t) = \sum_{n=-\infty}^{\infty} T(n;\ t)\, f(x-n), \tag{8.96}$$

where $T(n;\ t) = e^{-t} I_n(t)$ is the discrete analogue of the Gaussian kernel, and I_n are the modified Bessel functions of integer order (Abramowitz and Stegun 1964). Equivalently, this scale-space family can be defined in terms of a semi-discretized version of the diffusion equation

Consider the scale-space representation of a signal generated by a random noise signal. The probability that a point at a certain scale is say a local maximum point is equal to the probability that its value is greater than (or possibly equal to)[7] the values of its nearest neighbours:

$$\begin{aligned} &P(x_i \text{ is a local maximum at scale } t) \\ &\qquad = P((L(x_i;\ t) \geq L(x_{i-1};\ t)) \wedge (L(x_i;\ t) \geq L(x_{i+1};\ t))). \end{aligned} \tag{8.97}$$

If we assume that the input signal f is generated by a stationary normal process then also L will be a stationary normal process and the distribution of any triple $(L_{i-1}, L_i, L_{i+1})^T$, from now on denoted by $\xi = (\xi_1, \xi_2, \xi_3)^T$, will be jointly normal, which means that its statistics will be completely determined by the mean vector and the autocovariance matrix. Trivially, the mean of ξ is zero provided that the mean of f is zero.

[7] Although there are several ways to define a local extremum of a discrete signal using different combinations of ">" and "≥", these definitions will yield the same result with respect to this application.

Since the transformation from f to L is linear, the autocovariance C_L for the smoothed signal L will be given by

$$C_L(\cdot;\ t) = T(\cdot;\ t) * T(\cdot;\ t) * C_f(\cdot) = T(\cdot;\ 2t) * C_f(\cdot), \tag{8.98}$$

where C_f denotes the autocovariance of f. In the last equality we have made use of the semigroup property $T(\cdot;\ s) * T(\cdot;\ t) = T(\cdot;\ s+t)$ for the family of convolution kernels. If the input signal consists of white noise, then C_f will be the discrete delta function and $C_L(\cdot;\ t) = T(\cdot;\ 2t)$. Taking the symmetry property $T(-n;\ t) = T(n;\ t)$ into account, the distribution of ξ will be jointly normal with mean vector m_{3D} and covariance matrix C_{3D} given by:

$$m_{3D} = \begin{pmatrix} 0 \\ 0 \\ 0 \end{pmatrix}, \quad C_{3D} = \begin{pmatrix} T(0;\ 2t) & T(1;\ 2t) & T(2;\ 2t) \\ T(1;\ 2t) & T(0;\ 2t) & T(1;\ 2t) \\ T(2;\ 2t) & T(1;\ 2t) & T(0;\ 2t) \end{pmatrix}. \tag{8.99}$$

By introducing new variables $\eta_1 = \xi_2 - \xi_1$ and $\eta_2 = \xi_2 - \xi_3$, it follows that $\eta = (\eta_1, \eta_2)^T$ will be jointly normal and its statistics completely determined by

$$m_{2D} = \begin{pmatrix} 0 \\ 0 \end{pmatrix}, \quad C_{2D} = \begin{pmatrix} a_0(t) & a_1(t) \\ a_1(t) & a_0(t) \end{pmatrix}. \tag{8.100}$$

From well-known rules for the covariance $C(\cdot,\cdot)$ of a linear combination it follows that

$$\begin{aligned} a_0(t) &= C(\eta_1, \eta_1) = C(\eta_2, \eta_2) \\ &= 2(T(0;\ 2t) - T(1;\ 2t)), \end{aligned} \tag{8.101}$$

$$\begin{aligned} a_1(t) &= C(\eta_1, \eta_2) = C(\eta_2, \eta_1) \\ &= T(0;\ 2t) - 2T(1;\ 2t) + T(2;\ 2t). \end{aligned} \tag{8.102}$$

From $a_0(t) - a_1(t) = T(0;\ t) - T(2;\ t)$ and the unimodality property of T ($T(i;\ t) \geq T(j;\ t)$ if $|i| > |j|$) it follows that $a_0(t) > a_1(t)$ and trivially $a_0(t) > 0$ for all t. Now $p_d(t)$ can be expressed in terms of a two-dimensional integral

$$p_d(t) = \int\int_{\{\eta=(\eta_1,\eta_2):\ (\eta_1 \geq 0)\wedge(\eta_2 \geq 0)\}} \frac{1}{\sqrt{(2\pi)^2 |C_{2D}|}} e^{-\eta^T C_{2D}^{-1} \eta/2}\, d\eta_1\, d\eta_2. \tag{8.103}$$

After some calculations (Lindeberg 1991: appendix A.5.4) it follows that

$$p_d(t) = \frac{1}{4} + \frac{1}{2\pi} \arctan\left(\frac{a_1(t)}{\sqrt{a_0^2(t) - a_1^2(t)}} \right). \tag{8.104}$$

Observe that for any $a_0(t)$ and $a_1(t)$ this value is guaranteed to never be outside the interval $[0, \frac{1}{2}]$. With the expressions for $a_0(t)$ and $a_1(t)$, given by smoothing with the discrete analogue of the Gaussian kernel, the maximum value over variations in t is obtained for $t = 0$:

$$p_d(0) = \frac{1}{3}. \tag{8.105}$$

PROPOSITION 8.8. (DENSITY OF LOCAL EXTREMA;1D)
In the scale-space representation (8.96) of a one-dimensional discrete signal generated by a white noise stationary normal process, the expected density of local maxima (minima) in a smoothed signal at a certain scale t is given by (8.104) *with $a_0(t)$ and $a_1(t)$ according to* (8.101) *and* (8.102).

It is interesting to compare the discrete expression (8.104) with the earlier continuous result (8.92). The scale value where the continuous estimate gives a density equal to the discrete density at $t = 0$ is given by the equation $p_c(t) = p_d(0)$, that is by

$$\frac{1}{2\pi}\sqrt{\frac{3}{2}}\frac{1}{\sqrt{t}} = \frac{1}{3} \tag{8.106}$$

which has the solution

$$t_{c\text{-}d} = \frac{27}{8\pi^2} \approx 0.3420 \tag{8.107}$$

This corresponds to a σ-value of about 0.5848. Below this scale value the continuous analysis is, from that point of view, definitely not a valid approximation of what will happen to discrete signals. By combining proposition 8.8 with the effective scale concept we get:

COROLLARY 8.9. (EFFECTIVE SCALE FOR DISCRETE SIGNALS; 1D)
For discrete one-dimensional signals the effective scale parameter τ_d as function of the ordinary scale parameter t is given by

$$\tau_d(t) = A'' + B'' \log\left(\frac{4\pi}{3\pi + 6\arctan\left(\frac{a_1(t)}{\sqrt{a_0^2(t)-a_1^2(t)}}\right)}\right) \tag{8.108}$$

for some arbitrary constants A'' and $B'' > 0$ with $a_0(t)$ and $a_1(t)$ given by (8.101) *and* (8.102).

When defining the effective scale τ_d for discrete signals it is natural to let $t = 0$ correspond to $\tau_d = 0$. In that case A'' will be zero. Without loss of generality, we will from now on set $A'' = 0$ and $B'' = 1$.

8.7.3. *Asymptotic behaviour at fine and coarse scales*

A second order MacLaurin expansion of $p_d(t)$ (Lindeberg 1991: appendix A.5.5) yields

$$p_d(t) = \frac{1}{3} - \frac{1}{2\sqrt{3}\pi}t + \frac{1}{6\sqrt{3}\pi}t^2 + O(t^3). \tag{8.109}$$

This means that the effective scale $\tau_d(t)$ can be MacLaurin expanded (Lindeberg 1991: appendix A.5.5)

$$\tau_d(t) = \log\left(\frac{p_d(0)}{p_d(t)}\right) = \frac{\sqrt{3}}{2\pi}t + \left(\frac{1}{2\sqrt{3}\pi} + \frac{3}{8\pi^2}\right)t^2 + O(t^3), \tag{8.110}$$

showing that *at fine scales the effective scale τ for one-dimensional discrete signals is approximately an affine function of the ordinary scale parameter t.* On the other hand, a Taylor expansion of $p_d(t)$ at coarse scales gives

$$p_d(t) = \frac{1}{2\pi}\sqrt{\frac{3}{2}}\frac{1}{\sqrt{t}}\left(1 + \frac{1}{8t} + O(\frac{1}{t^2})\right) \tag{8.111}$$

which asymptotically agrees with the continuous result in (8.92). By inserting this expression into the expression for effective scale and by using $p_d(0) = \frac{1}{3}$ it follows that

$$\tau_d(t) = \log\left(\frac{p_d(0)}{p_d(t)}\right) = \log\left(\frac{2\pi}{3}\sqrt{\frac{2}{3}}\right) + \frac{1}{2}\log(t) + \log\left(1 - \frac{1}{8t} + O(\frac{1}{t^2})\right).$$

Hence, *at coarse scales the effective scale τ for one-dimensional discrete signals is approximately (up to an arbitrary affine transformation) a logarithmic function of the ordinary scale parameter t.*

The term $\log(1 - \frac{1}{8t} + O(\frac{1}{t^2}))$ expresses how much the effective scale derived for discrete signals differs from the effective scale derived for continuous signals, provided that the same values of the (arbitrary) constants A and B are selected in both cases.

8.7.4. *Comparisons between the continuous and discrete results*

As an illustration of the difference between the density of local maxima in the scale-space representation of a continuous and a discrete signal, we show the graphs of p_c and p_d in figure 8.14 (linear scale) and figure 8.15 (log-log scale). As expected, the curves differ significantly for small t and approach each other as t increases.

Numerical values quantifying this difference for a few values of t are given in table 8.1. It shows the ratio

$$\tau_{diff}(t) = \frac{\tau_d(t) - \tau_c(t)}{\tau_c(2t) - \tau_c(t)} = \frac{\tau_d(t) - \tau_c(t)}{\log(2)/2}, \tag{8.112}$$

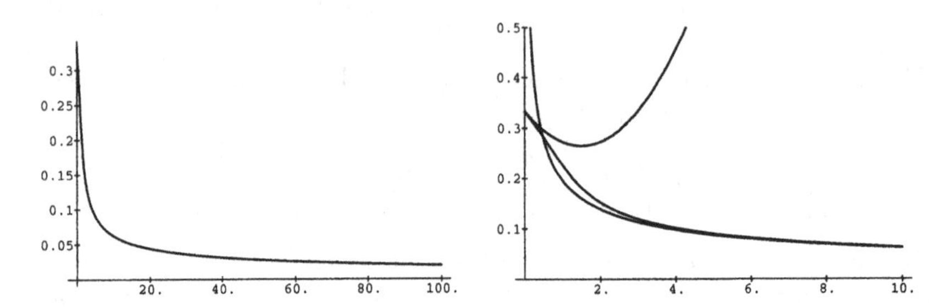

Figure 8.14. The density of local maxima of a discrete signal as function of the ordinary scale parameter t in linear scale. (a) Graph for $t \in [0, 100]$. (b) Enlargement of the interval $t \in [0, 10]$. For comparison the graphs showing the density of local extrema for a continuous signal $p_c(t)$ and the second order Taylor expansion of $p_d(t)$ around $t = 0$ have also been drawn. As expected, the continuous and discrete results differ significantly for small values of t but approach each other as t increases. The MacLaurin expansion is a valid approximation only in a very short interval around $t = 0$.

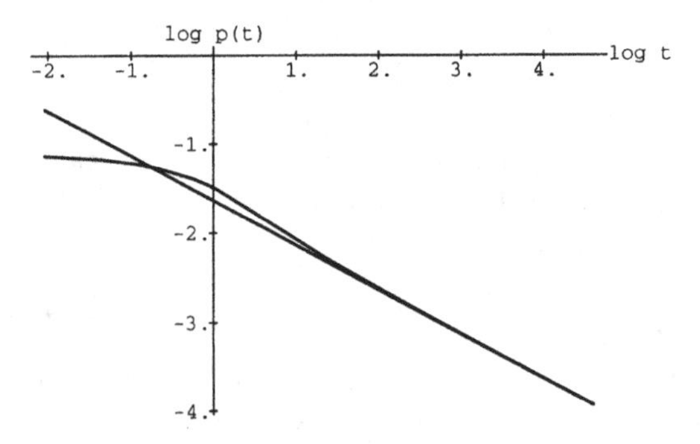

Figure 8.15. The density of local maxima of a continuous and a discrete signal as function of the ordinary scale parameter t in log-log scale ($t \in [0, 100]$). The straight line shows $p_c(t)$ and the other curve $p_d(t)$. It can be observed that p_c and p_d approach each other as the scale parameter increases. When t tends to zero, $p_c(t)$ tends to infinity while $p_d(t)$ tends to a constant ($\frac{1}{3}$).

t	$\tau_{diff}(t)$
0	∞
0.0625	250.30 %
0.25	67.46 %
1.0	-41.82 %
4.0	-10.47 %
16.0	-2.32 %
64.0	-0.56 %
256.0	-0.14 %
∞	0

Table 8.1. Indications about how the effective scale obtained from a discrete analysis differs from the effective scale given by the continuous scale-space theory. The quantity $\tau_{diff}(t)$ expresses the difference between $\tau_d(t)$ and $\tau_c(t)$ normalized such that one unit (100 %) in $\tau_{diff}(t)$ corresponds to the increase in τ_c induced by an increase in t with a factor of two.

which is a natural measure for how much the effective scale obtained from a continuous analysis differs from a discretely determined effective scale. The quantity is normalized so that one unit in τ_{diff} corresponds to the increase in τ_c induced by an increase in t with a factor of two.

8.7.5. Extension to two dimensions

The same type of analysis can, in principle, be carried out also for two-dimensional discrete signals. The probability that a specific point at a certain scale is a local maximum point is again equal to the probability that its value is greater than the values of its neighbours. Depending on the connectivity concept (four-connectivity or eight-connectivity on a square grid) we then obtain either a four-dimensional or an eight-dimensional integral to solve. However, because of the dimensionality of the integrals, no attempts have been made to calculate explicit expressions for the variation of the density as function of scale. Instead, for the purpose of implementation, the behaviour over scale has been simulated for various uncorrelated random noise signals (see section 7.7.1). From those experiments it has been empirically demonstrated that the $t^{-\alpha}$ dependence (with $\alpha \approx 1.0$) constitutes a reasonable approximation at coarse levels of scale.

The reason the exponent α changes from 0.5 to 1.0 when going from one to two dimensions can intuitively be understood by a dimensional analysis: Assume (as in appendix 7.7.1) that the standard deviation of the Gaussian kernel, $\sigma = \sqrt{t}$, can be linearly related to a characteristic length, x, in the scale-space representation of an N-dimensional signal at scale t. Moreover, assume that a characteristic distance d between the local extrema in that signal is linearly related to x. Then, the density of local extrema will be proportional to $d^{-N} \sim x^{-N} \sim \sigma^{-N}$, that is to $t^{-N/2}$.

8.8. Summary

We have analysed the behaviour of critical points in scale-space and shown that non-degenerate critical points will in general form regular curves across scales. Along those we have provided generally valid estimates of the drift velocity. At degenerate critical points the behaviour is more complicated and bifurcations may take place. For one-dimensional signals, the only bifurcation events possible when the scale parameter increases, are annihilations of pairs of local maxima and minima, while for two-dimensional signals both annihilations and creations of pairs of local extrema and saddle points can occur. Applied to grey-level and scale-space blobs only annihilations and merges will take place in the one-dimensional case, while the list of possibilities in two-dimensions comprises four types: annihilations, merges, splits and creations.

Finally, it should be pointed out that this analysis has been mainly concerned with the scale-space concept for continuous signals. When implementing this theory computationally it is obvious that one has to consider sampled (that is, discrete) data. At coarse scales, when a characteristic length of features in the image can be regarded as large compared to the distance between adjacent grid points, it seems plausible that the continuous results should constitute a reasonable approximation to what will happen in the scale-space representation of a discrete signal and vice versa. However, as indicated in section 8.7 this similarity will not necessarily hold[8] at fine scales. In those cases, a genuinely discrete theory might be needed. A thorough understanding of what happens to continuous signals under scale-space smoothing constitutes a first step towards this goal.

---

[8]Some conceptual complications arise in this context (for instance, what should be meant by drift velocity for discrete signals). It seems difficult to estimate such a quantity accurately, especially at fine scales, since in the discrete case local extrema will not move continuously, but rather in steps from one pixel to the next. Thus, one cannot talk about velocity, but rather about how long it takes until an extremum point moves one pixel. An alternative approach to this problem would be to analyse the feature points with sub-pixel accuracy (although this idea has not been carried out). Other conceptual problems concern what should be meant by singularities or degenerate and nondegenerate critical points in the discrete case. One possibility is to define these in terms of adjacent pixels having equal values, or by transitions (say e.g., blob bifurcations). But, will the classification of possible blob events still be valid in the discrete case?

9

Algorithm for computing the scale-space primal sketch

When building a representation like the scale-space primal sketch, several computational aspects must be considered. Algorithms are needed for smoothing grey-level images, detecting grey-level blobs in smoothed grey-level images, registering bifurcations, linking grey-level blobs across scales into scale-space blobs, and computing the scale-space blob volumes. In this chapter how this can be done will be briefly described. Before starting, it should be remarked that some algorithmic descriptions will, by necessity, be somewhat technical. Therefore, the reader is recommended to keep the overview description from chapter 7 in mind when penetrating this material. It should also be emphasized that, although the treatment deals mainly with the two-dimensional case, the ideas behind it are general and can (with appropriate modifications) be applied both in lower and higher dimensions.

9.1. Grey-level blob detection

To start with, an algorithm will be outlined for detecting grey-level blobs in a grey-level image. Let us consider the case with bright blobs on a dark background. The case with dark blobs on a bright background can be solved by applying the bright-blob detection algorithm to the inverted grey-level image.

9.1.1. One-dimensional signals

Grey-level blob detection in a one-dimensional discrete signal is trivial. In this case it is sufficient to start from each local maximum point and initiate a search procedure in each of the two possible directions (see figure 9.1). Each search procedure continues until it finds a local minimum point, i.e., as long as the grey-level values are decreasing. As soon as a minimum point has been found the search procedure stops, and the grey-level value is registered. The base-level of the blob is then given by the maximum value of these two registered grey-levels. From this information the grey-level blob is given by those pixels that can be reached from the local maximum point without descending below the base-level.

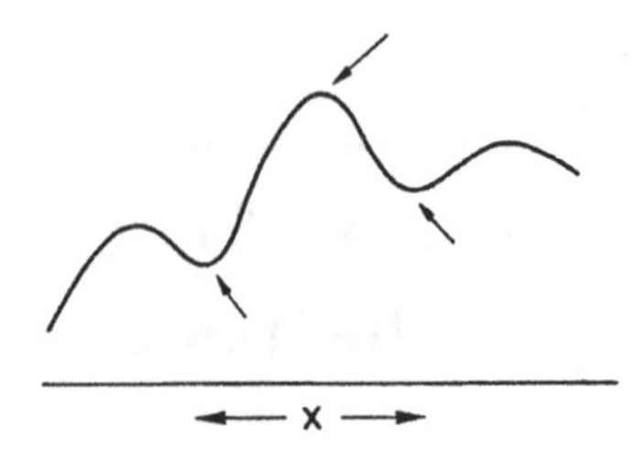

Figure 9.1. The blob detection algorithm for a one-dimensional discrete signal is trivial. The base-level of a bright blob is equal to the maximum value of the grey-levels in the two local minimum points surrounding the local maximum point of the blob.

The two-dimensional case is more elaborate, since the search then may be performed in a variety of directions. In section 9.1.4 a methodology will be described that avoids the search problem and instead performs a global blob detection based on a pre-sorting of the grey-levels.

9.1.2. Discrete local extremum: Generic vs. non-generic signals

A very important aspect when constructing algorithms for detecting grey-level blobs in discrete data is that of generic signals and non-generic signals. For generic signals (signals for which all pixel values are different) extrema will be isolated points. For non-generic signals, where connected pixels may have equal values, the situation is more complicated. Then, there are several possible ways to define a local extremum using different combinations of "$<$" and "$\leq$." For example, given a discrete function $g\colon \mathbb{Z}^N \to \mathbb{R}$ and a neighbourhood concept, a point $\xi \in \mathbb{Z}^N$ can be denoted:

- *weak local maximum:*
  if $g(\xi) \leq g(x)$ holds for all neighbours ξ of x,

- *strong local maximum:*
  if $g(\xi) < g(x)$ holds for all neighbours ξ of x,

- *semi-weak local maximum:*
  if $g(\xi) \leq g(x)$ holds for all neighbours ξ of x, and in addition $g(\xi) < g(x)$ for at least some neighbour ξ of x,

- *region-based local maximum:*
  if $g(\xi) \leq g(x)$ holds for all neighbours ξ of x , and in addition, by following connected points having the same grey-level value as x, it is impossible to reach a neighbour having a higher grey-level value.

For generic signals, all these formulations are equivalent, while they differ in situations where adjacent pixels have equal values (see figure 9.2 for an example). It is clear that when dealing with degenerate data, as is the case when grey-level values have been quantized, the *region-based local maximum* is the appropriate definition to work with.[1,2]

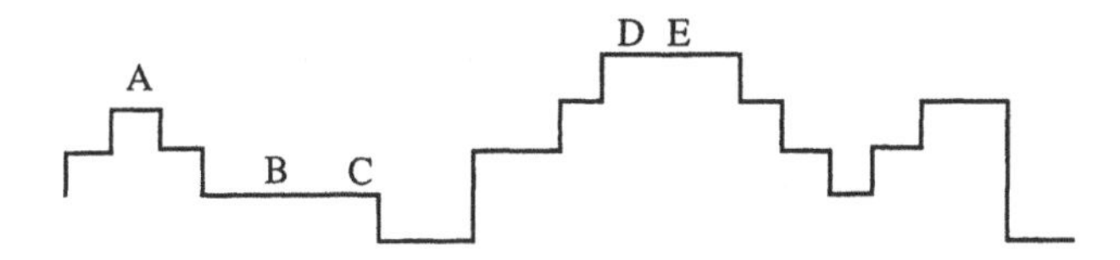

Figure 9.2. For non-generic signals, special care must be taken when defining the concept of a maximum point. In this figure, point A represents a maximum point as it would appear in a generic signal. Accordingly, it satisfies all the definitions of a maximum; it is a weak maximum, a strong maximum, a semi-weak maximum, as well as a region-based maximum. On the other hand, B is a weak maximum, while C is both a weak maximum and a semi-weak maximum. D is both a weak maximum, a semi-weak maximum and a region-based maximum, while E is both a weak maximum and a region-based maximum. The only definition of these that gives reasonable results in all cases is the region-based maximum.

9.1.3. Grey-level blob detection invariants

From the definition of a grey-level blob, it is easily realized that the following basic properties hold in the classification of the bright blobs of a discrete signal (see figure 9.3). To simplify the presentation, a temporary notation has been introduced: "region" denotes a set of connected pixels having the same grey-level value, "higher-neighbour" stands for a "neighbour region having a higher grey-level value" and "background" denotes a region that has been classified as not belonging to a blob.

a. If a region has no higher-neighbour, then it is a local maximum and will be the seed of a blob.

b. Else, if it has at least one higher-neighbour which is background, then it cannot be part of any blob and must be background.

[1]When implementing this concept computationally, one possibility is to pre-process the input data with the connected-component-labelling-algorithm, which assigns a unique identifier to each set of connected pixels having the same grey-level. Then, comparisons can be easily made between neighbouring *regions* consisting of connected pixels having the same grey-level.

[2]Note that the problem of non-generic data will be reduced when detecting local extrema in data that has been subject to scale-space smoothing, provided that the output results are stored in floating point precision with sufficient accuracy.

c. Else, if it has more than one higher-neighbour and if those higher-neighbours are parts of different blobs, then it cannot be a part of any blob, but must be background.[3]

d. Else, it has one or more higher-neighbours, which are all parts of the same blob. Then it must also be a part of that blob.

Starting from these properties sequential or parallel blob detection algorithms can be easily constructed.

$$\text{(a)}\quad \begin{matrix} - & - & - \\ - & X & - \\ - & - & - \end{matrix} \;\Rightarrow \text{max} \qquad\qquad \text{(b)}\quad \begin{matrix} \cdot & B^+ & \cdot \\ \cdot & X & \cdot \\ \cdot & \cdot & \cdot \end{matrix} \;\Rightarrow B$$

$$\text{(c)}\quad \begin{matrix} \cdot & \cdot & 1 \\ 2 & X & \cdot \\ \cdot & \cdot & \cdot \end{matrix} \;\Rightarrow B \qquad\qquad \text{(d)}\quad \begin{matrix} 1 & 1 & 1 \\ 1 & X & - \\ - & - & - \end{matrix} \;\Rightarrow 1$$

Figure 9.3. Illustration of the grey-level blob invariants labelled (a)–(d) above. In these figures, the symbol "X" denotes the central region to be classified, "–" a region having a lower grey-level than the central region, "B" a region classified as background, "B^+" a background region with a higher grey-level than the central region, "1" and "2" regions classified as belonging to regions labelled 1 and 2 respectively, and "·" an arbitrary region.

9.1.4. Sequential grey-level blob detection algorithm

The idea behind the algorithm is to initiate a blob seed in every (region-based) local maximum, and then let each maximum region grow until it meets with some other maximum region. If the growth procedure is performed in *descending grey-level order,* then it is guaranteed that no maximum region will grow too much. The case where adjacent points have equal values might lead to some practical problems, which however can be avoid by pre-processing the image with the connected-component-labelling-algorithm. Hence, given a finite discrete real-valued or integer-valued image perform the following steps:

1. Execute the connected-component-labelling-algorithm on the grey-level image in order to group connected points with equal values into regions.[4] After this step every set of connected pixels having same grey-level has been assigned the same unique region label.

[3] In fact, this region is a delimiting saddle region associated with the blob. (Note that according to the discrete definition of grey-level blob, the delimiting saddle point(s) of a grey-level blob is not included in its support region. The major reason for this is that we would like the support regions of different blobs to be disjunct regions.

[4] This step can be omitted if it is known in advance that no two connected pixels have equal values. Then, every pixel can be regarded as a region in the following description.

2. Sort the regions with respect to their grey-levels. For integer-valued images this may be done efficiently by indexing.

3. For each region, create a list of its neighbour regions having a higher grey-level.

4. Group the regions into blobs, i.e., for each region in descending grey-level order: Count how many references it has to neighbour regions with a higher grey-level.

   a. If the region has no such neighbours, then it is a local maximum point and will be the seed of a blob. Set a flag allowing the blob to grow, and store the grey-level of the region as the maximum grey-level of the blob.

   b. Else, if the region has a neighbour region with higher grey-level, which has been classified as background, then the current region cannot be a part of a bright blob and must also be classified as background.

   c. Else, if the region has more than one higher neighbour region, and if those neighbour regions are not parts of the same blob, then the region cannot be a part of a blob and must be set to background. For every blob, containing any of the neighbour regions, carry out the following:

      If the blob is still allowed to grow, then clear the flag that allows it to grow and store the current grey-level as the base-level of the blob. Store this region as one[5] saddle region associated with the blob.

   d. Else, if none of the previous conditions are true then the neighbour regions having a higher grey-level than the current region are all parts of the same blob. If that blob is still allowed to grow then the current region should be included as a part of that blob. Otherwise the region should be set to background.

For data given by scale-space smoothing it can in general be assumed that adjacent pixels in fact have different values, provided that the computations are carried out in floating point precision and that the output image is stored on that format. Therefore, this step will in some situations be superfluous. Other possible ways to ensure that the input data is non-degenerate is by modifying the least significant bits in the floating point numbers such that no pair of neighbouring pixels have equal values (the effects of such a modification should be negligible if the error introduced by this operation is kept below the numerical error in the implementation of the scale-space smoothing) or by introducing a pseudo-ordering on pixels having the same value. Such pseudo-ordering can, for example, be constructed from the coordinate values.

[5]Observe that for degenerate signals a grey-level blob can be delimited by more than one saddle point (where all such saddle points have the same grey-level), and that these "saddle points" in turn can be regions.

5. Create a blob image where all pixels in a region classified as blob are given the same (unique) label of the blob.

6. Traverse the grey-level image and the blob image simultaneously, and compute the contrast, area and volume for each blob. Store these values in a data structure together with the extremum regions and the saddle regions associated with the blobs.

With slight modifications this algorithm can be extended to computing a grey-level blob tree at a single level of scale, similar to the relational tree proposed by Ehrich and Lai (1978). Instead of letting a blob stop growing when a saddle point has been reached (step 4(c)), simply initiate a new seed at that saddle point. For example, if a saddle point S is encountered that delimits the extent of two grey-level blobs, G_1 and G_2, then it should be registered in a look-up-table that G_1 and G_2 are subarchs to the arch given by S. Then, in all further considerations, pixels labelled as belonging to S, G_1, and G_2 should be treated as equivalent when counting the number of references to regions with a higher grey-level. Moreover, the label associated with S should be used for labelling the pixels classified as belonging to this equivalence class. In this way, the relational tree can be stored using just one label image and one look-up-table for the subarch relations. A representation like Blom's (1992) nested hierarchy of level curves through saddle points can be obtained if in addition "the case with a dark blob contained in a bright blob" (see figure 7.3) is treated.

9.2. Linking grey-level blobs into scale-space blobs

Linking blobs across scales could be a potential source to difficult matching problems, since blobs can move, disappear, merge, split, or be created when the scale parameter changes. However, the notion of a scale-space with a *continuous* scale parameter provides a simple way to circumvent these problems in many cases, since the scale step may be varied at will. If confronted with a problematic matching situation, the matching difficulties can often be avoided by a refinement of the scale sampling. If the scale step is *adaptively* made just fine enough, then it should be trivial to judge which grey-level blobs belong to the same scale-space blob.

9.2.1. Basic idea: Adaptive scale sampling

According to the classification for continuous signals (see section 8.4), there are four possible types of (generic) blob events in scale-space: annihilation, merge, split, and creation. Assuming that this property transfers to the discrete case, all one must look for are moving blobs, and bifurcations of these four types.

If a situation is encountered with, say, three blobs at a fine scale, seeming to all belong to the same coarse scale blob, then the situation can (under the assumption of generic signals) be resolved into bifurcations of these four types by a sufficient number of refinements of the sampling along the scale direction. This idea, combined with the notion of a scale-space representation with a continuous scale parameter, constitutes the basic principle behind the *adaptive scale linking algorithm,* which essentially refines[6] the scale sampling until all relations between blobs at adjacent levels of scales can be decomposed into these primitive transitions.

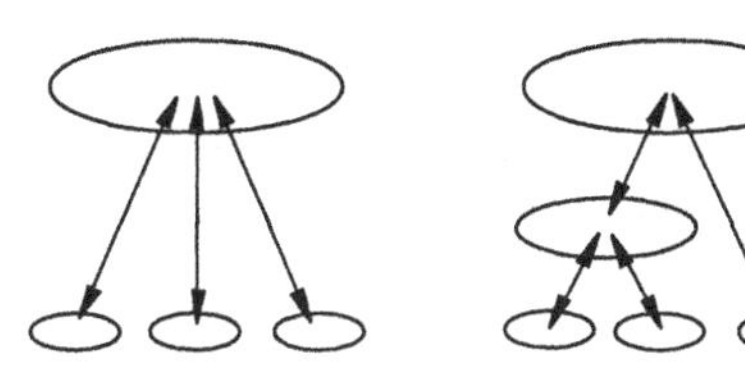

Figure 9.4. Generically, an encountered situation with, say, three blobs at a fine scale that all seem to belong to the same coarse scale blob can be resolved into a sequence of two successive blob merges by a refinement of the scale sampling. (Observe that only the support regions of the blobs are shown in this figure.)

9.2.2. Blob-blob matching

Based on this idea and using a heuristics that it should be unlikely[7] that a blob moves outside its support region, the blob linking between two adjacent scales can be performed based on spatial coincidence. A straightforward strategy is to start with a relatively fine initial sampling in scale, and for each pair of scale levels investigate if there are blobs with overlapping support regions. If so, they are registered as a matching candidates of each other:

DEFINITION 9.1. (BLOB-BLOB MATCHING CANDIDATE)
Let S_F and S_C be the support regions of two grey-level blobs G_F and G_C existing at two adjacent scale levels t_F and t_C respectively, where $t_F < t_C$. G_F is said to be a blob-blob matching candidate from above of G_C, denoted

$$G_F \swarrow_{b-b} G_C, \tag{9.1}$$

if there exists some pixel in S_C contained in S_F. Similarly, G_C is said to be a blob-blob matching candidate of G_F from below, denoted

$$G_C \nwarrow_{b-b} G_F, \tag{9.2}$$

[6] Note that this refinement principle cannot be applied as easily in pyramid representations, in which the scale levels have been set *in advance* and there is a fixed scale step beyond which refinements are not possible.

[7] Compare with observation 8.4 in chapter 8.

if there exists some pixel in S_F contained in S_C.

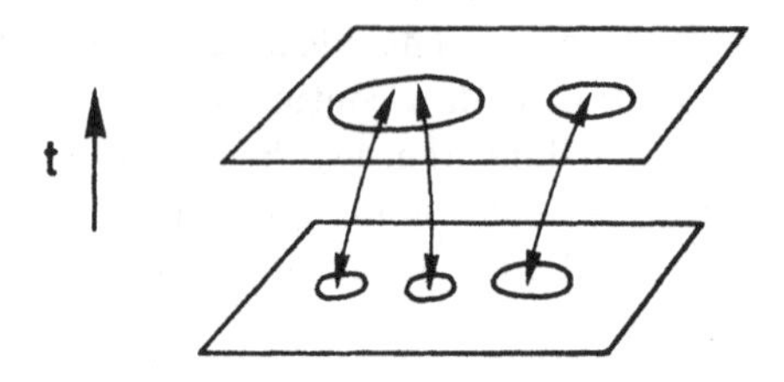

Figure 9.5. The blob linking between scale levels is based on spatial coincidence, i.e., if two grey-level blobs at adjacent scales have a spatial point in common, then they are registered as matching candidates of each other. In this example, the left situation is registered as a candidate blob merge, while the right situation is a candidate plain link within the same scale-space blob.

Obviously, matching candidates of this type are bidirectional:

$$(G_F \swarrow_{b-b} G_C) \Longleftrightarrow (G_C \nwarrow_{b-b} G_F). \tag{9.3}$$

Given such relations between the grey-level blobs from the scale-space representations at two adjacent levels of scale, the following primitive types of elementary matching situations can be discerned; candidate link within a scale-space blob, candidate annihilation, candidate merge, candidate split and candidate creation (see also figure 9.6).

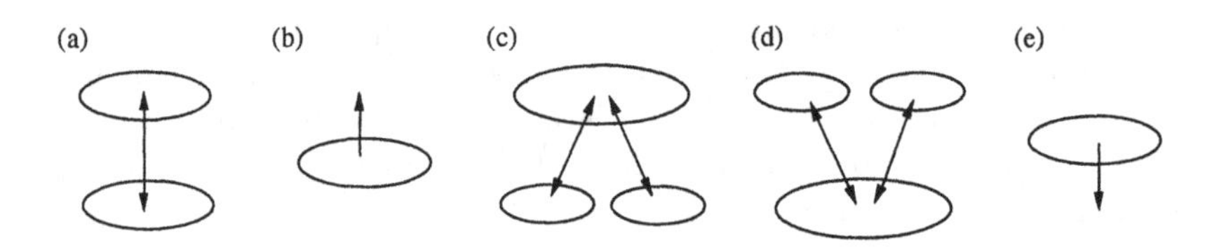

Figure 9.6. Elementary matching situations given by matching relations between blobs at different scales; (a) plain link, (b) annihilation, (c) merge, (d) split, and (e) creation.

As indicated above, the idea behind the scale linking algorithm is that it should be possible to decompose all relations between blobs at adjacent levels of scale into primitive relations of these five types, simply by refinements of the scale sampling. For example, if a blob has more than two matching candidates then a refinement should be made.

9.2.3. Blob-extremum matching

There are, however, situations where this methodology may lead to an unnecessarily large number of refinements. Consider, for example, a pair of neighbouring blobs, that are two blobs sharing the same delimiting saddle point, which slowly drift with the scale-space smoothing (see figure 9.7(a)). Then, a very large number of refinements may actually be needed in order to resolve this situation into a series of plain links.

The efficiency in such situations can be substantially improved by allowing for blob-extremum matching, that is by testing for the inclusion of the extremum points at one level scale in the grey-level blobs at the other level of scale:

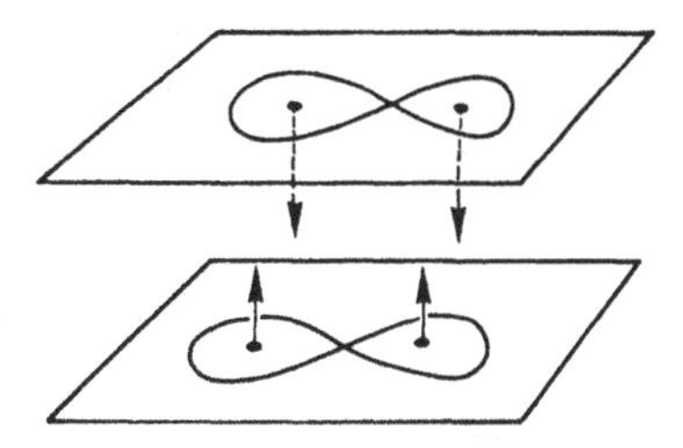

Figure 9.7. Basing the matching just on blob-blob matching candidates might lead to an unnecessarily large number of refinements for configurations with two neighbouring blobs that drift slowly due to the scale-space smoothing. In such situations, blob-extremum matching can be used for improving the matching (especially at coarser levels of scale). The idea is to gather additional matching candidates based on inclusion of the local maximum points at one level of scale in the grey-level blobs at the other level of scale. If the matching candidates are mutual and unique, then a match is accepted without refinement.

DEFINITION 9.2. (BLOB-EXTREMUM MATCHING CANDIDATE)
Let $t_F < t_C$ be two scale levels, let G_F be a grey-level blob at scale t_F with extremum point E_F and support region S_F, and let G_C a grey-level blob at scale t_C with support region S_C and extremum point E_C. Then, G_F is said to to be an blob-extremum matching candidate from above of G_C, denoted

$$G_F \swarrow_{b-e} G_C, \tag{9.4}$$

if E_C is contained in S_F. Similarly, G_C is said to be an blob-extremum matching candidate of G_F from below, denoted

$$G_C \nwarrow_{b-e} G_F, \tag{9.5}$$

if E_F is contained in S_C.

It is clear that these matching relations are not necessarily bidirectional. The idea behind this construction is that if the blob-extremum matching candidates resolve a situation with a pair of double candidates, then the situation can be registered as a pair of candidate plain links. It turns out that these situations are rather common at coarser scales. Compare this with figure 7.10, where two blobs "hang together" but drift slowly due to the smoothing. As we shall see later, these types of blob-extremum relations can also be used for stating stronger matching conditions than the blob-blob coincidence requirements.

9.2.4. Registering bifurcations in scale-space

What remains to be decided is when a blob match should be accepted. In the current implementation a scale refinement is performed, in principle, each time an unclear matching situation occurs, and matches are accepted in principle only when all blob events between the two scale levels can be classified as belonging to either one of the primitive cases: plain link within a scale-space blob, blob annihilation, blob merge, blob split, or blob creation. There are several possible ways to define situations that are candidates of being bifurcation situations. To enable a clear statement of this, let use introduce a notation for the number of matching relations associated with a certain grey-level blob G:

- $\sharp \swarrow_{b-b} (G)$ denotes the total number of blob-blob matching relations *from above, starting* at G, that G is involved in.
- $\sharp \searrow_{b-b} (G)$ denotes the total number of blob-blob matching relations *from above, ending* at G, that G is involved in.
- $\sharp \swarrow_{b-e} (G)$ denotes the total number of blob-extremum matching relations *from above, starting* at G, that G is involved in.
- $\sharp \searrow_{b-e} (G)$ denotes the total number blob-extremum matching relations *from above, ending* at G, that G is involved in.

Similarly, $\sharp \nwarrow_{b-b} (G)$, $\sharp \nearrow_{b-b} (G)$, $\sharp \nwarrow_{b-e} (G)$ and $\sharp \nwarrow_{b-e} (G)$ denote the number of matching candidates *from below* associated with a certain grey-level blob G.

9.2.5. Weak conditions for bifurcation situations

Given these relations between blobs at different scales, candidate situations for being links within the same scale-space blob, bifurcation situations, or complex situations to be subject for further refinements can be formalized as in the following definitions. (Grey-level blobs at the finer one of the two involved scales are denoted G_F, G_{F1}, and G_{F2}, while grey-level blobs at the coarser scale level are written G_C, G_{C1}, and G_{C2}):

DEFINITION 9.3. (WEAK LINK CANDIDATE)
$\{G_F, G_C\}$ are said to form a weak link situation between t_F and t_C if

$$(\sharp \nwarrow_{b-b} (G_F) = 1) \wedge (\sharp \swarrow_{b-b} (G_C) = 1) \wedge (G_C \nwarrow_{b-b} G_F). \quad (9.6)$$

DEFINITION 9.4. (WEAK ANNIHILATION CANDIDATE)
$\{G_F\}$ is said to form a weak annihilation situation between t_F and t_C if

$$(\sharp \nwarrow_{b-b} (G_F) = 0). \tag{9.7}$$

DEFINITION 9.5. (WEAK MERGE CANDIDATE)
$\{G_{F1}, G_{F2}, G_C\}$ are said to form a weak merge situation between t_F and t_C if

$$(\sharp \nwarrow_{b-b} (G_{F1}) = 1) \wedge (\sharp \nwarrow_{b-b} (G_{F2}) = 1) \wedge (\sharp \swarrow_{b-b} (G_C) = 2) \wedge (G_C \nwarrow_{b-b} G_{F1}) \wedge (G_C \nwarrow_{b-b} G_{F2}). \tag{9.8}$$

DEFINITION 9.6. (WEAK SPLIT CANDIDATE)
$\{G_F, G_{C1}, G_{C2}\}$ are said to form a weak split situation between t_F and t_C if

$$(\sharp \nwarrow_{b-b} (G_F) = 2) \wedge (\sharp \swarrow_{b-b} (G_{C1}) = 1) \wedge (\sharp \swarrow_{b-b} (G_{C2}) = 1) \wedge (G_F \swarrow_{b-b} G_{C1}) \wedge (G_F \swarrow_{b-b} G_{C2}). \tag{9.9}$$

DEFINITION 9.7. (WEAK CREATION CANDIDATE)
$\{G_C\}$ is said to form a weak creation situation between t_F and t_C if

$$(\sharp \swarrow_{b-b} (G_C) = 0). \tag{9.10}$$

9.2.6. Strong conditions for bifurcation situations

So far, neither the locations of the extremum points nor the relations between delimiting saddle points at blob merges and blob splits have been used. By taking these into account, the following can be defined (below $S_{delimit}(G)$ denotes the delimiting saddle point of a grey-level blob G and *nonshared*(S) means that the saddle point S is non-shared):

DEFINITION 9.8. (STRONG LINK CANDIDATE)
$\{G_F, G_C\}$ are said to form a strong link situation between t_F and t_C if they form a weak link situation between t_F and t_C, and in addition

$$(G_C \nwarrow_{b-e} G_F) \wedge (G_C \swarrow_{b-e} G_F) \wedge (\text{nonshared}(S_{delimit}(G_F))) \wedge (\text{nonshared}(S_{delimit}(G_C))). \tag{9.11}$$

DEFINITION 9.9. (STRONG MERGE CANDIDATE)
$\{G_{F1}, G_{F2}, G_C\}$ are said to form a strong merge situation if between t_F and t_C they form a weak merge situation between t_F and t_C, and in addition

$$(G_C \nwarrow_{b-e} G_{F1}) \wedge (G_C \nwarrow_{b-e} G_{F2}) \wedge (S_{delimit}(G_{F1}) = S_{delimit}(G_{F2})). \tag{9.12}$$

DEFINITION 9.10. (STRONG SPLIT CANDIDATE)
$\{G_F, G_{C1}, G_{C2}\}$ are said to form a strong split situation between t_F and t_C if they form a weak split situation between t_F and t_C, and in addition

$$\begin{aligned}
&(G_F \swarrow_{b-e} G_{C1}) \wedge (G_F \swarrow_{b-e} G_{C2}) \wedge \\
&(S_{delimit}(G_{C1}) = S_{delimit}(G_{C2})).
\end{aligned} \tag{9.13}$$

Note that it cannot be required that the extremum point of the coarser scale blob involved in a blob merge should necessarily belong to some of the blobs at the finer scale, or that any such relation should hold in a blob split. When formalizing the matching criterion for pairs of double candidates described in figure 9.7 we get:

DEFINITION 9.11. (STRONG DOUBLE LINK CANDIDATE)
$\{G_{F1}, G_{F2}, G_{C1}, G_{C2}\}$ are said to form a strong double link situation between t_F and t_C if

$$\begin{aligned}
&(\sharp \nwarrow_{b-b} (G_{F1}) = 2) \wedge (\sharp \nwarrow_{b-b} (G_{F2}) = 2) \wedge \\
&(\sharp \swarrow_{b-b} (G_{C1}) = 2) \wedge (\sharp \swarrow_{b-b} (G_{C2}) = 2) \wedge \\
&(G_{C1} \nwarrow_{b-b} G_{F1}) \wedge (G_{C1} \nwarrow_{b-b} G_{F2}) \wedge \\
&(G_{C2} \nwarrow_{b-b} G_{F1}) \wedge (G_{C2} \nwarrow_{b-b} G_{F2}) \wedge \\
&(G_{C1} \nwarrow_{b-e} G_{F1}) \wedge (G_{C2} \nwarrow_{b-e} G_{F2}) \wedge \\
&(G_{F1} \swarrow_{b-e} G_{C1}) \wedge (G_{F2} \swarrow_{b-e} G_{C2}) \wedge \\
&(S_{delimit}(G_{F1}) = S_{delimit}(G_{F2})) \wedge (S_{delimit}(G_{C1}) = S_{delimit}(G_{C2})).
\end{aligned} \tag{9.14}$$

When this condition is satisfied G_{C1} is regarded as belonging to the same scale-space blob as G_{F1}, and G_{C2} is regarded as belonging to the same scale-space blob as G_{F2}.

To express similar stronger conditions for blob annihilations and blob creations is not as easy, since in this case we have to ensure that we have not failed to find a matching candidate that should have been registered. Of course, one could require the delimiting saddle point to be non-shared, but for small blobs this condition will be far from sufficient. In particular blobs covering just one pixel are difficult, since the heuristics stating that it should be unlikely for a blob to move outside its support region is definitely not valid.

9.2.7. *Extended neighbourhood search*

For tracking small size blobs the previously mentioned criteria will not be sufficient. For example, the drift of a blob covering just one pixel will be impossible to capture unless some additional gathering of matching candidates is carried out. This means that a situation that should have been registered as a plain link could give rise to one annihilation and one creation unless some precautions are taken. Therefore, in the current implementation, an extended neighbourhood search is performed in a region (of width one pixel) around every point involved in a weak creation situation. The purpose is to investigate if there are other blobs nearby which are involved in weak annihilation situations. A blob creation is accepted only if no such blobs can be found, and, in addition, the same conclusion holds through a small number of refinements.

9.2.8. Bidirectional matching

In contrast to many matching algorithms (in, for example, motion analysis), where the matching is performed only in one direction (that is with increasing time), this matching procedure is purely geometric and bidirectional. The matching candidates are always registered from both directions. Therefore, the scheme can *equivalently* be started from a fine scale or a coarse scale. The first approach can be advantageous if the scale-smoothing is implemented as cascade smoothing. The second approach could on the other hand have advantages when focusing the attention on significant image structures detected at a coarse scale.

9.2.9. Delimiting the refinement depth

The decomposition property, meaning that relations between grey-level blobs at different levels of scale can be resolved into relations of the five primitive types (shown in figure 9.6), is guaranteed to hold only for generic signals. Therefore, to avoid a possible infinite number of refinements in situations when the algorithm is presented with degenerate data, an upper bound is introduced on the number of refinements allowed to take place. If this number is reached, then a *complex* blob event is registered. Although this situation has not been encountered in any realistic images, it has occurred sometimes in highly regular and noise-free synthetic data.

9.2.10. Scale levels and computation of refinement scale

The algorithm is initiated with a relatively fine sampling along the scale direction, corresponding to about $\frac{1}{3}$ - $\frac{1}{2}$ octave in t. At coarse scales the scale levels are distributed such that the scale step measured in effective scale is approximately constant. The maximum scale[8] is determined from the size of the image (the outer scale) and the minimum scale is set to a low value[9] (the inner scale). When refinements are needed, the refinement

[8] For an image of size 256×256, the maximum scale has usually been set to $t = 1024$, a scale at which there is usually only one blob in the image, and the boundary effects have started to become important.

[9] This scale value may be zero, but because of computational aspects it might be practical to use a higher value. In the experiments is has throughout been set to either 0, 1, or 2.

scale is computed from the existing scale levels t_1 and t_2 based on the notion of effective scale

$$t_{refine}(t_1, t_2) = \tau_{eff}^{-1}((\tau_{eff}(t_1) + \tau_{eff}(t_2))/2), \tag{9.15}$$

where τ_{eff} denotes the transformation function from the ordinary scale parameter to the effective scale parameter and τ_{eff}^{-1} its inverse. The function values are computed from interpolation in a table with simulation data accumulated from white noise images (see section 7.7.1).

9.3. Basic blob linking algorithm

To summarize, an algorithm for linking grey-level blobs across scales into scale-space blobs can be based on the following steps. (The treatment below is based on the weak matching relations only to illustrate the idea. It should be obvious how the stronger criteria can be incorporated in an analogous manner.)

1. Determine an initial set of scale levels, from some minimum scale t_{min}, given by the inner scale of the image, to some maximum scale t_{max}, given by the outer scale of the image. Distribute the intermediate scale levels such that the scale step, measured in effective scale is approximately constant. At coarse scales this means that the ratio between successive scale values will be about constant. At fine scales instead the differences between successive scale values will be approximately equal. Push these scales onto a stack of scale levels to be processed later. In a hardware implementation, dedicated hardware can be used for each scale level.

2. Extract the grey-level blobs from the image at the finest scale using the grey-level blob detection algorithm.

3. Get the next scale level from the stack of scale levels.

   a. Extract the blobs at the current level of scale.

   b. For each grey-level blob at the current scale level, determine how many matching candidates it finds at the previous scale level. Similarly, for each grey-level blob at the previous scale level, determine how many matching candidates it finds at the current scale level.

   c. If some grey-level blob has more than two match candidates, then the matching is non-trivial. Similarly, if there is a pair of double candidates,[10] i.e., if there is a blob having two matching

[10]Many situations of this type can be resolved with an extended blob-extremum matching—especially at coarser levels of scale. Then, the refinement step will not be necessary.

candidates, and if one of the matching candidates in turn also has two possible match candidates, then the matching is also difficult. In these cases perform a refinement:

i. Push the current scale level into the set of scale levels to be computed.
ii. Compute a refinement scale between the current scale level and the previous one.
iii. Continue with step 3.

d. Else, if some grey-level blob at the coarser level does not find a match candidate at the finer level, the situation is more complicated. According to scale-space theory this situation may in fact occur (but not very often). There could also be some other natural explanations as to why the algorithm fails to find match candidates:

i. The blob may have moved outside the spatial region it covered at the previous level of scale. This phenomenon applies mostly to blobs with small areas—in particular blobs consisting of only one pixel. If such blobs move, they will always be lost if the matching is based on only spatial overlap.
ii. Numerical errors may have violated the scale-space properties.

In this implementation an extended search in a small neighbourhood (of distance 1) around the coarse-scale blob in order to gather more matching candidates. If exactly one such candidate has been found, and if that blob has no other match candidates, then a blob match will be accepted, and the two grey-level blobs will be linked into the same scale-space blob. Otherwise, a refinement will be performed. However, if the refinement depth is too deep, then a blob creation will instead be registered.

e. Else, each blob has either one or two matching candidates, and the matching candidates will be accepted.

i. If a blob at the fine scale has exactly one match candidate at the coarse scale, and if that candidate in turn has exactly one match candidate to the fine scale, then link the grey-level blobs into the same scale-space blob.
ii. If a grey-level blob at the finer level does not have a match candidate at the coarser level, then register a blob annihilation.
iii. If a blob at the coarser level finds two match candidates at the finer level and if these blobs in turn have exactly one

match candidate each at the coarser level, then register a blob merge.

iv. If a blob at the finer level finds two match candidates at the coarser level and if these blobs in turn have one match candidate each, then register a blob split.

4. Store the registered relations between grey-level blobs at different scales. Then, traverse all the scale levels and compute the scale-space blob volume and the scale-space lifetime for each scale-space blob (see section 9.4 for details).

9.3.1. *Blob linking vs. extremum linking*

It should be stressed that grey-level blobs are much easier to track across scales than are local extrema, since the blob concept *associates a region* with every local extremum point. If instead the multi-scale scale analysis would have been based on local extrema only, then the matching problem would often be more difficult, since local extrema can move much faster than the blobs. Ambiguous situations could easily occur. Bifurcation situations especially would be harder to identify. If at some level of scale the algorithm loses the track of a local extremum point, then it is hard to say if it is because the extremum point has moved much faster than expected or because it has disappeared (been annihilated). It is in this context the blob regions are important, since they give natural spatial regions in which there are no other local extrema. They also define natural regions to search for blobs in at the next level of scale (compare with observation 8.4 in chapter 8).

9.4. Computing scale-space blob volumes

Once the scale linking has been performed and the bifurcations have been registered, it is straightforward to compute the scale-space blob volumes. First, every grey-level blob volume, as computed according to (7.4), is transformed according to (7.21) and (7.22). To reduce the sensitivity of V_m and V_σ to the actual scaling of the grey-level values in the image, the tabulated values are rescaled with a uniform scaling factor determined from a least squares fit between the experimental values and the tabulated values at the finest levels of scale.

Second, given these normalized grey-level blob volumes, the normalized scale-space blob volumes are computed according to (8.7). Assuming that a scale-space blob is represented by a number of grey-level blobs at scales $t_1 < t_2 < \ldots < t_n$, and assuming that the scale-space blob does not exist at the nearest finer and coarser scales, $t_0 < t_1$ and $t_{n+1} > t_n$ respectively, the integral is discretized using the trapezoid rule of integration

$$S_{vol,norm}(r) \approx \frac{1}{2} \sum_{i=0}^{n} (V(t_i) + V(t_{i+1})) \, (\tau_{eff}(t_{i+1}) - \tau_{eff}(t_i)). \quad (9.16)$$

The bifurcation scales, t_0 and t_{n+1}, are localized to the mean values (computed in effective scale) of the nearest coarser and finer scales around the bifurcation. The grey-level blob volume at a bifurcation is set to zero for annihilations and creations as well as for the smaller blobs involved in merges and splits. For the larger blob involved in a blob merge or a blob split, the grey-level blob volume estimated by the volume of the grey-level blob at the nearest scale level included in that scale-space blob.

9.5. Potential improvements of the algorithm

The major design criterion behind this implementation has been to compute the representation as accurately as possibly. Very little work has been spent on the computational efficiency, since the main object has been to investigate what type of information can be extracted from a representation of the proposed type. In this section, some obvious improvements will be described, which can be used for improving the performance.

9.5.1. Local refinements

The refinements are currently made globally. In other words when a difficult situation has been encountered, for which refinements are necessary, *all* grey-level blobs at the involved scales are subject to this process. Obviously, there would be an improvement to perform the refinements *regionally*, such that only the grey-level blobs involved in the ambiguous situation are affected. Then, only a window rather than the entire image needs to be processed.

9.5.2. Drift velocity estimates

The success of the linking algorithm depends very much on a fine sampling in scale. A natural extension is to incorporate drift velocity estimates of the blobs to get further verification of the bifurcation situations and the blob matches; compare also with section 9.2.7. Such an estimate could also give more precise information about how dense the scale sampling really needs to be, possibly implying that a fewer number of scale levels needs to be treated. Observe that some discrete aspects may have to be introduced if such an approach is taken.

Another possible way of improving the performance could be by analysing the variation of the volume and contrast of the grey-level blob and compare with analytical results as those derived in section 8.5.3. One could also use drift velocity estimates as those derived in section 8.1.1, or build up a model of the motion of the extremum point as function of scale. No such methods have, however, been implemented.

9.5.3. Approximate description

As said earlier, the main objective throughout this work has been to try to reduce the computational and numerical errors as far as possible when computing the representation. However, it often turns out that many of the situations that lead to refinements correspond to structures that are later on rejected as being non-significant. Therefore, it seems plausible that the performance could be improved if those refinements could be avoided (in other words, if just an approximate description could be computed). Selecting a proper trade-off between the degree of approximation introduced errors may, however, require extensive experimentation, and has not been carried out.

9.5.4. Subsampling at coarse scales

The representations at coarser levels of scale are highly redundant. Another approximation would be to subsample the images at coarser scales, as is done in pyramids to reduce the number of pixels to be processed. An important issue to consider if such an approach is taken is to ensure that "no severe discontinuities" are introduced in the scale direction (compare also with section 4.5.4).

9.6. Data structure

To give a rough idea of what information can be available in a data structure representing the scale-space primal sketch, a brief description is given of what kinds of objects could be defined in an actual implementation of this concept, and what types of data can be stored in those (see also figure 7.13).

GREY-LEVEL BLOB:

polarity: bright or dark
scale level: pointer
extremum point: pointer
delimiting saddle point: pointer (list for degenerate signals)
support region: pointer
grey-level blob volume

EXTREMUM POINT:

position: pixel coordinates (several pixels for degenerate signals)
grey-level value
grey-level blob: pointer to the grey-level blob to which this saddle point serves as the seed

SADDLE POINT:

position: pixel coordinates (several pixels for degenerate signals)
grey-level value
grey-level blobs: pointers to the grey-level blobs to which this saddle point serves as a delimiting saddle point

SUPPORT REGION:

extreme coordinates: delimiting (rectangular) box
blob area: number of pixels in the region
first order moments: giving the center of gravity
second order moments: allowing for an ellipse approximation giving the major and minor axes, which in turn give the blob orientation
pixel representation: can be encoded either as a bit map, or in a more compact form as e.g. run-length coding row by row
boundary flag: telling whether the region belongs to the image boundary or not

SCALE-SPACE BLOB:

polarity: bright or dark
significance: normalized scale-space blob volume
bifurcation event at the appearance scale: pointer
bifurcation event at the disappearance scale: pointer
grey-level blobs: pointers to all the grey-level blobs the scale-space blob consists of
selected scale level: pointer
grey-level blob at the selected scale level: pointer
boundary flag: telling whether there is a grey-level blobs belonging to the image boundary or not

BIFURCATION EVENT:

type of bifurcation: can be either

- one of the generic bifurcation situations: annihilation, merge, split, creation,
- a non-generic complex bifurcation, with more than three scale-space blobs involved, which cannot be resolved into primitive transformations of the previously listed types,
- a flag indicating that the minimum or the maximum scale of the analysis has been reached.

participants from above: pointers to the scale-space blobs at the coarser scale that are involved in the bifurcation
participants from below: pointers to the scale-space blobs at the finer scale that are involved in the bifurcation

spatial position
bifurcation scale: scale value (or interval)

SCALE LEVEL:
scale value
smoothed grey-level image: pointer
bright blob image: all bright grey-level blobs coded as labels
dark blob image: all dark grey-level blobs coded as labels
bright grey-level blobs: pointers to all bright grey-level blobs at this scale
dark grey-level blobs: pointers to all dark grey-level blobs at this scale
next coarser scale level: pointer
next finer scale level: pointer

Finally, it is convenient to create an object that can serve as a handle to all these sub-objects:

SCALE-SPACE PRIMAL SKETCH:
scale levels: it can be useful to represent all the scale levels accessed by the refinement algorithm both as a linked list and as a refinement tree
bright scale-space blobs: pointers to all the bright scale-space blobs
dark scale-space blobs: pointers to all the scale-space blobs
bright bifurcations: pointers to all the bifurcations in which bright scale-space blobs are involved
dark bifurcations: pointers to all the bifurcations in which dark scale-space blobs are involved

Of course, depending on the actual application it will in some situations be computationally more efficient not to compute all this information when building the data structure, and due to memory considerations, pieces of information may have to be thrown away during the process. For example, regarding grey-level blobs it is often sufficient to save only those blobs that correspond to the selected scale of a scale-space blob.

Part III

The scale-space primal sketch: Applications

10

Detecting salient blob-like image structures and their scales

A major motivation for this research has been to investigate whether or not the scale-space model allows for determination and detection of stable phenomena. In this chapter it will be demonstrated that such determination and detection are indeed possible, and that the suggested representation can be used for extracting regions of interest with associated stable scales from image in a solely bottom-up data-driven way. The treatment is based on the following *assumption:*

> *Structures, which are significant in scale-space, are likely to correspond to significant structures in the image.*

This statement has been expressed on a general form, since it can be speculated that the approach can be applied also structures[1] other than the blobs considered here. More precisely, since the primitives that will be used are scale-space blobs, the heuristic selection method is formulated as follows:

ASSUMPTION 10.1. (RANKING OF BLOB STRUCTURES ON SIGNIFICANCE) *In the absence of other evidence, a scale-space blob having a large normalized scale-space blob volume in scale-space is likely to correspond to a relevant blob-like region in the image.*

A scale-space blob will, in general, exist over some range of scales in scale-space. When there is a need for reducing the amount of data represented, and to select a single scale and region as representative of the scale-space blob, the following postulates are suggested.

ASSUMPTION 10.2. (SCALE SELECTION FROM MAXIMUM BLOB RESPONSE OVER SCALES) *In the absence of other evidence, the scale at which a scale-space blob assumes its maximum normalized grey-level blob volume over scales is likely to be a relevant scale for representing that blob.*

[1]For example, it seems plausible that the lifetime of an edge in scale-space is an important property for measuring significance. As will be demonstrated in section 11.3, a multi-scale blob detection approach can be useful in junction detection, if applied to curvature data computed from the multi-scale N-jet representation.

ASSUMPTION 10.3. (SELECTION OF SPATIAL REPRESENTATIVE)
In the absence of other evidence, the spatial extent of a scale-space blob can be represented by the support region of the blob at the scale level selected according to assumption 10.2.

The ranking on significance depends on the actual scaling of the four coordinates in the scale-space representation. Therefore, the extraction method implicitly relies upon the assumption that it should be sufficient to transform the coordinates once and for all as was done in section 7.7.

ASSUMPTION 10.4. (NORMALIZATION FROM REFERENCE DATA)
The coordinate axes in the scale-space representation can be normalized based on the behaviour in scale-space of reference data.

Next, experimental results will be given demonstrating how these assumptions can be used for segmenting out intuitively reasonable regions from various types of imagery. First, however, motivations will be given to why these assumptions have been stated.

10.1. Motivations for the assumptions

A central problem in low-level vision concerns what should be meant by image structure. In other words, which features in an image should be regarded as significant, and which ones can be rejected as insignificant or as due to noise. As we discussed in the introduction, this problem seems impossible if stated as a pure mathematical problem, as is the segmentation problem if seen in isolation. Nevertheless, since there are indications that biological vision systems are able to perform natural pre-attentive groupings in images (Witkin and Tenenbaum 1983; Lowe 1985), one may speculate whether there are any inherent properties in data that can be used for defining such groupings.

The scale-space primal sketch constitutes an attempt to express such groupings for blob-like structures using a formal mathematical framework, which constructs primitives from the scale-space representation (solely in terms of the singularities that occur in scale-space).

10.1.1. Stability in scale-space: Salience

When Witkin (1983) coined the term scale-space, he observed a marked correspondence between perceptual salience and stability in scale-space:

> ... intervals that survive over a broad range of scales tend to leap out to the eye ...

Assumption 10.1 constitutes an extension of this observation to a heuristic principle for extracting blob-like image structures. The significance measure is, however, not based on the scale-space lifetime alone, since as mentioned in section 7.6, blobs due to noise can survive over large ranges of scale, if located in regions with slowly varying grey-level.

Observe how this measure of significance relates to a definition of structure in terms of *transformational invariance.* If a feature is to be useful for recognition, it must necessarily be stable with respect to small disturbances. Otherwise it would hardly be useful, since it would be impossible to compute it accurately. Here, this stability requirement is used for actually formulating an operational method for detecting image structures—by subjecting the image to *systematic parameter variations*, and explicitly measuring the stability of the structures (here, blobs) under parameter variations (here, scale variations).

In line with this idea, assumption 10.1 states that the scale-space blobs that are the most stable ones under variations of the scale parameter in scale-space are the most likely to correspond to significant image structures. Of course, the reverse statement does not hold. There are many other sources of information (e.g. lines in line drawings), which are not captured by a blob concept and scale-space smoothing.

Note, that this use of transformational invariance is different than what is usually meant by invariance in an algebraic or a geometric sense; the transformational invariance of the scale-space blobs concerns local topological properties that are stable over finite intervals of parameter variations.

10.1.2. Reduction of the representation: Abstraction

Because of complexity arguments, the entire parameter variation information from the low-level modules cannot be transferred to the higher-level modules in a vision system.

Assumption 10.2 and assumption 10.3 express such a desire to represent a scale-space blob with a grey-level blob at a single level of scale, to give a more compressed representation—an *abstraction*—for further processing.

The motivation for selecting the scale at which the maximum of the normalized grey-level blob volume is assumed is that it should reflect the scale where the blob response is maximally strong. It turns out that this scale will often be close to the appearance scale of the scale-space blob, except at blob splits and blob creations, for which the grey-level blob volume may[2] be zero at the appearance scale.

It is worth noting that assumption 10.2 implies a projection from a four-dimensional scale-space blob to a three-dimensional grey-level blob,

[2]More precisely, at blob creations the grey-level blob volume of the new blob is always zero. At blob splits, the grey-level blob volume of the blob associated with the new local extremum is zero, while the volume of the other blob involved in the bifurcation may be non-zero (see the detailed analysis of the fold singularity in sections 8.5.3–8.5.3).

and that assumption 10.3 implies a projection from that grey-level blob to its two-dimensional support region.

10.2. Basic method for extracting blob structures

The basic methodology for extracting significant blobs from an image should now be obvious from the previous presentation.

- Generate the suggested scale-space primal sketch, where blobs are extracted at all levels of scale, and linked across scales into scale-space blobs.
- Compute the normalized scale-space volume for each scale-space blob based on the notion of effective scale and effective grey-level blob volumes.
- Sort the scale-space blobs in descending significance order, i.e., with respect to their normalized scale-space blob volumes.
- For each scale-space blob determine the scale where it assumes its maximum grey-level blob volume, and extract the support region of the grey-level blob at that scale.

10.3. Experimental results

Figure 10.1 and figure 10.3 show the result of applying this procedure to two images, one with a telephone and a calculator, and the other with a set of blocks. The reader is encouraged to study these images carefully.

For display purposes, the N most significant dark scale-space blobs have been extracted. Each blob is displayed at its representative scale (that is the previously mentioned scale at which the scale-space blob assumes its maximum grey-level blob volume). The spatial representative of each blob (which is the blob support region of the grey-level blob at the representative scale) is marked in a binary image, where black indicates the existence of a significant dark blob, and white represents background. To avoid overlap, the display routine shifts to a new fresh image each time the addition of a new blob would imply overlap between two blobs.

We can see that in the block image, the individual blocks are extracted. Also, at coarser scales, adjacent blocks are grouped into coarser scale units, and the imperfections in the image acquisition near the boundaries are pointed out. In the telephone scene, the buttons, the keyboard, the calculator, the cord, and the receiver are detected as single units.

As an illustration of the spatial relations between the blobs at different scales, figure 10.2 and figure 10.4 show the blob boundaries superimposed onto each other. More experimental results, including bright blobs, are presented in following sections.

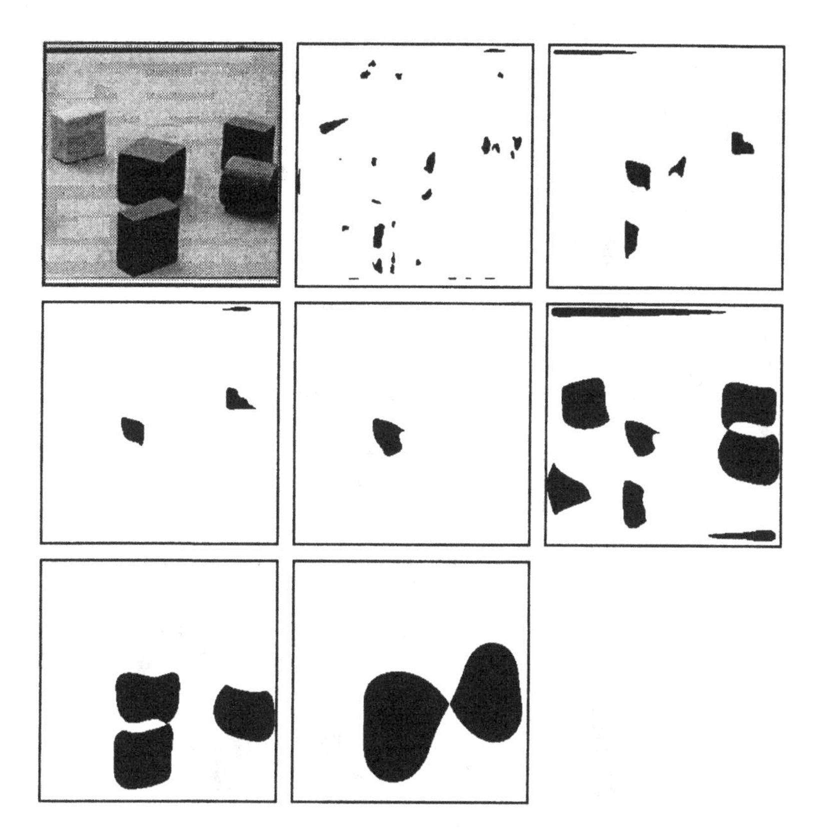

Figure 10.1. The 50 most significant dark blobs from a block image. (Note how these images have been produced—they are not just blob images at a few levels of scale. Instead every blob has been marked at its selected scale. Finally, the blobs have been drawn in different images to avoid overlap.)

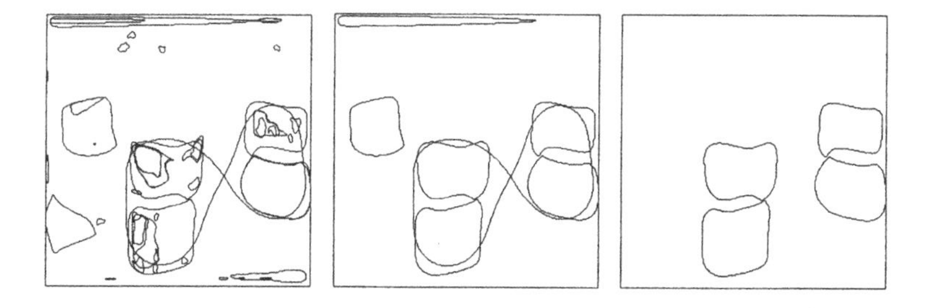

Figure 10.2. Boundaries of the dark scale-space blobs extracted from the block image in figure 10.1; (left) the 50 most significant dark blobs, (middle) low threshold on the significance measure set in one of the "gaps" in the sequence of significance values (between 74 and 131), and (right) high threshold on the significance measure set in another "gap" (between 298 and 591). (The significance values are shown in table 10.1.)

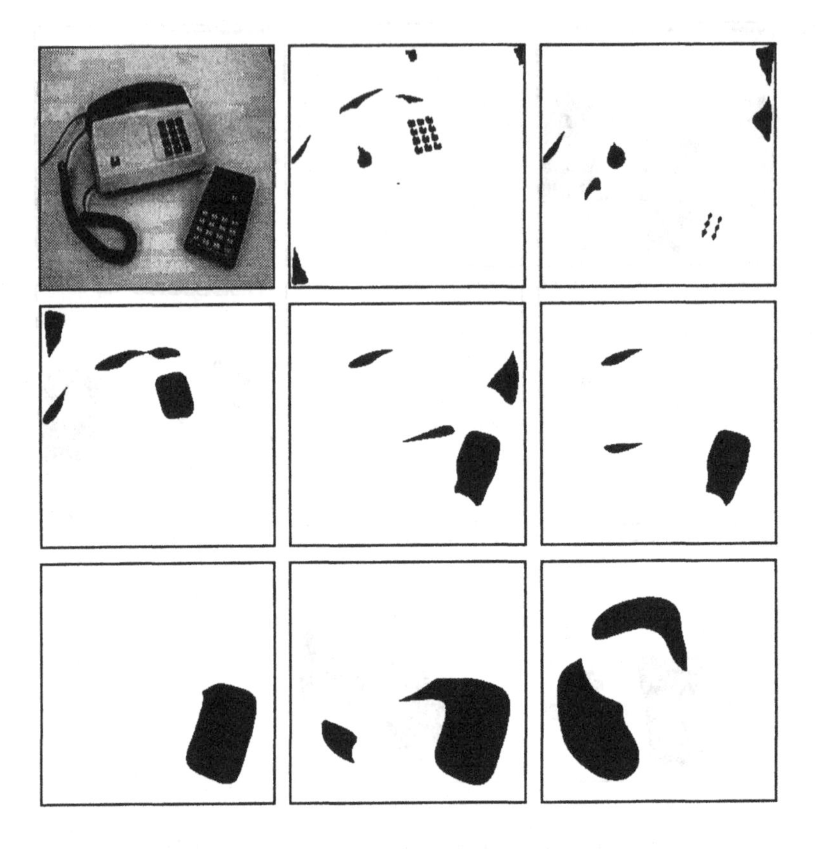

Figure 10.3. The 50 most significant dark blobs from a telephone and calculator image.

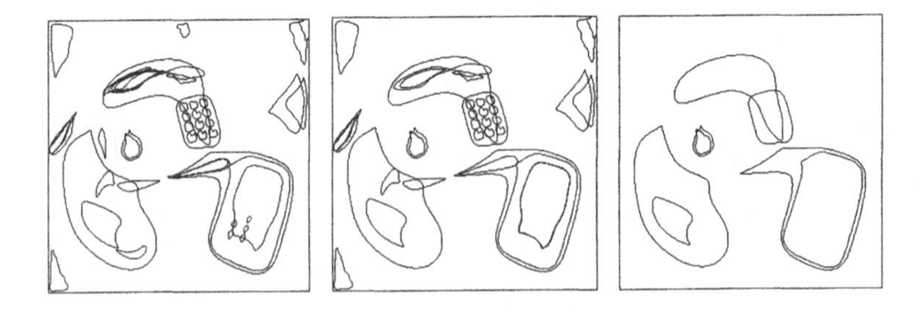

Figure 10.4. Boundaries of the dark scale-space blobs extracted from the telephone and calculator image in figure 10.3; (left) the 50 most significant dark blobs, (middle) low threshold on the significance measure set in one "gap" in the sequence of significance values, (right) high threshold on the significance measure set in another "gap."

Significance	Scale	Blob label
1450.55	32.00	1760
1266.43	64.00	1767
1030.53	50.80	1764
591.16	80.60	1768
297.60	812.90	1770
284.72	645.10	1769
150.64	45.25	1761
131.99	28.51	1758
73.69	45.25	1763
63.51	35.91	1065
35.92	28.51	1759
35.42	22.65	1753
20.45	8.00	1703
17.43	8.99	1702
12.84	11.99	1723
9.94	28.51	1757
6.84	4.00	1256
6.20	9.53	1708
5.33	14.25	1725
5.10	2.00	1440
4.85	1.40	1471
4.03	2.87	1610
3.84	16.00	1731
2.41	4.73	1679
2.30	1.00	1078
2.27	1.10	1265
2.21	10.10	1713
2.21	8.99	1706
2.07	1.22	251
2.02	1.00	1072
1.98	1.00	1070
1.96	1.00	1187
1.95	2.65	1243
1.95	5.00	1686
1.93	1.05	1371
1.87	6.40	1611
1.83	1.00	1286
1.83	1.00	1183
1.79	1.00	1083
1.75	1.00	1336
1.72	1.10	1393
1.71	1.00	212

Table 10.1. Significance values and selected scale levels for the 40 most significant scale-space blobs computed from the block image. Note that a few blobs have significance values clearly standing out from the others.

Figure 10.5. Significance values of the 50 most significant blobs from the block image. The significance value of each blob has been marked with bold vertical line along a horizontal logarithmic scale. The thin vertical lines indicate the manually selected thresholds used in figure 10.2.

Figure 10.6. Corresponding significance values and manually selected thresholds for the blobs from the telephone and calculator image in figure 10.4.

Let us conclude by stressing that *the intrinsic shape of the grey-level landscape is extracted in a completely bottom-up data-driven way without any assumptions about the shape of the primitives* (except for the fact that the scale-space smoothing favours blob-like objects, since it is equivalent to correlation with a Gaussian-shaped kernel).

A segmentation is obtained, which is coarse in the sense that the localization of object boundaries may be poor, due to the natural distortions of shape which occur in scale-space. However, the segmentation is safe in the sense that those regions, which are given by the scale-space blobs with large scale-space volume, really serve as landmarks of significant structure in the image, with information about

- the *approximate location and extent* of relevant regions in the image.
- an *appropriate scale* for treating those regions.

This is exactly the kind of coarse information[3] that is necessary for many higher-level processes (see chapter 11).

10.4. Further treatment of generated blob hypotheses

The number of scale-space blobs selected for display above is, of course, rather arbitrary. Note, however, that there is a well-defined ranking between the blobs. If one studies their significance values (see table 10.1, figure 10.2, and figure 10.6), one can observe that those blobs we regard as the most significant have significance values standing out from the significance values of the others. Hence, it seems plausible that a few re-

[3]The scale-space primal sketch contains much more information than is presented in this rudimentary output. For example, the registered blob bifurcations in scale-space have not been illustrated. Nor have the hierarchical relations induced by the blob events between blobs at different levels of scale been shown. This information is however explicit in the computed representation.

gions can be extracted based on this observation alone. In more general situations, there is a need for feedback and reasoning.

The output information from this representation should not be overestimated. Since it is a low-level processing module, the output should be interpreted mainly as indicators signalling that "there might be something there of about that size—now some other module should take a closer look." From this viewpoint it can be noted how well the extracted blobs describe blob-like features in the previous images, considering that the blobs have been extracted almost without any *a priori* information.
In principle, a reasoning process working on the output from the scale-space primal sketch can operate in either of two possible modes:

- Use a threshold on the significance measure. In a real system, such a threshold may in some situations be set from given context information and expectations.

- Evaluate the generated hypotheses in decreasing order of significance, i.e., try first to interpret the first hypothesis in a feedback loop, then consider the second one, etc. Continue as long as the hypotheses deliver meaningful interpretations for the higher-level modules.

An inherent property of this representation is that it does not have any limiting requirement that there is just one possible interpretation of a situation. Instead it generates a variety of hypotheses. Given some region in space, several hypotheses may be active for it (or parts of it) concerning different structures at different scales.

10.5. Properties of the scale selection method

In this section, some consequences of the suggested scale selection method will be described. The presentation is not intended to be theoretically rigorous, but rather to convey an intuitive understanding for what qualitative properties the stated assumption leads to.

10.5.1. Relations between object size and selected scale

The scale value given by assumption 10.2 does not necessarily reflect the size of the blob region in the image. Although large values of the scale parameter in general will lead to images with large size features, there is no *direct* relationship between the size of an object and its selected scale. In certain situations large size objects may in fact be assigned relatively small scale values (although the opposite situation can be expected not to occur). The scale value given by assumption 10.2 should therefore rather be interpreted as an *abstract scale parameter*, or as reflecting *the smallest amount of smoothing* for which a region in the image manifests itself as a single blob entity.

Consider for example an image with, say, a few squares of fixed size. The scale value, where for the first time one of the squares appears as a blob, can vary substantially depending on the noise level in the image and on where the squares are located relative to each other. In the ideal noise-free and texture-free case, i.e., when there are no interfering fine-scale structures present, the selected scale for each one of the squares will be zero. Only for coarse scale structures, which only exist as groupings of other primitive fine scale structures, the selected scale will be non-zero in the ideal noise-free case (for example, a letter formed by arranging the blocks in a certain pattern with some spacing between them).

10.5.2. Partial ordering

Hence, these scale values do not induce any total ordering of regions with respect to their relative size, but rather a *partial ordering.* By and large the following property holds: If two structures overlap, i.e., if a fine scale structure is superimposed onto a coarser scale structure, then the coarser scale structure will be assigned a larger scaler value than its superimposed fine scale structure. On the other hand, if similar structures are located sufficiently far apart from each other in an image, then the reverse relation may actually hold.

However, the situation is even further complicated. At blob splits, the blob existing after the bifurcation will be larger than the blobs existing before the bifurcation. Therefore, the scale values given by assumption 10.2 give reliable information about the relative size[4] of two objects only when the objects overlap, and in addition the blobs can be related to each other through a series of bifurcations free from blob splits and blob creations.

10.5.3. Multiple blob responses

As can be seen in figure 10.3, *multiple blob responses* can be obtained from an image region. This is a common phenomenon in the scale-space primal sketch that arises because a large (significant) blob merges with a small (insignificant) blob and forms a new scale-space blob. From the definition of scale-space blob it follows that every time a bifurcation takes place, the (involved) grey-level blobs existing before the bifurcation will be treated as belonging to different scale-space blobs than the (involved) grey-level blobs existing after the bifurcation.

[4]On the other hand it is not even clear that it is desirable to use the scale values for size comparisons, since the size of a region can be easily estimated from the size of its blob support region.

10.6. Additional experiments

More experimental results demonstrating the properties of the scale-space primal sketch and the suggested way to extract image structures from this representation[5] are given in figures 10.7–10.27; see also section 11.4 for an application to the analysis of aerosol images, and section 11.5 for examples with textures.

Figure 10.7 and figure 10.9 show an indoor table scene and the 50 most significant bright and dark blobs extracted from the grey-level image. In figures 10.8–10.12 the boundaries of these blobs are displayed and also the results of superimposing the blob boundaries onto the original image. To give a rough idea of the significance values, thresholds have been selected manually in "gaps" in the sequences of significance values. In this scene, most of the meaningful objects are brighter than the background and that these objects are found. In addition, the bright blobs respond also to illumination phenomena on the table, in the background as well as to specularities. The detected dark blobs correspond to the two background regions and various shadows due to the objects on the table.

For one object, the curved pipe in the right part of the image, only the specularity on its surface is detected. The object fails to stand out as a single blob unit. This illustrates a characteristic property of the representation: a region, which borders upon both a darker region and a brighter region, cannot be expected to be detected as a single blob region by this method. According to the blob definition (compare with figure 7.1 and figure 7.6), only regions that are brighter or darker than their background will be treated as "blobs," (see figure 10.13). To be able to detect regions also of this latter type it seems necessary to include more information into the analysis (e.g., derivatives).

Figures 10.14–10.17 show similar results from a scattered office scene, where most of the important objects are darker than their background. The handle of the hammer, the heap of screws, the black tape reel, the label of the hammer, and some other dark regions are all detected as dark blobs. The grey tape reel is not detected as a blob, since it is a neighbour of both a darker region and a brighter region. It can be observed that the blob corresponding to the handle of the hammer spreads relatively far from the boundary of the actual object. This phenomenon occurs for isolated objects far away from other competing blobs of the same polarity.

[5] All these experiments have been performed with images of size 256×256 pixels. The scale-space convolutions were carried out with floating point calculations, and the image boundaries were treated in the following way: When detecting *bright* blobs on dark background the image was extended using the *minimum* grey-level value. Conversely, when looking for *dark* blobs on bright background the image was extended with the *maximum* grey-level value. The infinite support convolution kernel (the discrete analogue of the Gaussian kernel) was truncated at the tails such that the truncation error ϵ was guaranteed not to exceed 0.0005.

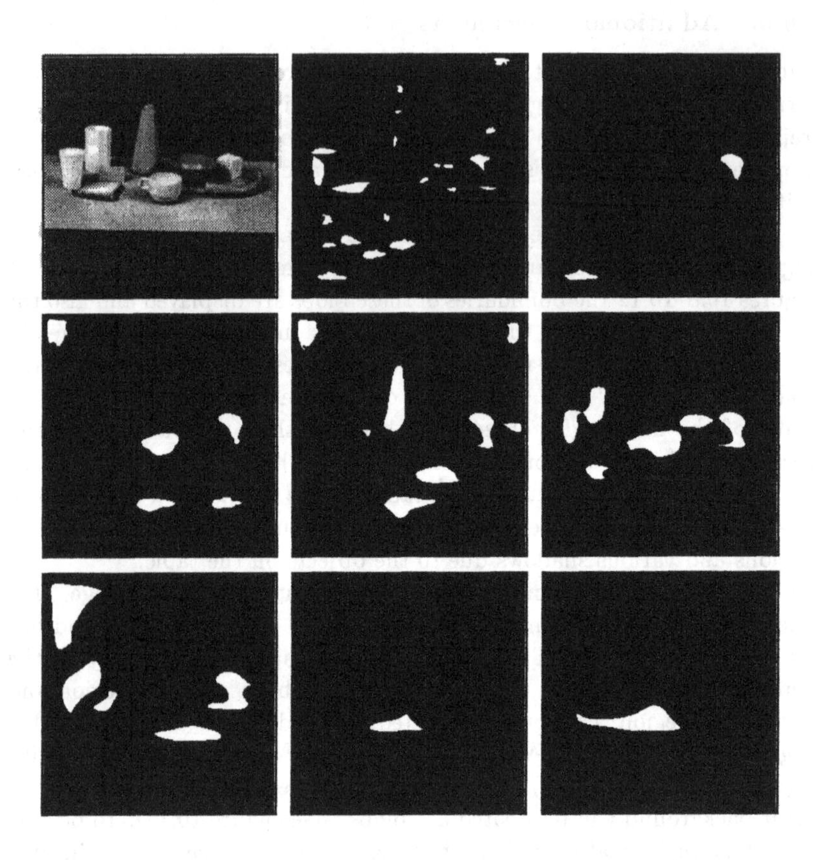

Figure 10.7. The 50 most significant bright blobs from a table scene. (One blob corresponding to the entire image has been suppressed.)

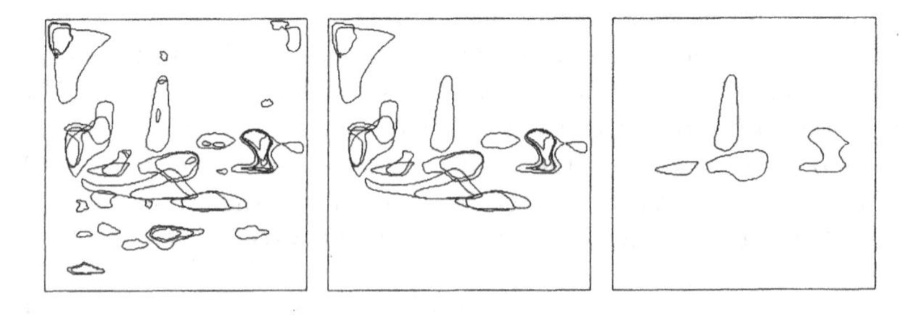

Figure 10.8. Boundaries of the bright blobs extracted from the table scene; (left) the 50 most significant bright blobs, (middle) low threshold on the significance measure set in one of the "gaps" in the sequence of significance values, (right) high threshold on the significance measure set in another "gap."

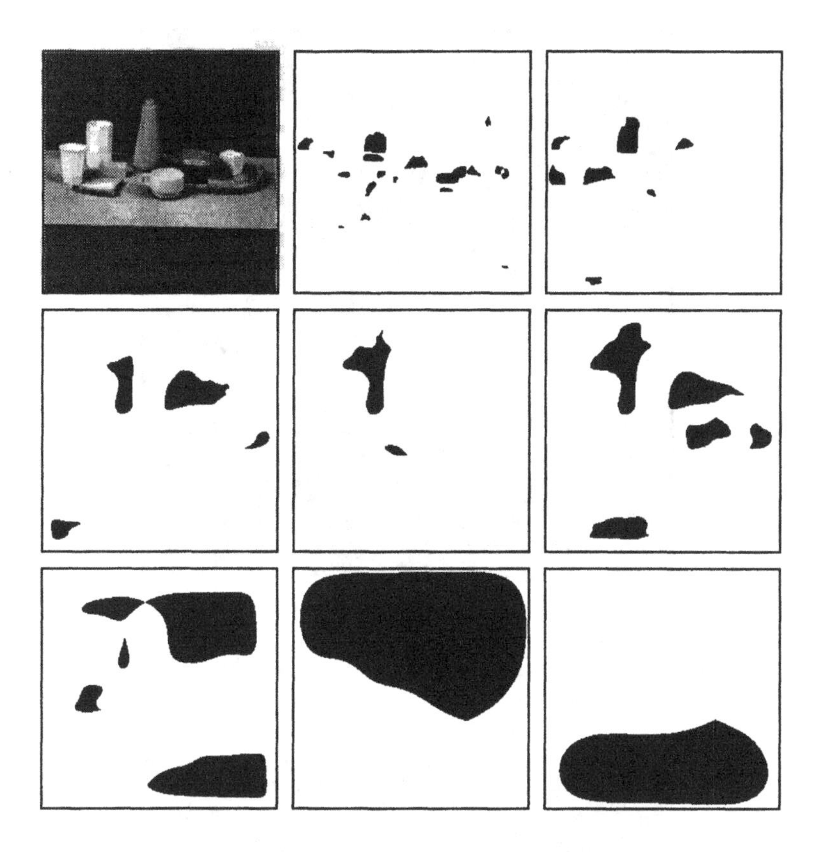

Figure 10.9. The 50 most significant dark blobs from a table scene. (About 10 multiple blob responses have been suppressed in order to save space. All blob boundaries are, however, shown in figure 10.10(a) below.)

Figure 10.10. Boundaries of the dark blobs extracted from the table scene; (left) the 50 most significant dark blobs, (middle) low threshold on the significance measure set in one of the "gaps" in the sequence of significance values, (right) high threshold on the significance measure set in another "gap."

Figure 10.11. Boundaries of the extracted bright blobs superimposed onto a bright copy of the original grey-level image. The lower threshold has been used on the significance values.

Figure 10.12. Boundaries of the extracted dark blobs superimposed onto a bright copy of the original grey-level image. The lower threshold has been used on the significance values.

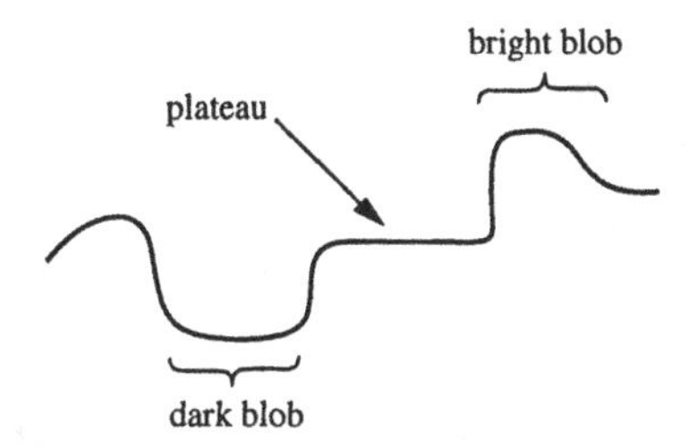

Figure 10.13. According to the definition of grey-level blob, only regions that are brighter or darker than their background will be classified as blobs. A region that borders upon both a darker and a brighter region does not satisfy the blob definition, which means that the plateau in the figure will not be detected as one unit by the algorithm. To extract such regions it seems necessary to include some kind of gradient information into the analysis.

However, this can be compensated for (for instance, when matching blobs to edges). See section 11.1.2 for a description. The bright blobs correspond to the holes in the two tape reels as well as various regions on the table.

Figure 10.18 and figure 10.20 display the extracted dark and bright blobs from an outdoor image of a house, a scene where there are both dark and bright objects with meaningful interpretation. Figure 10.19 and figure 10.21 show the boundaries of these blobs. It can be observed that the windows of the house are detected as well as various parts of the wall, the sky and parts of the tree.

Finally, as a test of the stability of the representation, figures 10.22–10.25 display the results from another image of the same telephone and calculator as in figure 10.3, where the background has been changed to textured cloth, and the camera and the objects in the scene have been moved. Clearly, the important regions (the receiver, the cord, the keyboard, the buttons and the calculator) are still being found. The bright blobs respond to the telephone, the hole in the cord, other regions delimited by dark objects, various illumination phenomena in the background, the bright buttons of the calculator, and the regions between the buttons of the telephone.

For comparison, corresponding results for the original telephone and calculator image are shown in figures 10.26–10.27. Similar types of regions are extracted in the two cases, both for the dark and the bright blobs, although the interference effects for the coarse scale blobs are different, mainly because the distance between the telephone and the calculator has been changed.

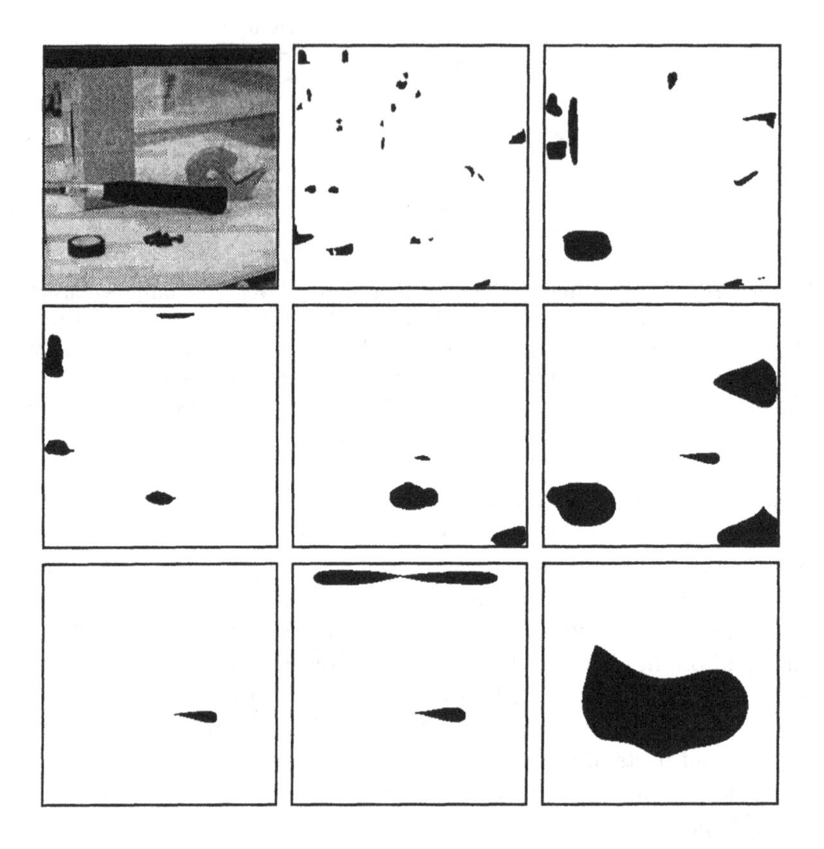

Figure 10.14. The 50 most significant dark blobs from a scattered office scene.

Figure 10.15. Boundaries of the dark blobs extracted from the scattered office scene. (a) The 50 most significant blobs. (b) Low threshold on the significance measure set in one of the "gaps" in the sequence of significance values. (c) High threshold on the significance measure set in another "gap,"

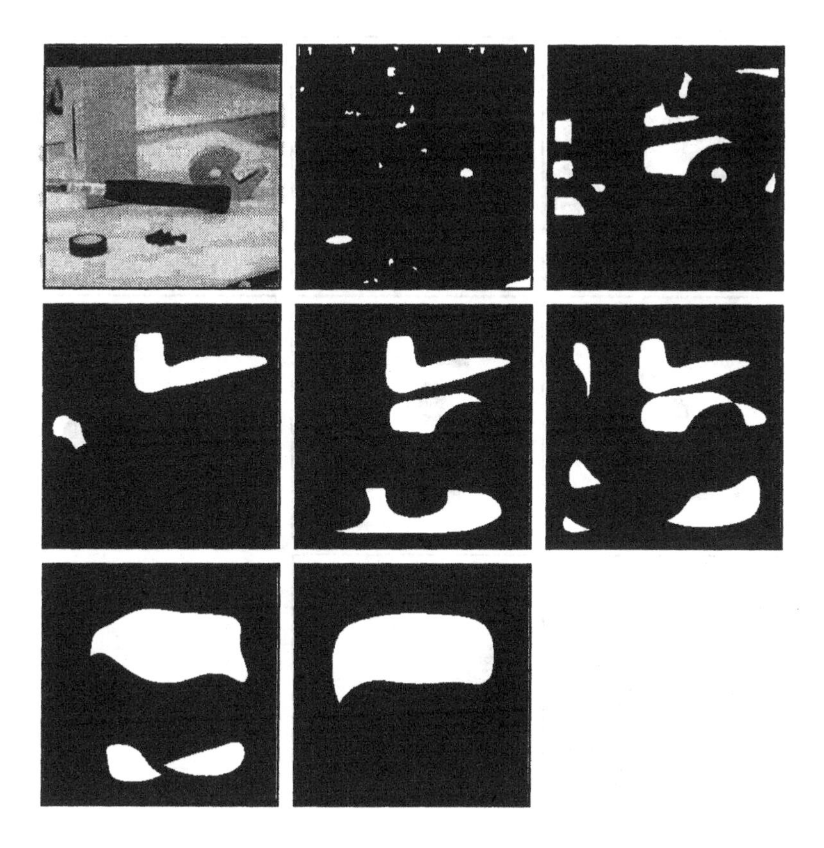

Figure 10.16. The 50 most significant bright blobs from a scattered office scene.

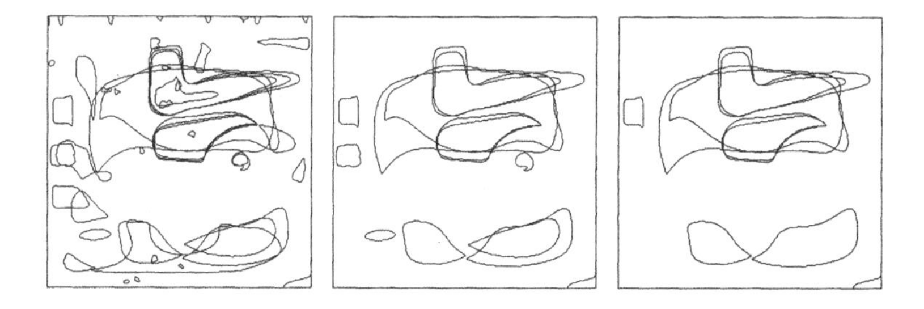

Figure 10.17. Boundaries of the bright blobs extracted from the scattered office scene; (left) the 50 most significant blobs. (middle) low threshold on the significance measure set in one of the "gaps" in the sequence of significance values, (right) high threshold on the significance measure set in another "gap."

Figure 10.18. The 50 most significant dark blobs from an image of the Godthem Inn at Djurgården, Stockholm.

Figure 10.19. Boundaries of the dark blobs extracted from the Godthem Inn image; (left) the 50 most significant blobs. (middle) low threshold on the significance measure set in one of the "gaps" in the sequence of significance values, (right) high threshold on the significance measure set in another "gap."

Figure 10.20. The 50 most significant bright blobs from an image of the Godthem Inn at Djurgården, Stockholm. (In the last case, the entire image has been classified as one bright blob.)

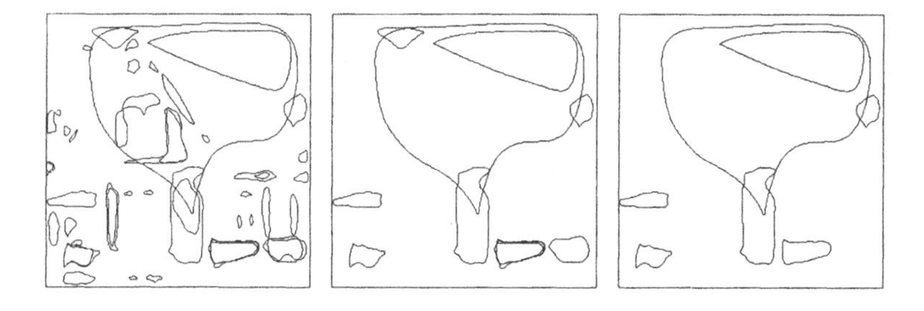

Figure 10.21. Boundaries of the bright blobs extracted from the Godthem Inn image; (left) the 50 most significant blobs, (middle) low threshold on the significance measure set in one of the "gaps" in the sequence of significance values, (right) high threshold on the significance measure set in another "gap."

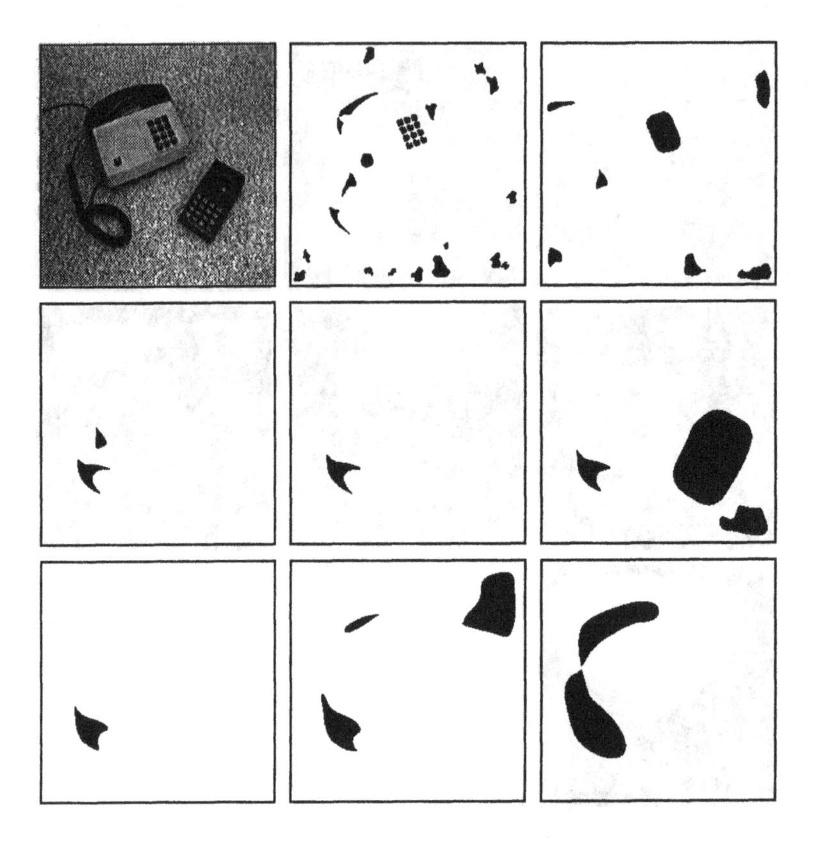

Figure 10.22. The 50 most significant dark blobs from a telephone and calculator image. The background is textured.

Figure 10.23. Boundaries of the dark blobs extracted from the telephone and calculator image with textured background, (left) the 50 most significant blobs, (middle) low threshold on the significance measure set in one of the "gaps" in the sequence of significance values, (right) high threshold on the significance measure set in another "gap."

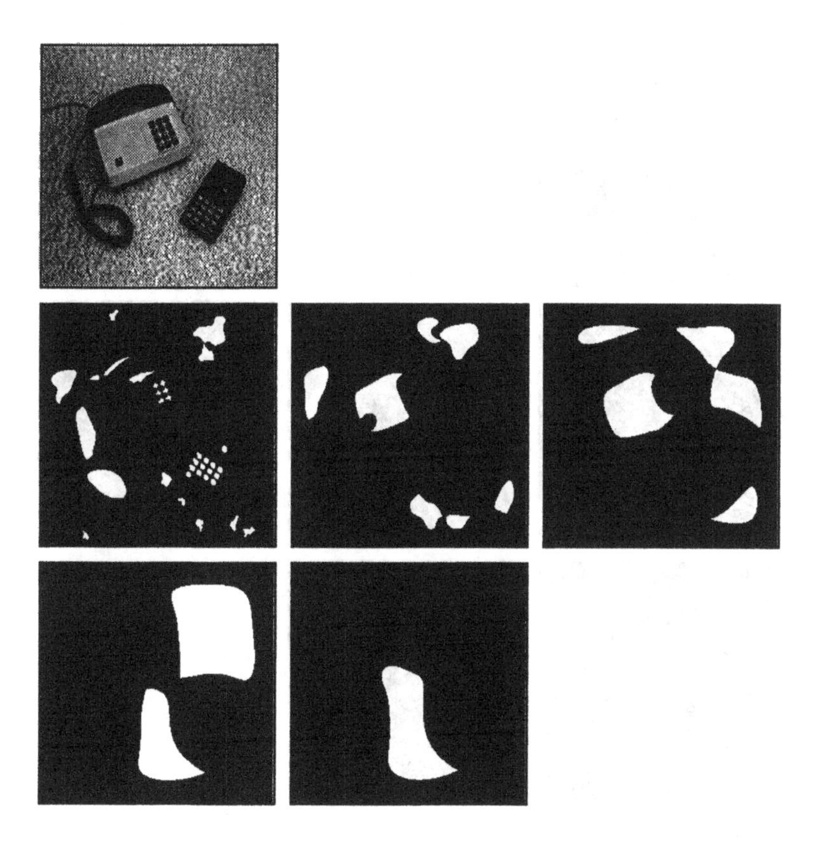

Figure 10.24. The 50 most significant bright blobs from a telephone and calculator image. The background is textured.

Figure 10.25. Boundaries of the bright blobs extracted from the telephone and calculator image with textured background, (left) the 50 most significant blobs, (middle) low threshold on the significance measure set in one of the "gaps" in the sequence of significance values, (right) high threshold on the significance measure set in another "gap."

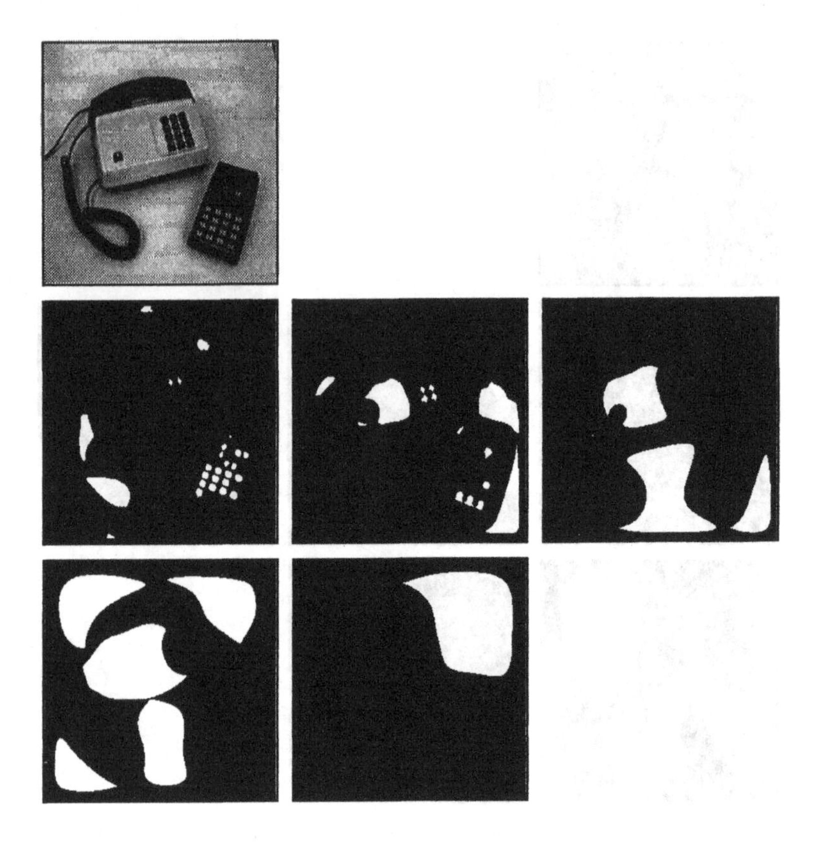

Figure 10.26. The 50 most significant bright blobs from the telephone and calculator image with smooth background. (The dark blobs were shown in figure 10.1.)

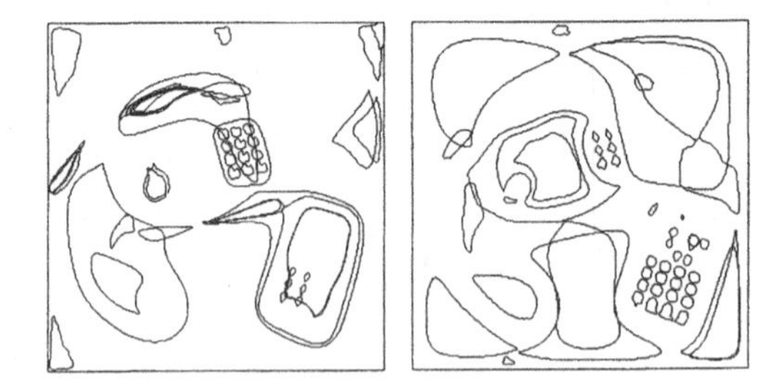

Figure 10.27. Boundaries of the dark and bright blobs extracted from the telephone and calculator image with smooth background, (left) the 50 most significant dark blobs, (right) the 50 most significant bright blobs.

11

Guiding early visual processing with qualitative scale and region information

Many methods in computer vision and image analysis implicitly assume that the problems of scale detection and initial segmentation have already been solved. One example is in edge detection, where the selection of step size for the gradient computations leads to a trade-off problem. A small step size gives a small truncation error, but the noise sensitivity might be severe. Conversely, a large step size will in general reduce the noise sensitivity, but at the cost of an increased truncation error. In the worst case a slope of interest can be missed and meaningless results obtained, if the difference quotient approximating the gradient is formed over a wider distance than the size of the object in the image. The problem originates from a basic scale problem, namely that the issue of scale must be considered when selecting a mask size for computing spatial derivatives.

Other examples can be obtained from most "shape-from-X" methods, which in general assume that they are applied to a domain in the image where the underlying assumptions are valid (e.g., corresponding to a region in the image corresponding to one facet of a surface, etc.).

A principle that will be advocated in this chapter is that the *qualitative scale and region information* extracted from the scale-space primal sketch can be useful for *guiding other visual processes,* and will *simplify their tasks.* More specifically, it will be proposed that in the absence of other evidence, scale levels for further processing can be selected from the scale-space primal sketch, and that the blob support regions can provide coarse size information to other algorithms. It is suggested that this type of information can be used for delimiting the search space for further processing, for example, such that the processing can be performed selectively, and matching can be carried out *regionally* in a neighbourhood of a blob instead of globally over the entire image.

Of course, the scale-space smoothing also leads to shape distortions. In particular, the amplitude of spatial derivatives can be expected to decrease by the scale-space smoothing. Therefore, the actual numerical values of derivatives computed from the coarse scale representations cannot be expected to be *quantitatively* accurate. However, for finding *qualitative* features, not depending on the actual scaling of the intensity, such as edges, local extrema, singularities etc., the *detection* step can be carried out at a coarse scale. Then, once the existence of a feature has been established, if precise numerical values are required, it should be possible to compute those in a second step, e.g., by fitting an appropriate model to the original data. This notion will be made more precise in chapters 13–14, where a genuine two-stage approach is developed for detecting structures at coarse scales and then localizing them to finer scales.

11.1. Guiding edge detection with blob information

As a first application of the suggested methodology, an integration with an edge detection method known as *edge focusing* (Bergholm 1987) is described. The main idea is to use the scale and region information for guiding an edge detection scheme working at an adaptively determined coarse level of scale. It will be demonstrated that this task can be simplified, and that thresholding on gradient magnitude can be avoided. Given a significant scale-space blob, edge detection is performed at the selected scale of the blob scale-space blob. Then, a matching step is carried out between the support region of the blob and the edges in a neighbourhood of the blob. Finally, the matched edges are tracked to finer scales to improve the localization (see figure 11.1 for a schematic overview).

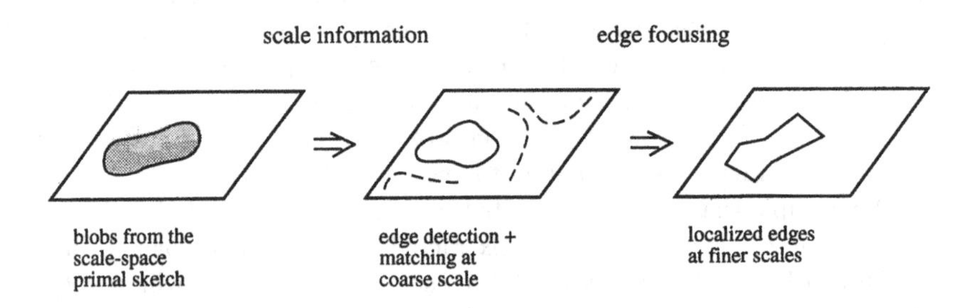

Figure 11.1. Schematic overview of the proposed integration of the scale-space primal sketch module with edge focusing.

The underlying motivation for this approach is to circumvent the inherent conflict between noise suppression and localization in *simultaneous* detection and localization of image structures, which has been discussed by, e.g, Canny (1986). By detecting structures at a coarse scales, the detection problem will be simplified at the cost of poor localization. The localization is then improved in a second stage by tracking the edges to finer scales in scale-space.

It is not maintained that the presented scheme describes any "optimal way" to solve every occurring subproblem. Instead, the intention is to exemplify how a connection between the scale-space primal sketch and other modules can be done, and to support the claim that the type of qualitative scale and region information extracted by the scale-space primal sketch really is useful as input to other visual processes.

11.1.1. Edge detection at a coarse scale

A rather simple edge detector is used deliberately. The image is smoothed to the scale associated with the scale-space blob. Then, derivatives along the two coordinates are estimated by difference approximations, and a non-maximum suppression step (Canny 1986; Korn 1988) is performed (without thresholding on gradient magnitude) to obtain thin edges.[1] To suppress spurious noise points, only edge segments of length exceeding, say two pixels, are accepted.

11.1.2. Matching blobs to edges

Associating blobs with edges leads to a matching situation. The matching procedure used for associating blobs with edges is based on spatial coincidence, and is a combination of the following three criteria:

Geometric coincidence. The edge segment should "encircle," "be included in," or "intersect" the blob. A convenient way to express such a criterion is by requiring it to be *impossible to draw a straight line separating the edge from the blob,* (see figure 11.2).

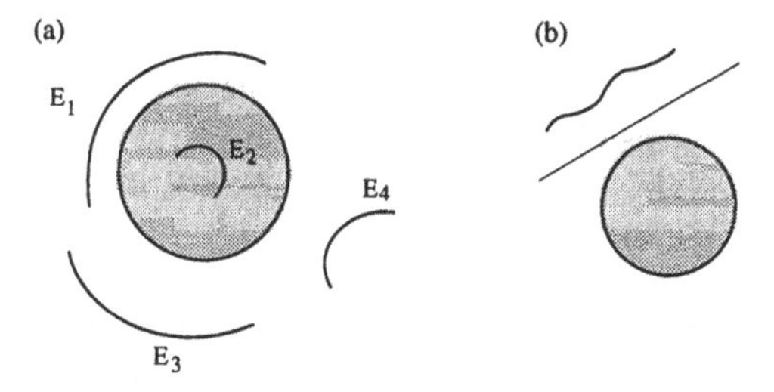

Figure 11.2. The geometric coincidence condition means that the edge should either surround the blob, be included in it, or intersect the blob—it should be impossible to draw a straight line separating the edge from the blob (b). In example (a) edges E_1 and E_2 are treated as matching candidates of the blob, while edges E_3 and E_4 are not.

A simple way to approximate this criterion computationally is as follows: Let $B \subset \mathbb{R}^2$ be the set of points contained in the support region of a blob, and let $E \subset \mathbb{R}^2$ be the set of points covered by an edge segment.

[1]Edges are defined as the ridges of the gradient magnitude map, i.e., the points for which the gradient magnitude assumes a maximum in the gradient direction.

Further, given any region R and any arbitrarily rotated coordinate system $(\xi, \eta) \in \mathbb{R}^2$, define the extreme coordinate values $\xi_{min}, \xi_{max} \in \mathbb{R}$ by

$$\xi_{min}(R) = \min_{(\xi,\eta)\in R} \xi; \quad \xi_{max}(R) = \max_{(\xi,\eta)\in R} \xi. \tag{11.1}$$

Now, an edge segment E is regarded as a matching candidate of a blob B if

$$\xi_{min}(E) \leq \xi_{max}(B), \quad \xi_{max}(E) \geq \xi_{min}(B) \tag{11.2}$$

hold in a sufficiently large number of directions. For practical implementation, this condition is required to hold along both the coordinate directions of a standard Cartesian coordinate system, and in a corresponding coordinate system rotated by 45 degrees.

Proximity. The edge segment should *not be too far away from the blob boundary.* In other words, the edge segment should comprise at least some point located near the boundary ∂B of the blob. This condition can be stated as

$$\min_{x_E\in E;\ x_B\in\partial B} ||x_E - x_B||_2 \leq d(t)/2, \tag{11.3}$$

where $d(t)$ represents a characteristic length[2] at scale t. The purpose of this criterion is to prevent (interior and exterior) edges far away from the blob boundary from being associated with the blob; see figure 11.3(a).

Voronoi diagram of the grey-level blob image. The edge segment should not be strongly associated with other blobs. A natural way to express such a criterion is in terms of a Voronoi diagram of the grey-level blob image at the selected scale. The image plane is divided into a number of labelled regions. The Voronoi region associated with a blob B is defined as those points $x \in \mathbb{R}^2$ for which B is the nearest region to P. An edge segment is then regarded as a Voronoi matching candidate of a blob if it has at least one point in common with the Voronoi region associated with the grey-level blob. This condition prevents edges that are closely related one particular blob from being associated with other blobs; see figure 11.3(b). In fact, this type of criterion turns out to be useful also to several other matching problems.

Composed matching procedure. For an edge segment to be accepted as a matching candidate of a blob, it must satisfy all these criteria. Hence, the matching is relatively restrictive. It is also improved by the fact that

[2]For implementational purpose, this characteristic length is determined as the square root of an experimentally determined blob area $A_m(t)$ at scale t. It is accumulated in the same way as the statistics of the grey-level blob volume $V_m(t)$.

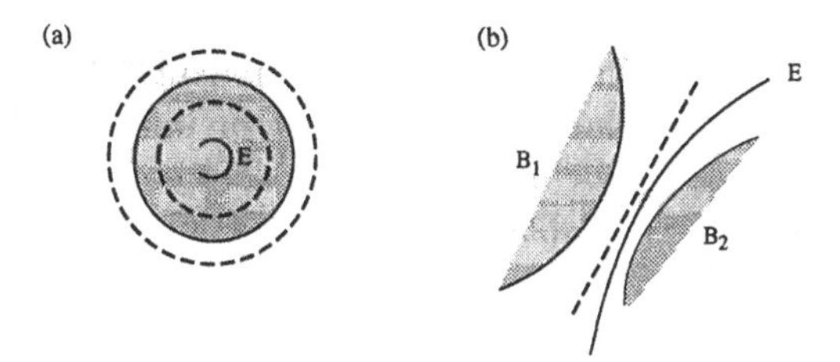

Figure 11.3. (a) The purpose of the proximity criterion is to prevent edges far away from the blob boundary from being associated with the blob. (b) The purpose of the Voronoi region matching is to prevent edges strongly related to one blob from being associated with other nearby blobs.

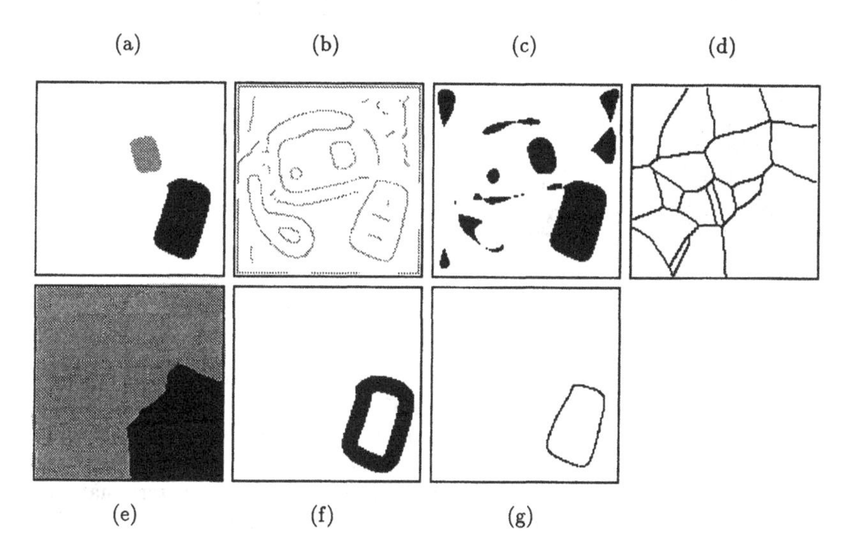

Figure 11.4. The matching procedure between blobs and edges for one blob from the telephone and calculator image. (a) The support region of a dark scale-space blob (black). (b) Edges detected at the scale given by the blob. (c) All grey-level blobs at the same level of scale. (d) Voronoi diagram of the grey-level blob image. (e) The Voronoi region corresponding to the given blob. (f) The proximity stripe around the blob edge. (g) Resulting edges matched to the blob.

it is performed at a scale at which a blob has manifested itself. Once it is known that a spatial region has given rise to a large blob at some level of scale, it seems unlikely that conflicting edges should appear at the same scale, since most interfering structures ought to be suppressed by the scale-space smoothing.

Figures 11.4–11.5 illustrate two such matching situations from the block image and the telephone and calculator images respectively. The figures show the blob to be matched, edges detected at the scale given by the blob, the grey-level blobs at the same level of scale, the Voronoi diagram of the support regions of the grey-level blobs, as well as the matched edges. It can be seen that this matching criterion selects reasonable subsets of edges to that may be associated with the blobs.

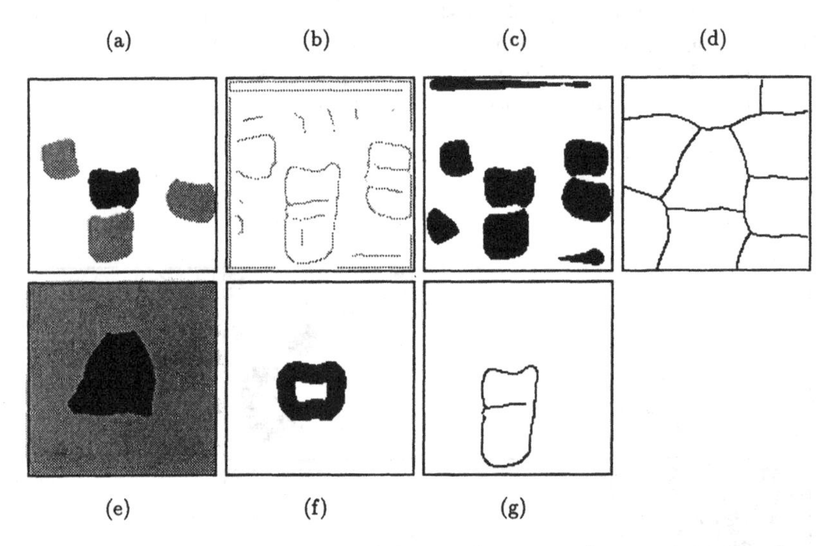

Figure 11.5. Similar illustration of the matching procedure as above, but for a blob from the block image. Note that one of the edge segments spreads far away from the blob, since the matching algorithm does not include any mechanism for breaking up long edge segments into shorter ones. However, we will demonstrate below that the focusing procedure itself provides a cue for such determination—at finer scales the elongated edge will split into two well separated sets of edges (see figure 11.10).

The main problem with this matching procedure is that it does not include any mechanism for splitting long edge segments into shorter ones. Hence, certain edge segments can be very long and spread far away from the blob boundary; see the example in figure 11.8.

Improving the localization of the blob boundary. For a single isolated blob, the proximity matching criterion in its original formulation can lead to problems. In such cases the boundary of the grey-level blob support region can spread far away from the boundary of the actual "object" in the image, since there will be no competing blob that delimit the extent of the blob support region. Hence, the blob might extend far beyond the "actual boundary," but with a relatively flat intensity slope (see figures 11.6–11.7).

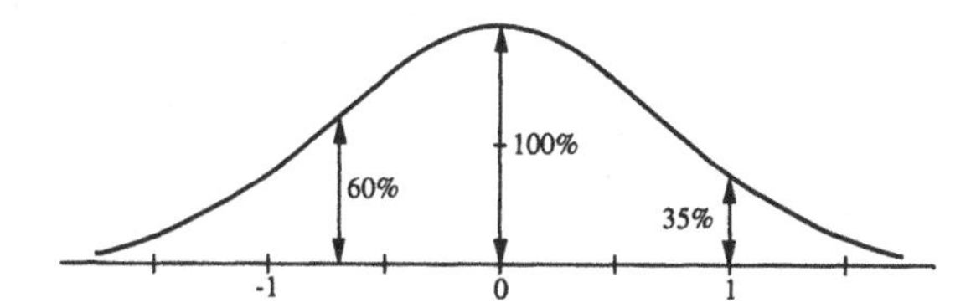

Figure 11.6. A single isolated blob has an infinite support region. To obtain a more useful spatial representative for the purpose of matching, the blob boundary is modified in the proximity step. A straightforward approach is to clip the blob at a grey-level corresponding to the edge. For a Gaussian intensity profile, edges given by non-maximum suppression correspond to a clipping level of about 61% ($\approx e^{-1/2}$). In the current implementation, a clipping level of 35% ($\approx e^{-1}$) is used. Observe that the clipping is used *only* in the proximity matching step.

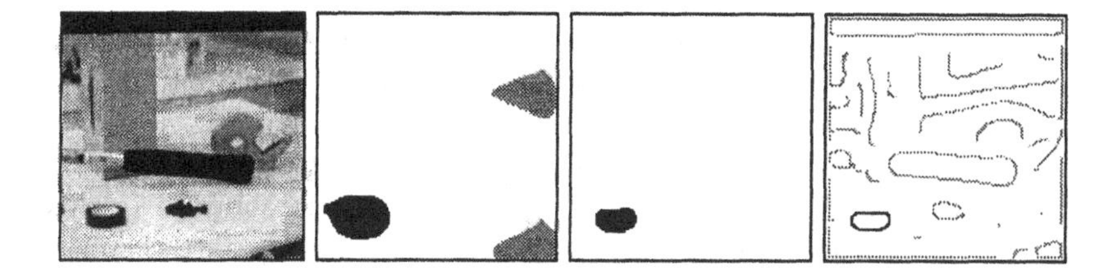

Figure 11.7. Example of the effect of clipping: (left) original grey-level image, (middle left) the (unclipped) blob corresponding to the tape reel in the lower left corner, (middle right) the effect of clipping that blob, (right) Edges matched to the clipped blob.

11.1.3. Blob-initiated edge focusing

Edge focusing (Bergholm 1987, 1989) is a method for following edges through scale-space. The basic principle is to detect edges at a coarse scale, where the detection problem can be expected to be easier, and then track the edges to a finer scale, to improve the localization, which can be very poor at coarse scales. In this sense, the method tries to circumvent the conflict between noise suppression and edge localization that has been discussed by Canny (1986) among others.

It has been shown (Bergholm 1987, 1989) that if the focusing procedure is performed such that the scale step $\Delta\sigma$, expressed in $\sigma = \sqrt{t}$, is less than $\frac{1}{2}$, then for most common edge configurations the edges are guaranteed to move not more than one pixel from one level to the next. When this property holds the matching will be trivial—in order to find the corresponding edges at the finer level of scale, it suffices to perform edge detection in a one-pixel neighbourhood around the edges at the coarser scale. (Obviously, there are situations where such a fixed scale sampling can lead to problems; see the analysis of drift velocities in scale-space in section 8.1.1.)

In this application, the focusing procedure is initiated from several scale levels, since the significant blobs from the scale-space primal sketch manifest themselves at different scales. Hence, the blobs are first pre-sorted in decreasing scale order. The procedure starts with the coarsest scale blob, detects edges at that scale, and matches those to the blob. This gives the input data for the edge focusing procedure, which follows these edges to the scale given by the second blob. Then, the edge detection and matching steps are repeated at the new scale level, and the resulting edges are added to the output from the previous focusing step. This new edge image serves as input for another focusing procedure, which traces the edges to the next finer scale etc.

Figure 11.8 illustrates some steps from this composed procedure applied to the telephone and calculator image. To reduce the number of blob hypotheses treated, a threshold has been introduced on the significance value. The "final result" is shown in the lower right corner.

Observe that this method, called *blob-initiated edge focusing,* is not just another edge detector. Instead, the edge elements obtained in this way are more likely to correspond to meaningful entities, since they are *explicitly grouped* into edge segments, and are associated with blobs and explicit scale information. Note that label information for the edge segments can be easily inherited during the edge focusing process.

Figure 11.8. (Opposite page): Illustration of the composed blob-edge focusing procedure for the telephone and calculator image. The left column shows the active blob hypothesis; its blob support region has been marked with black. The middle column shows the edge image at the level of scale given by the previous blob; matched edge segments are drawn black while the other edge pixels are grey. The right column shows the result after focusing, just before a new blob is considered. The image in the lower right corner displays "the final result," i.e. the edges that are related to the dark blobs in the image. The scale and significance values for the different blobs are from top to bottom (101.6, 14.1), (50.8, 252.8), (32.0, 11.4), (25.4, 660.9), (14.3, 40.8), (6.4, 63.6), and (1.3, 13.2) respectively.

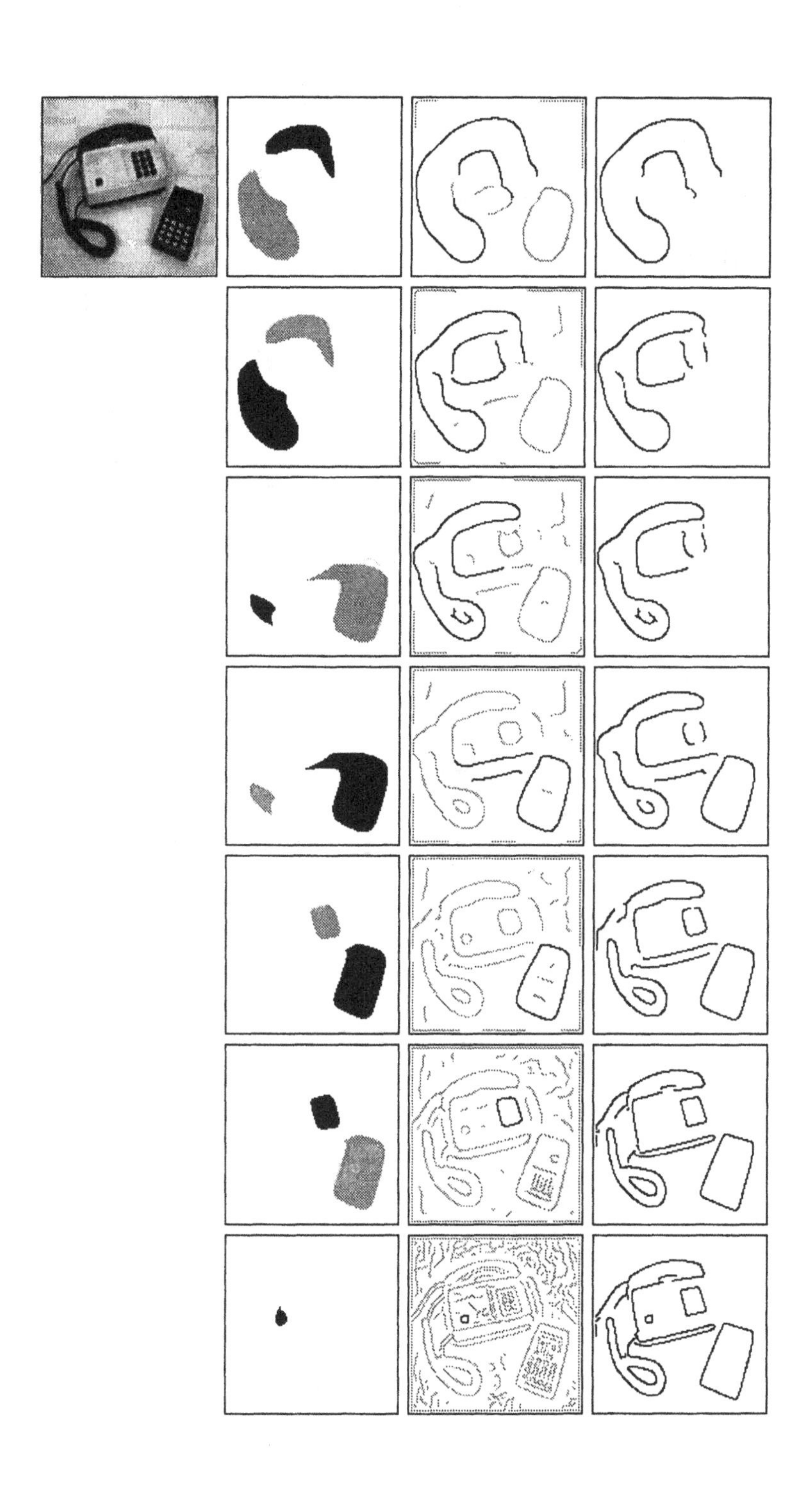

Figure 11.9. (Opposite page): Similar illustration of the composed blob-edge focusing procedure for the block image. The scale and significance values for the different blobs are from top to bottom (203.2, 15.5), (161.3, 10.7), (20.2, 20.3), (12.7, 52.2), (12.7, 4.6), (12.7, 35.7), and (8.0, 48.9) respectively.

A simpler way to implement the method is as follows: Instead of treating all the hypotheses generated by the scale-space primal sketch simultaneously during the edge focusing phase, they can of course also be treated separately. Each blob can start up its own focusing scheme, which processes the edge segments matching that blob independently of the edges corresponding to other blobs. Then, the relations between blobs and edges will be obvious and the interference effects between edge segments from different blobs during the edge focusing will be eliminated; see figure 11.10 for an example. This example also demonstrates another important property of the methodology. An edge segment that extends far outside a blob may split into several segments during the focusing procedure. By registering such bifurcations, there is a potential for reducing the number of edge segments associated with the blob.

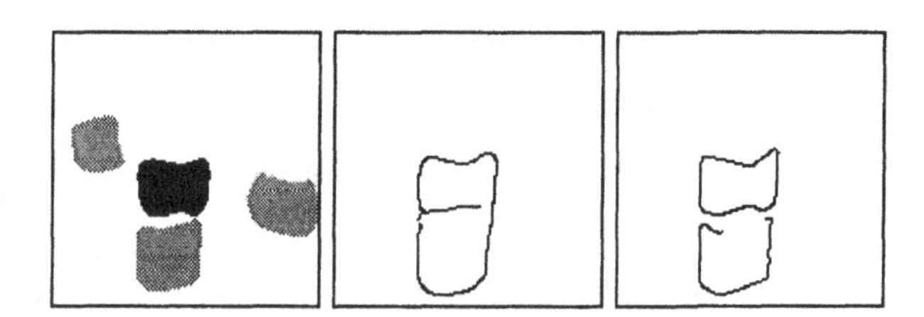

Figure 11.10. Individual blob hypothesis treatment for the a blob from the block image. (a) A blob from the scale-space primal sketch. (b) Matched edges. (c) After focusing to finer scales ($t = 1.0$) the edge segment splits into two separate edge segments, corresponding to the two adjacent blocks.

With this integration experiment, we have avoided manual setting of two of the tuning parameters in the edge focusing algorithm, the initial scale for detection, and the threshold on gradient magnitude. What remains undetermined is the stop scale down to which the edge focusing procedure should be performed. Here, it has been throughout set to $t = 1$, a scale where the sampling effects due to the discrete grid start to become important. It seems plausible that further guidance could be obtained by studying the behaviour of the focused edges in scale-space. One approach is developed in chapter 13 (see also Sjöberg and Bergholm (1988) and Zhang and Bergholm (1993)).

The integration of the two algorithms exemplifies the previously mentioned guidance of focus-of-attention. Note that the processing initiated by the scale-space primal sketch is performed only for a small subset of the image data. In this sense the approach bears similarity with the idea of a "focused beam" derived by Tsotsos (1990) from complexity arguments; see also the experimental work by Culhane and Tsotsos (1992). The scale-space primal sketch serves as a module for generating regions of interest. In those, selective and more refined processes can later be invoked.

11.2. Automatic peak detection in histograms

The scale-space primal sketch is well suited for automatic cluster detection, since it is designed for detection of bright blobs on dark background and vice versa. Hence, it lends itself as a natural module for peak detection in algorithms based on histogramming techniques. Although it is well-known that histogram-based segmentation can hardly be expected to work globally on entire images (due to illumination variations, interference because of many regions, etc.), such methods can often give useful results *locally* in small windows, where only a few regions of distinctly different characteristics (e.g., colour or grey-level) are present.

11.2.1. Experiments: Histogram-based colour segmentation

Figures 11.11–11.12 illustrate how the scale-space primal sketch can constitute a helpful tool in such histogram modality analysis of multi-spectral data. It shows histograms of the (two-dimensional) chroma[3] information from two real-world images, together with blobs detected by the scale-space primal sketch, as well as backprojections of the blobs to the original image plane. It can be seen that the extracted blobs induce a meaningful partitioning of the histogram, corresponding to regions in the image with distinctly different colours.

Of course, there is a decision finally to be made about which peaks in the histogram should be counted as being significant. However, it seems plausible that the significance values given by the scale-space blob volumes reflect the situation in a manner useful for such reasoning, especially since the regions around the peaks are extracted automatically. In these examples, (single) thresholds have been set manually in "gaps" in the sequences of significance values; see the captions of figures 11.11–11.12.

[3]The colour images have been converted from the usual RGB format to the CIEu^*v^* 1976 format (Billmeyer and Saltzman 1982), which aims at separating the intensity and the chroma information. The histograms are formed only over the (two-dimensional) chroma information, ignoring the (one-dimensional) intensity information.

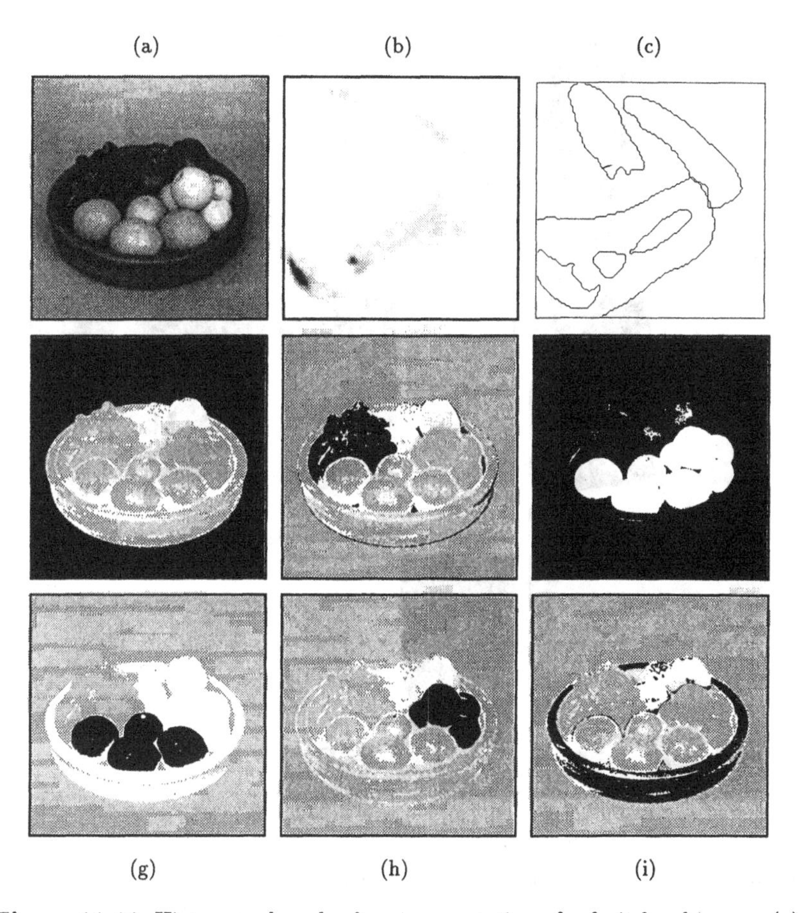

Figure 11.11. Histogram-based colour segmentation of a fruit bowl image: (a) Grey-level image. (b) Histogram over the chroma information. (c) Boundaries of the 6 most significant blobs detected by the scale-space primal sketch. (d)–(i) Backprojections of the different histogram blobs (in decreasing order of significance). The pixels corresponding to the various blobs have been marked in black. (The region in figure (f) is the union of the regions in figures (d), (e) and (i).) The significance values of the accepted blobs were 42.6 (background), 8.3 (grapes), 3.6 (oranges), 3.1 (apples), 3.0 (bowl), and for the rejected blobs 2.0 and less (2.0, 1.9, 1.8, 1.4, 1.3, 1.1, 1.1, 1.1, ...).

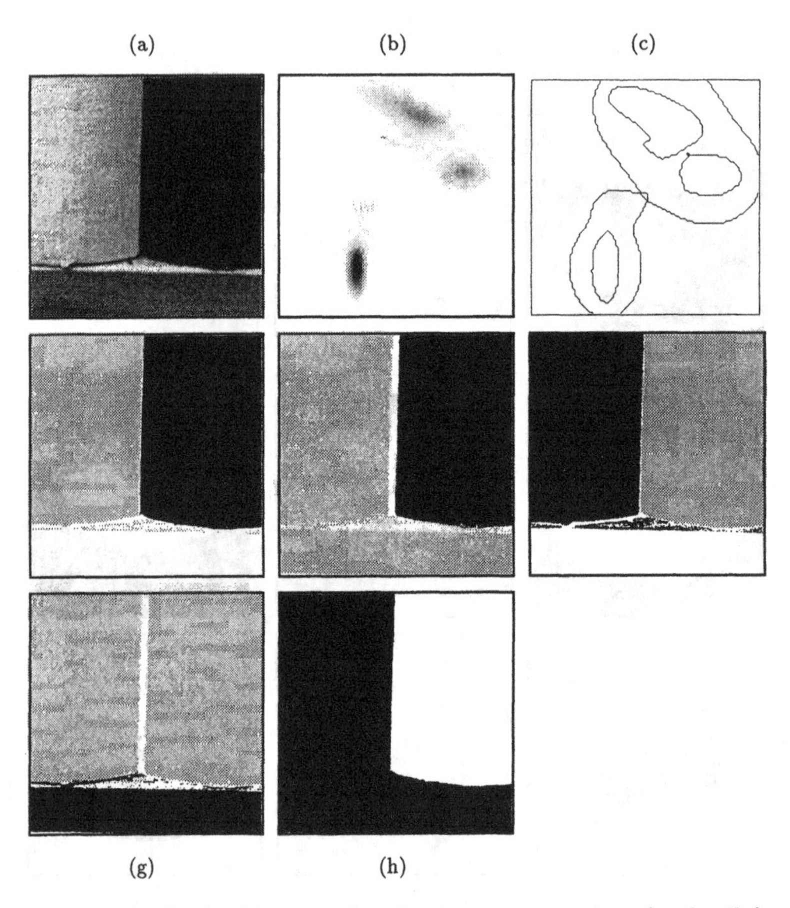

Figure 11.12. Similar histogram-based colour segmentation of a detail from an office scene. The image shows a small window from a bookcase with two binders (yellow and blue) on a shelf made of (yellowish) wood. The displayed blobs have significance 187.9 (blue binder, large blob), 173.7 (blue binder, small blob), 170.1 (yellow binder), 80.6 (shelf), and 66.7 (yellow binder and shelf). As can be seen, two blobs corresponding to the blue binder have been detected. This is a common phenomenon in the scale-space primal sketch, that arises because a large blob merges with a small (insignificant) blob and forms a new scale-space blob. Two such duplicate blobs corresponding to the yellow binder (significance 18.0) and the shelf (significance 17.9) have been suppressed. The remaining blobs had significance 2.5, 2.0, 2.0, 2.0, 1.2, 1.2, 1.2, 1.1, and less.

11.2.2. On the sensitivity to quantization effects

It can also be noted that this peak detection concept will be less sensitive to quantization effects in the histogram acquisition than many traditional peak detection methods. The problems due to too fine a quantization in the accumulator space will be substantially reduced, since the scale-space blurring will lead to a propagation of information between different accumulator cells. Thus, even though the original histogram may have been acquired using "too many and too small" accumulator cells, large scale peaks will be detected anyway, since the contents of their accumulator cells will merge to large scale blobs in scale-space after sufficient amounts of blurring.

Finding peaks in histograms is a problem that arises in many contexts. Let us point out that the case with colour-based histogram segmentation has been considered just as one possible application of the scale-space primal to histogram analysis. Because of the general purpose nature of this tool, there are potential applications for similar techniques such as Hough transforms, texture classification etc. in two as well as other dimensions. For related work, see (Carlotto 1987; Mokhtarian and Mackworth 1986).

11.3. Junction classification: Focus-of-attention

More generally, the scale-space primal sketch can serve as a primitive mechanism for focus-of-attention. As an illustration of this, an experimental work will be briefly described, where the scale-space primal sketch has been used for guiding the focus-of-attention of an active head-eye system applied to a specific test problem of classifying junctions.

The presentation is aimed at demonstrating how the suggested approach can be used when addressing some of the most fundamental problems in active analysis:

- How to generate hypotheses about the existence of objects?
- How to determine where to look?
- At what scale(s) to analyse image structures?

11.3.1. Background: Active analysis of local image and scene structure

It is often argued that human vision provides a two-legged proof of the feasibility of general purpose machine vision. Human vision is, however, a highly dynamic process working over time. Its input data change continuously, as well as its system parameters, which vary due to eye movements, fixation, foveation, etc. In fact, this is also the case when looking at static and monocular scenes, such as images. Therefore, if a vision system is restricted to the analysis of pre-recorded images, it is not clear that the machine vision problem is posed in a comparable context, or, put differently, that machine vision is given a fair chance.

A central theme in computational vision is the derivation of low-level cues from local features of image brightness. One may for example try to select an important subset of edges from a given image. However, the results obtained even from a first-rate method are usually inferior to those predicted by a human looking at the corresponding scene. Hence, subsequent high-level interpretations often face difficult problems due to shortcomings in the low-level features extracted.

Referring to these remarks, one can point out at least three reasons to this. One is that several edge detectors, like the Canny–Deriche (1987) edge detector, usually are based on fixed models (step edges with noise), and have no means of detecting that another model may be more appropriate, e.g., that of a corner. The second reason is that feature detectors usually operate in isolation, and are not integrated with other visual modules. For example, the contrast usually varies along an edge. It is even common that it changes polarity. Hence, in the lack of contrast between two regions, there may be no way to locally establish the existence of an edge at a certain point. The third reason is that a performance corresponding to a human looking at the actual scene can hardly be expected due to limits in resolution, and the lack of other visual cues that are available in active analysis, like depth from focus, vergence, etc.

Standard camera systems usually have a visual angle of about 50 degrees and give image matrices of say 512×512 to 1000×1000 pixels. This could be compared to the 2 degrees in foveal vision, which in view of visual acuity can be said to correspond to an image resolution of about 200×200 pixels. Obviously, this difference implies that multi-resolution processing, like done in pyramids, cannot be seen as analogous to foveation. The images coming out of a standard camera system are too much limited by resolution, at least if the images are *overview* images of, say, a large number of real-world objects.

Here, such tasks will be considered in a framework that does not have such limitations. The idea is that visual tasks can be made more tractable by using *focus-of-attention* and *foveation.*[4] In other words, what we aim at is to acquire images with a sufficiently high resolution, such that the phenomena under study can be *clearly resolved.* This implies that some mechanism is needed for controlling the image acquisition process—the visual system must be *active.*

An approach like this raises computation problems different from those appearing in methods for analysing local structure in normal overview images. Obviously, a method is needed for deciding where to focus the attention. Moreover, the increased resolution is likely to enhance the

[4]Here, foveation means acquisition of new images with locally increased resolution.

noise. This may call for different types of algorithms for analysing local structure. There is a problem of detecting structure when the resolution and window size are dynamically varied. What will be used here is a method of stability of responses, closely related to the transformational invariance used for detecting image structures in scale-space based on their stability over scales.

11.3.2. Test problem: Active detection and classification of junctions

In this section, this general approach and in particular the idea of foveating interesting structure will be applied to the test problem of detecting and classifying junctions. A motivation for selecting this problem as test problem is that junctions provide important cues to three-dimensional structure (Malik 1987). For example, T-junctions generically indicate interposition, and hence depth discontinuities.

It is well-known that edge detectors have problems at junctions. Zero-crossings of a function used for defining edges (e.g., the second-order directional derivative in the gradient direction; see section 6.2.1) cannot correctly divide three regions at a 3-junction. If first-order derivatives are used, other problems arise (e.g., in the computation of the gradient direction). To overcome these problems, direct methods for junction detection have been proposed (Kitchen and Rosenfeld 1982; Dreschler and Nagel 1982; Förstner and Gülch 1987; Koenderink and Richards 1988; Noble 1988; Deriche and Giraudon 1990; Florack *et al.* 1992). While such methods can be used for *finding* junction candidates, they usually give no information about what *type* of junction has been found.

Brunnström *et al.* (1990) have demonstrated that a classification of junctions can be performed by analysing the modalities of local intensity and directional histograms during a simulated foveation process. The ba-

Intensity	Edge direction	Classification hypothesis
unimodal	any	noise spike
bimodal	unimodal	edge
bimodal	bimodal	L-junction
trimodal	bimodal	T-junction
trimodal	trimodal	3-junction

Table 11.1. Basic classification scheme for local intensity and directional distributions around a candidate junction point. These specific junction types arise from a simple world model with polygon-like objects, in which shadows, surface markings, and accidental alignments are assumed not to occur. The extension to higher-order configurations, like a shadow-edge crossing a physical edge, is straightforward. (Adapted from Brunnström *et al.* 1990).

sic principle of the method is to accumulate local histograms of the grey-level values and the directional information around candidate junction points, which are assumed to be given by some interest point operator. Then, the numbers of peaks in the histograms can be related to the type of junction according to table 11.1.

The motivation for this scheme is that in the neighbourhood of a point where three edges join, there will generically be three dominant grey-level peaks corresponding the three surfaces that meet at that point. If the point is a 3-junction, an arrow-junction, or a Y-junction, then the edge direction histogram will (generically) contain three major peaks, while two directional peaks can be expected at a T-junction. Similarly, at an L-junction there will in general be two intensity and two directional peaks. Noise spikes and edges must be considered, since interest point operators tend to give false alarms at such points. Situations with more than three peaks in either the intensity or the directional histogram are treated as non-generic, or as corresponding to surface markings or shadows.

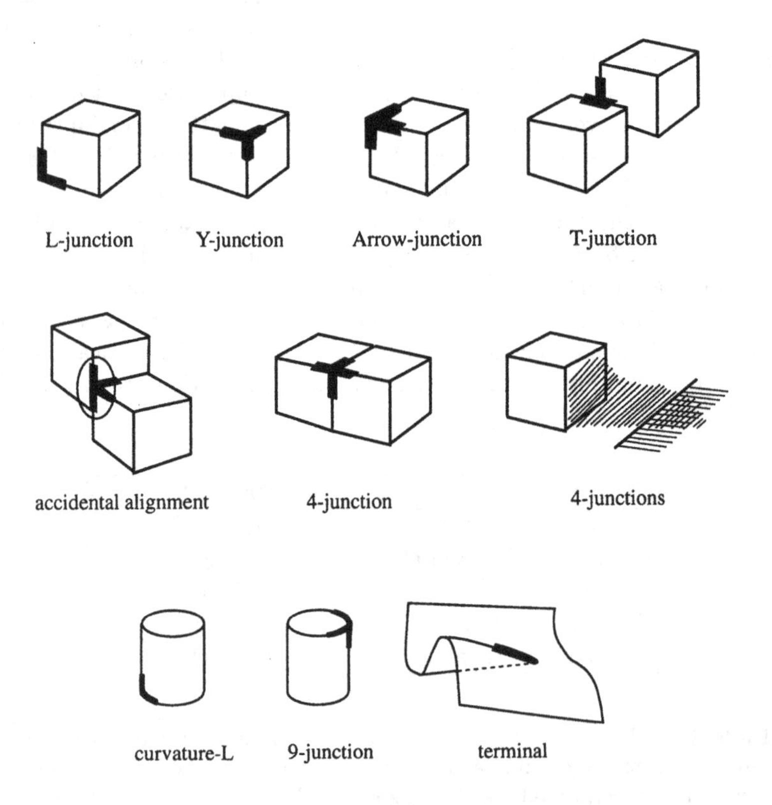

Figure 11.13. Basic junction types under different assumptions about the world; (top row) an ideal block world, (middle row) accidental alignemens, surface markings and shadows, (bottom row) curved surfaces.

Of course, the result from such a histogram analysis cannot be regarded as a final classification, since the spatial information is lost in the histogram. A hypothesis is obtained, which must be verified in some way, e.g. by backprojection into the original data. Therefore, the algorithm is embedded in a classification cycle, see Brunnström *et al.* (1990) for further information.

11.3.3. Context information required for histogram classification

Taking such local histogram properties as the basis for a classification scheme leads to two obvious questions: Where should the window for accumulating the statistics be located, and how large should it be?

The scale-space primal sketch approach can provide valuable clues for both these tasks. Here, it will be used for the following purposes:

- For determining coarse regions of interest constituting hypotheses about the existence of objects, or parts of objects, in the scene, and for selecting scale levels for analysing image structures in those regions further.

- For detecting candidate junction points in curvature data, and for providing information about window sizes for histogram accumulation.

Let us now motivate these choices. To estimate the number of peaks in the histogram, some minimum number of samples will be required. With a precise model for the imaging process as well as the noise characteristics, one could conceive deriving bounds on the resolution, at least in some simple cases. However, as will be developed below, direct setting of a single window size immediately valid for correct classification seems to be a very difficult or even an impossible task.

11.3.4. The scale problem in junction classification

If the window is too large, then other structures than the actual corner region around the point of interest may be included in the window, and the histogram modalities would be affected. Conversely, if it is too small, then the histograms, in particular the directional histogram, could be severely biased and deviate far from the ideal appearance in case the physical corner is slightly rounded—a scale phenomenon that seems to be commonly occurring in realistic scenes. (This effect does not occur for an ideal (sharp) corner, for which the inner scale is zero.) A too small a window might also fall outside the actual corner if the interest point is associated with a localization error. An example illustrating these effects for a rounded corner of a plastic detail is shown in figure 11.14.

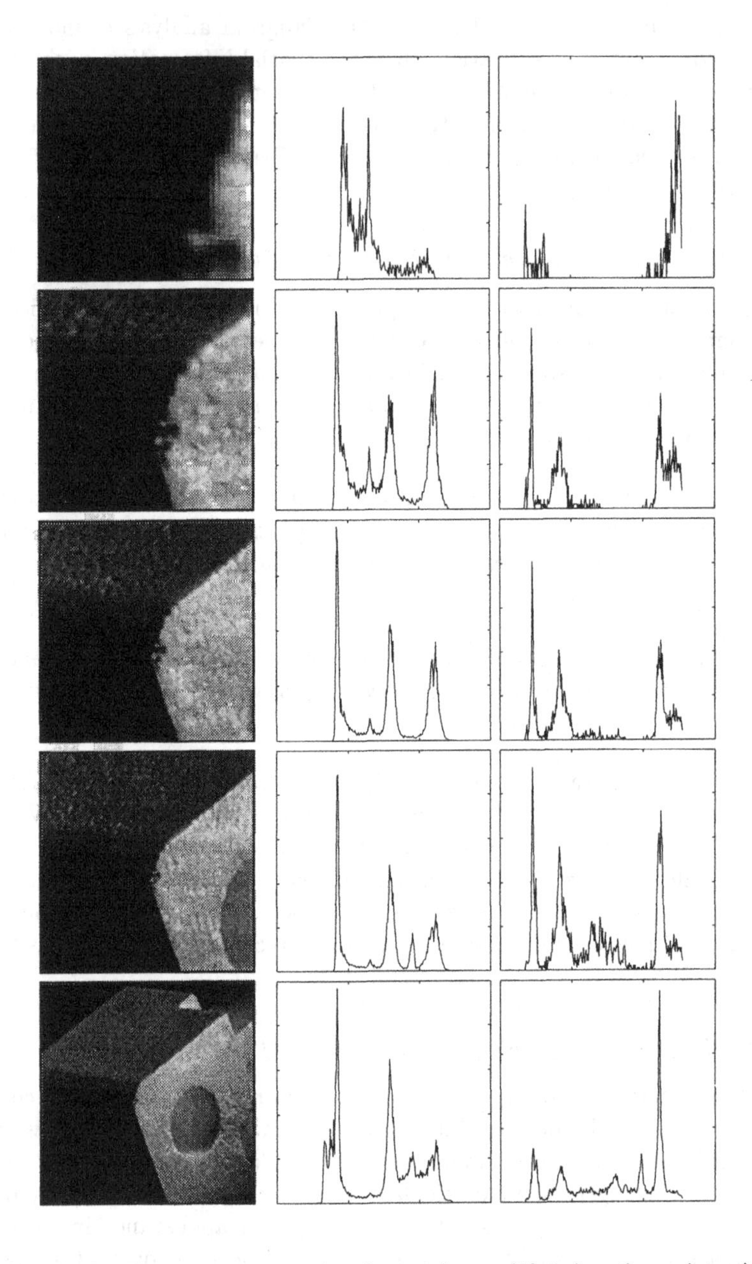

Figure 11.14. Histograms accumulated around a candidate junction point using windows of different size. (left) Selected window. (middle) Intensity histogram (accumulated only for non-edge pixels). (right) Directional histogram (accumulated only for edge pixels). Observe that a correct classification based on histogram modalities can be obtained only over a certain range of window sizes.

11.3.5. Foveation: Acquiring new data with locally increased resolution

Therefore, the methodology that has been adopted is to first determine a coarse estimate of a natural window size around the junction candidate, and then invoke the process of *simulated foveation.*

Foveation means that new image data are acquired in a region around the candidate junction, such that the region of interest is resolved by a sufficient number of samples (here, for accumulating sufficiently dense histograms). With a foveated sensor this can be achieved simply by directing the fovea towards the region of interest. Lacking a foveated sensor, this process is simulated by using the zoom capability of a head-eye system (see the following).

Simulated foveation in turn means that the size of the region of interest (over which the histogram is accumulated) is varied locally in a *continuous* manner (even though the data still has to be sampled at discrete resolutions). In this way, the problem of selecting a proper window size can be circumvented. The method is based on the assumption that stable responses will occur for the models that best fit the data, which closely relates to the systematic parameter variation principle described in section 10.1.1. For each region of interest during this parameter variation, it is investigated whether the window covers a sufficient number of pixels in the image. If the number is below a certain threshold, typically 64×64 pixels, then a new image of higher resolution is acquired by increasing the amount of zooming.

11.3.6. Detecting candidate junctions at coarse scales

Several different types of corner detectors have been proposed in the literature (Kitchen and Rosenfeld 1982; Dreschler and Nagel 1982; Förstner and Gülch 1987; Koenderink and Richards 1988; Noble 1988; Deriche and Giraudon 1990; Florack *et al.* 1992). A problem that has not been very much treated though is that of at what scale(s) the junctions should be detected. Corners are usually regarded as pointwise properties, and are thereby treated as very fine scale features.

In this treatment a somewhat unusual approach will be taken. Corners will be detected at coarse scales by blob detection on curvature data. The motivation for this approach is that realistic corners from man-made environments are usually rounded, which means that small size operators will have problems in detecting those from the original image. Another motivation is to detect the interest points at coarser scales to simplify the detection and matching problems. Now, since corners are usually regarded as pointwise properties, smoothing the image before detecting such points may seem like a contradiction, because of the risk that important interest points disappear by this operation. However, as will be demonstrated below, this approach will be applicable for detecting coarse scale corners corresponding to the rough outline of say a polygon-like object, (provided that the contrast is sufficient between the different regions that meet in the junction).

 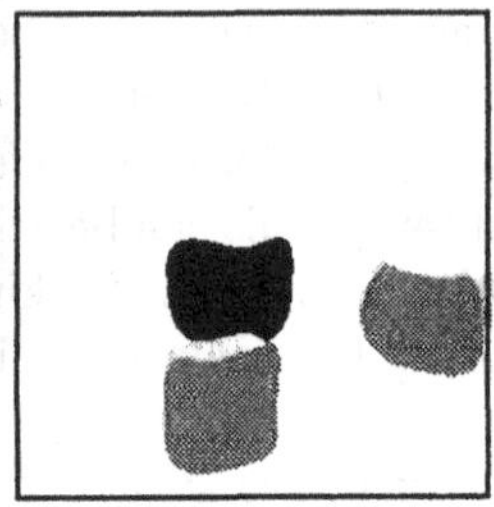 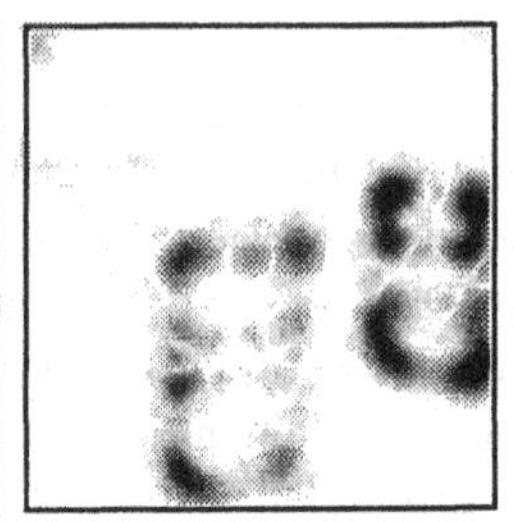

Figure 11.15. Computing the (absolute value of) the rescaled level curve curvature $\tilde{\kappa}$ at a coarse scale given by a significant blob from the scale-space primal sketch; (left) grey-level image. (middle) a significant dark scale-space blob extracted from the scale-space primal sketch (marked with black), (right) the absolute value of the rescaled level curve curvature computed at the scale of the scale-space blob. (Observe that the curvature data is intended to be valid only in a region around the scale-space blob that invoked the analysis.)

Figure 11.15(c) shows the result of computing the absolute value[5] of the rescaled level curve curvature (see section 6.2.2) on the block image at a scale given by a significant blob from the scale-space primal sketch. It can be observed that this operation gives strong response in the neighbourhood of corner points.

11.3.7. Regions of interest: Curvature blobs

The curvature information is, however, still implicit in the data. Simple thresholding on the magnitude of the response will in general not be sufficient for detecting candidate junctions. Therefore, to *extract* interest points from this output, it can be useful to perform blob detection on the curvature information using the scale-space primal sketch. Such blobs are termed *curvature blobs.*

Figure 11.16(b) shows the result of applying this operation to the data in figure 11.15(c). Note that a set of regions is extracted corresponding to the major corners of the block. Also note that the support regions of the blobs serve as natural descriptors for a characteristic size of a region around the candidate junction. Hence, this information is used for setting

[5]By taking the absolute value, positive and negative values are treated in the same way. This step is not necessary and should be omitted in situations where the sign information is important. Here it is used mainly in order for the algorithm to be able to handle all junctions with the same operation, and to simplify the presentation.

(coarse) upper and lower bounds on the range of window sizes for the focusing procedure.

Of course, a matching needs to be performed between the curvature blobs and the original blob that invoked the analysis. This can be done, basically, by testing the blobs for mutual inclusion. Figure 11.16(c) shows those curvature blobs that overlap spatially with the original scale-space blob, and in addition for which the selected scale values are smaller than the selected scale of the scale-space blob.

Figure 11.16. Multi-scale blob detection applied to the curvature data in figure 11.15; (left) raw grey-level blobs detected from the curvature image; note the high noise sensitivity in detecting blobs at a single level of scale, (middle) blob boundaries of the 50 most significant scale-space blobs detected from the same image (detected by applying the scale-space primal sketch to the curvature data), (right) blob boundaries of the blobs matched to the original scale-space blob that invoked the analysis.

A trade-off with this approach is that the estimate of the location of the corner will in general be affected by the smoothing operation. Therefore, it should be emphasized that the main goal of the first step is to *detect* candidate junctions at the possible cost of poor localization. A coarse estimate of the position of the candidate corner is given by the (unique) local maximum associated with the blob. Then, if improved localization is needed, it can be obtained using edge and curvature information at finer scales (see section 13.4).

11.3.8. Experimental methodology: Guiding the process

To summarize, the methodology proposed for classifying junctions can be described as follows:

1. Assume that the camera is positioned in a reference position. Take an overview image of the scene using a wide field of view.

2. Compute the scale-space primal sketch of the overview image for both dark and bright blobs to generate a set of regions of interest.

3. For each scale-space blob in decreasing order of significance:

   a. Take a closer look at the structure under consideration, i.e. acquire a new image in a window around the region of interest.

   b. Compute the (rescaled) level curve curvature at the scale given by the scale-space blob determining the current region of interest (multiplied by the magnification factor in the zoom step).

   c. Compute the scale-space primal sketch of that data. Match the blobs obtained from this curvature image to the previously obtained blob region determining the current region of interest. This generates a new set of smaller regions of interest, which are candidates for containing junction points.

   d. For each such candidate junction region (from now on denoted curvature blob) in decreasing order of significance:

      i. If necessary acquire a new image to guarantee a sufficient number of samples for the histogram accumulation.

      ii. Invoke the classification procedure to that junction. Direct the gaze towards the local maximum associated with the curvature blob. Set the minimum and maximum window size from the size of the support region of the curvature blob (e.g., by multiplication and division to some constant; typically 2 or 3).

      iii. Carry out the focusing procedure, in which local intensity and directional histograms are accumulated for a set of window sizes and their modalities determined.

      iv. Analyse the stability of the interpretations and select a representative window size as an abstraction.

      v. Verify the interpretation. Currently this is done by comparing two different groupings the distributions of the grey-level and directional information in the window selected as representative for the focusing process.[6]

For the purpose of experimentation, this process has been implemented using the KTH head-eye system developed by Pahlavan and Eklundh (1992), which allows for algorithmic control of the image acquisition.

Of course, there are several problem concerning control strategies of the reasoning and image acquisition processes associated with this type of analysis. There are no claims that the methodology used here is "op-

[6]The verification method used in these experiments consists of the following basic steps: Fit straight lines to the edges in the window around the candidate junction selected as representative for the process. From these edges, compute a partition of the window. Then, compare with the backprojections of the grey-level histogram peaks to the original data, and invoke a statistical test (see Brunnström 1993 for details).

timal" in any sense or a final solution. The blobs are currently selected in decreasing order of significance, and new images are accumulated if the number of pixels associated with the window size goes below a predetermined values. The presentation here is rather intended to exemplify how this type of data-driven analysis, combined with local qualitative reasoning and an active camera head, can be used in early vision.

11.3.9. Experiments: Fixation and foveation

Now, some experimental results will be described of applying the suggested methodology to a scene with a set of blocks. An overview of the setup is shown in figure 11.17(a). The blocks are made out of wood with weakly textured surfaces and rounded corners.

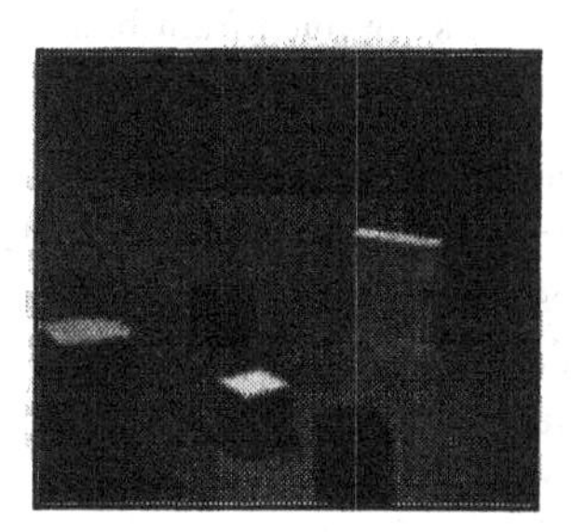

Figure 11.17. Overview image of a scene under study.

Figure 11.18. Boundaries of the 20 most significant dark and bright scale-space blobs; (left) dark blobs, (right) bright blobs.

Figure 11.18 illustrates the result of extracting dark and bright blobs from the overview image using the scale-space primal sketch. The boundaries of the 20 most significant bright and dark scale-space blobs have been displayed. Each such blob constitutes a hypothesis about the existence of an object, a face of an object, or some illumination phenomenon in the scene.

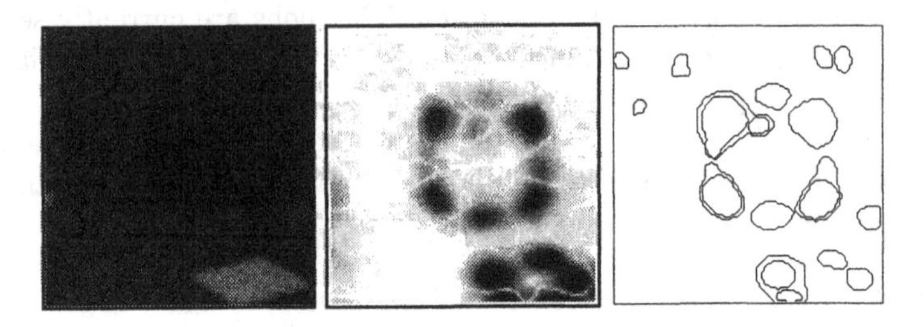

Figure 11.19. Zooming in to a region of interest given by a scale-space blob from the previous processing step; (left) a window around the region of interest, set from the location and the size of the blob, (middle) the rescaled level curve curvature computed at the scale of the blob, (right) the boundaries of the 20 most significant curvature blobs obtained from blob detection.

In figure 11.19 the cameras of the head-eye system have been redirected towards one of the dark blobs corresponding to the central block, and a new image of higher resolution has been acquired around the region of interest. This step simulates foveation. At the scale of the scale-space blob (transformed with respect to the increased sampling density), the level curve curvature is computed, and curvature blobs are detected using the scale-space primal sketch; see figure 11.19(c).

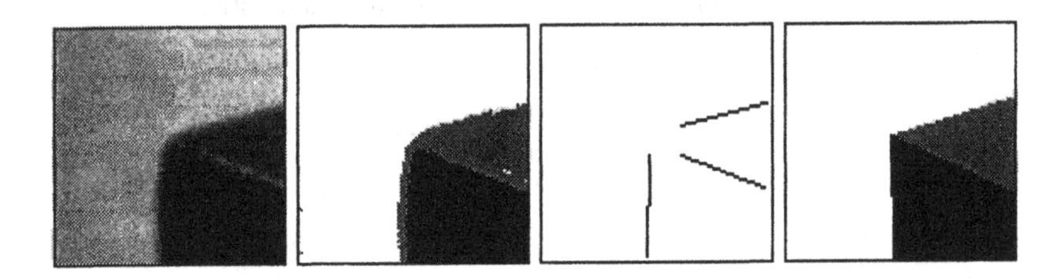

Figure 11.20. Zooming in to a junction candidate given by a curvature blob; (left) maximum window size for the focusing procedure set from the size of the curvature blob, (middle left) backprojected peaks from the intensity histogram, (middle right) lines computed from the backprojected peaks from the directional histogram, (right) schematic illustration of the classification result in which a simple junction model has been adjusted to the data.

In figure 11.20 the algorithm has zoomed in further to one of the curvature blobs, and invoked a histogram classification procedure tuned to the size of the curvature blob. This junction was classified as a 3-junction based on three peaks stable with respect to variations in window size, detected in the grey-level and directional histograms respectively. Figure 11.21 shows corresponding results for an L-junction. As an indication of the ability to suppress "false alarms," figures 11.22–11.23 show the results of applying the classification procedure to a point along an edge and a point in the background.

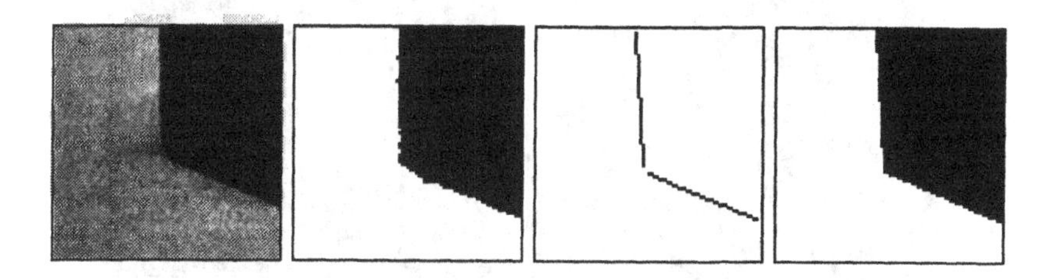

Figure 11.21. Similar classification result for an L-junction.

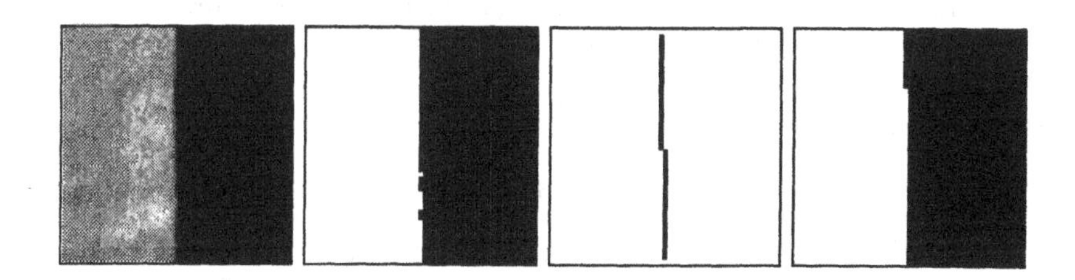

Figure 11.22. Similar classification result at an edge point.

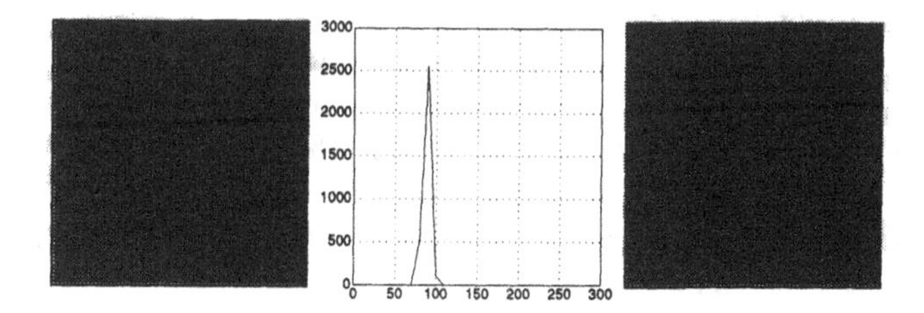

Figure 11.23. Applying the classification procedure to a point in the background. (a) Maximum window size. (b) Grey-level histogram of the maximum window. (c) Backprojected grey-level histogram peak at representative window size. Classification result: Noise spike.

Another example is shown in figure 11.24. Here, the classification results for a number of blobs have been backprojected as labels onto an overview image of the scene.

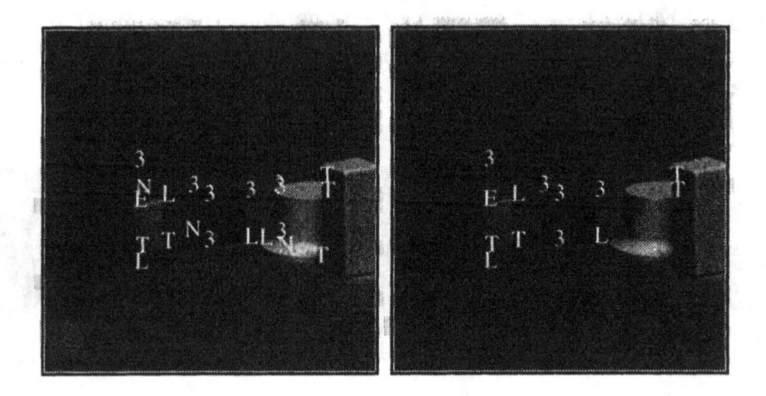

Figure 11.24. Classification result of a whole process backprojected onto an overview image. The left image shows the result after the histogram analysis. There is one label for each input candidate junction. In the right image only the verified junctions have been labelled. (The meaning of the labels are: E = edge-point, L = L-junction, T = T-junction, 3 = Y- or arrow-junction, N = non-generic.) (Adapted from Brunnström 1993.)

11.3.10. Additional cues from active analysis: Accommodation distance

The ability to control gaze and focus does also facilitate further feature classification, since the camera parameters, such as the focal distance and the zoom rate, can be controlled by the algorithm. This can for instance be applied to the task of investigating whether a T-junction in the image is due to a depth discontinuity or a surface marking in the world. This section demonstrates how such a classification task can be solved monocularly, using focus.

In figure 11.25(a)–(b) the system has zoomed in to a curvature blob associated with a scale-space blob corresponding to the bright block. The example demonstrates the effect of varying the focal distance and measuring the variation of a simple measure on image sharpness varies with the focal distance. This measure is the sum of the squares of the gradient magnitudes in a small window.

Two curves are displayed in figure 11.25(c); one with the window positioned at the left part of the approximately horizontal edge, and one with the window positioned at the lower part of the vertical edge. Clearly, the two curves attain their maxima for different accommodation distances. The distance between the peaks gives a measure of the relative depth between the two edges, which in turn can be related to absolute depth values by a calibration of the camera system.

Corresponding results for a T-junction due to surface markings are shown in figures 11.25(d)–(e). Here, the two graphs attain their maxima at approximately the same position, indicating that there is no depth discontinuity at this point.

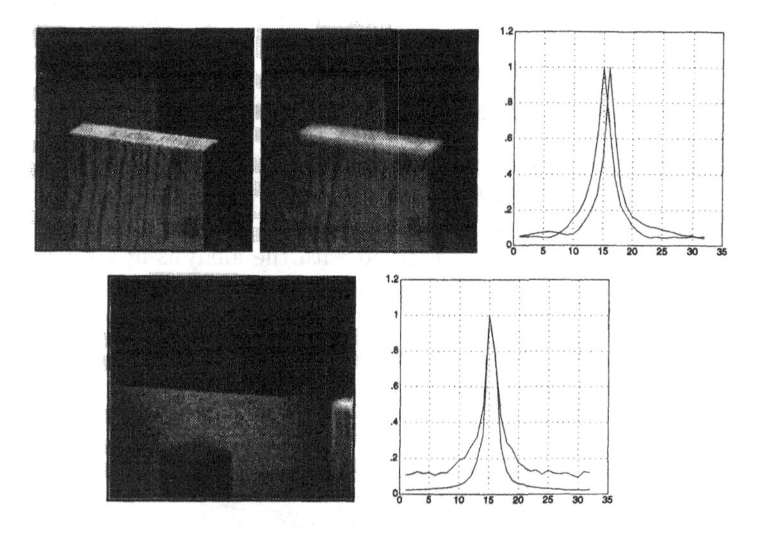

Figure 11.25. Depth from accommodation distance: This example shows the effect of varying the focal distance at two T-junctions corresponding to a depth discontinuity and a surface marking respectively. In the upper-left image the camera was focused on the left part of the approximately horizontal edge, while in the upper-middle image the camera was focused on the lower part of the vertical edge. In both cases the accommodation distance was determined from an auto-focusing procedure (implemented by Horii 1992), which maximizes a simple measure on image sharpness. The graphs on the upper-right display how this measure varies as function of the focal distance. The lower row shows corresponding results for a T-junction due to a surface marking. It can be observed that in the first case the two curves attain their maxima at clearly distinct positions (indicating the presence of a depth discontinuity), while in the second case the two curves attain their maxima at approximately the same position (indicating that the T-junction is due to a surface marking).

11.3.11. Summary and discussion: Junction classification

To summarize, this experiment indicates how the scale-space primal sketch can be used in dynamic situations like focus-of-attention. Such mechanisms are necessary if computer vision systems, are to perform their tasks in a complex and dynamic world. It should be emphasized that the treatment here describes on-going experimental work, and that there is still more work to be done concerning control strategies of the reasoning process. Nevertheless, the presentation illustrates some basic ideas of how the suggested approach can be used in an active vision situation, and specifically, how qualitative scale and region information can be used for guiding a junction detection module by detecting curvature blobs from grey-level data. While the presented classification methods are conceptually simple, a necessary prerequisite for them all is the ability of the visual system to foveate. This includes a mechanism for guiding the focus-of-attention.

11.4. Example: Analysis of aerosol images

As one example of how the scale-space primal sketch can be used for various image analysis tasks, a specific application will be considered in this section, which has arisen from a physical problem.

The treatment will be concerned with the analysis of a certain type of high-speed photographs of aerosols generated by nozzles for fluid atomization. A typical example of such an image is shown in figure 11.26. What one perceives are some kind of clusters in the drop distribution, seemingly periodically spaced in the spread direction of the aerosol.

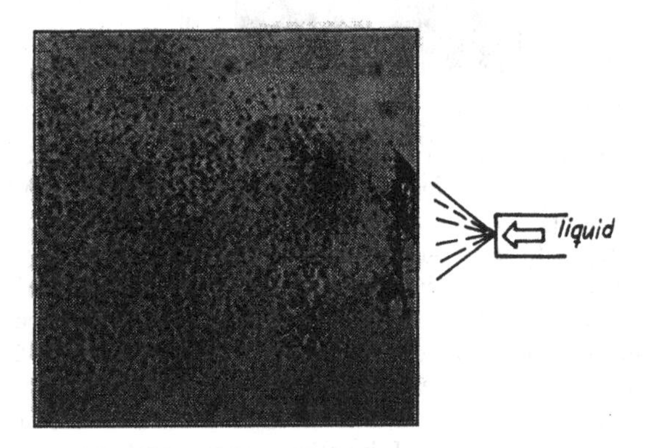

Figure 11.26. Photograph of an aerosol generated by a nozzle for fluid atomization (fuel injector). The time of exposure is approximately 20 nanoseconds.

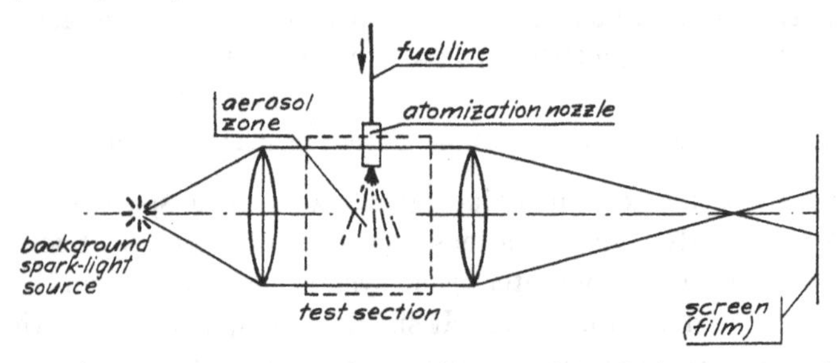

Figure 11.27. Schematic view of a shadow-graph optical system used in the physical experiments for acquiring the aerosol images. The fluid, subjected to a pressure of about 0.1 MPa, enters the nozzle and becomes atomized. The short exposure time is accomplished by performing the experiments in a dark room and illuminating the test section with a short flash.

If these events really exist, then the physical interpretation would be that there are periodical (or oscillatory) phenomena taking place in the fuel atomization process. This is a theoretically interesting question, important for the deeper understanding of the combustion processes in combustors.[7] Usually it is assumed that fuel injectors produce aerosols with a relatively uniform droplet distribution, but the high-speed photograph in figure 11.26 seems to indicate that this is not always the case. One may speculate that the occurrence of such non-uniformities could represent a possible driver for abnormal combustion events, which in turn could result in a deteriorated emission situation possibly affecting the exhaust production and the fuel consumption.

However, it is not easy to say directly that these periodic structures really are there, since they correspond to coarse scale phenomena while the dominating kind of objects in the image is small dark blobs, i.e., fine scale phenomena. Therefore it is of interest to develop objective methods for analysing these structures.

Here it will be demonstrated in a straightforward manner that: these structures can be *enhanced* by a scale-space representation of the image, and they can be *extracted automatically* with the scale-space primal sketch.

11.4.1. Experimental results

It should be stressed that these data are extremely irregular with a very high noise level. Figure 11.28 displays the intensity variations in a cross-section of the image along the spread direction of the aerosol. Therefore, conventional segmentation techniques can be expected to have severe problems when applied to this type of data.

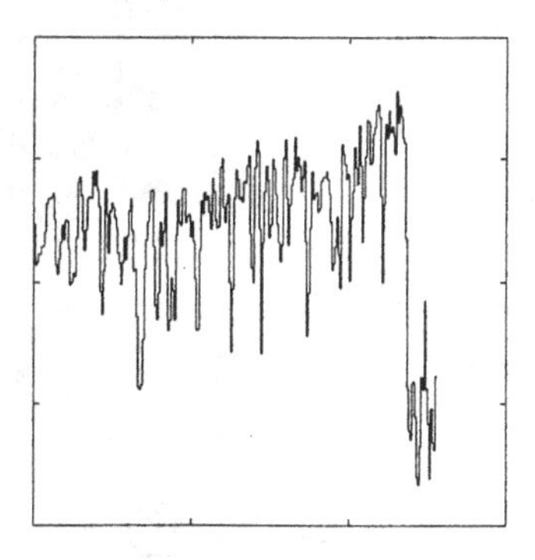

Figure 11.28. Intensity variations in a central cross-section along the spread direction of the aerosol.

[7]Further information about the physical background to the problem can be found in (Lindeberg and Eklundh 1991) and (Valdsoo 1989a,b,c).

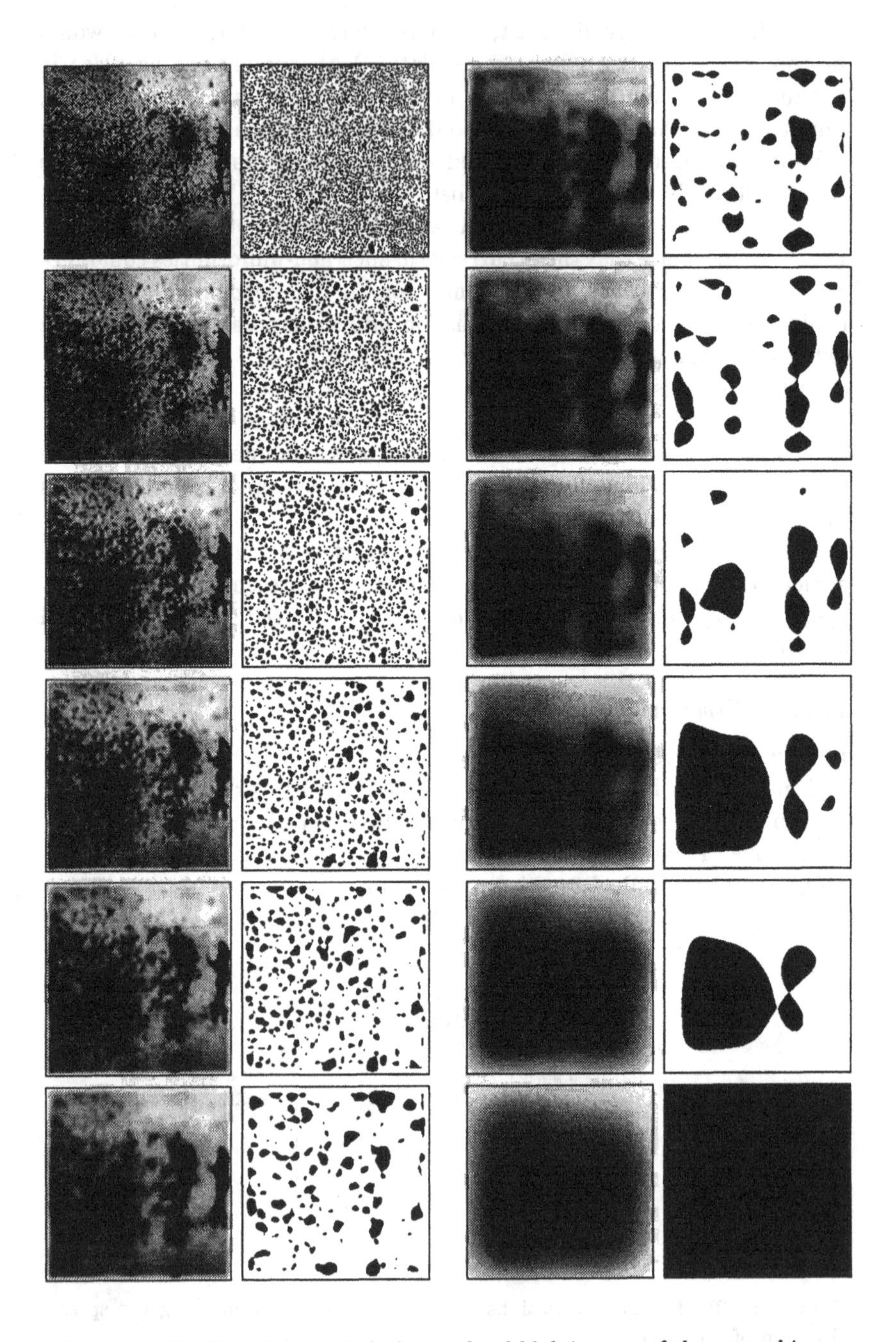

Figure 11.29. Grey-level and dark grey-level blob images of the aerosol image at scale levels $t = 0, 1, 2, 4, 8, 16, 32, 64, 128, 256, 512$, and 1024 (from top left to bottom right).

Figure 11.29 shows the resulting scale-space representations together with the extracted blobs for a set of (logarithmically distributed) scale levels. It can be seen that the drop clusters, which we earlier perceived as periodic structures in the original image, now appear as large dark blobs at the coarser levels of scale (t = 128, 256, 512). Although the scale-space representation enhances these clusters, one still relies on a visual and subjective observer in order to extract and verify the existence of these periodic phenomena. Some natural questions that were raised from the application point of view were the following: (i) Can any one(s) of these smoothed images be regarded as a proper description(s) of the original image? (ii) Which blobs can be regarded as significant structures in the image? Figure 11.30 displays the result of extracting the 50 most significant dark blobs from this image, and figure 11.31 shows the the boundaries of the significant blobs.

It can be observed that the periodically occurring drop clusters we perceived in the image are now detected as significant blob-like structures by the scale-space primal sketch. Since, in contrast to many other methods in image analysis, this method is essentially free from "tuning parameters," and arbitrarily selected error criterions or thresholds, one may argue that the features detected by this algorithm can be regarded as reflecting inherent properties of the image—they are not just enforced effects of the analysis method.

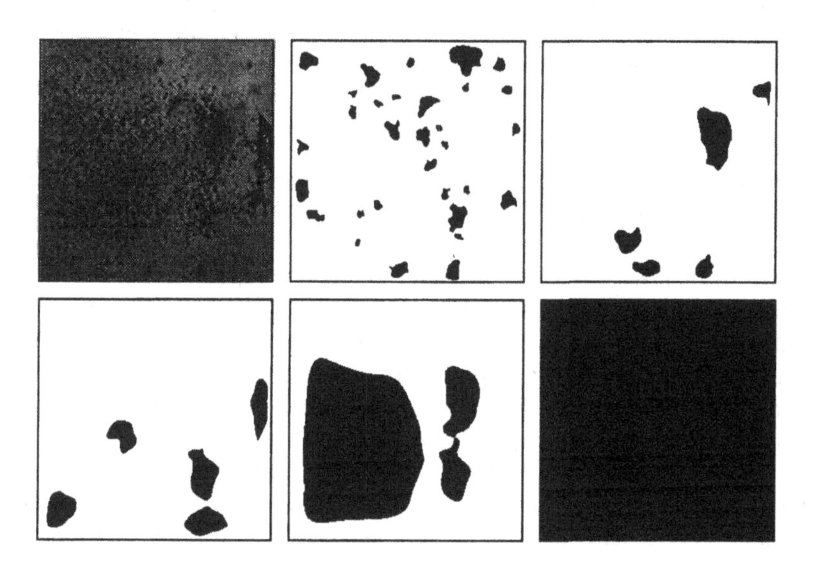

Figure 11.30. Original aerosol image and the 50 most significant dark blobs determined from the scale-space primal sketch.

Figure 11.31. Boundaries of the 50 most significant dark blobs.

11.4.2. Conclusions

The intention of this experiment has been to demonstrate that the scale-space primal sketch concept is a useful tool for automatic extraction of those periodic structures that were brought out by a scale-space representation of the aerosol image. Further experiments with this approach are given in (Lindeberg and Eklundh 1991).

This presentation is mainly intended to demonstrate the potential of using the scale-space primal approach as a primary tool in this kind of image analysis applications. Of course, more work needs to be done to arrive at a fully automated analysis method for this particular problem. Nevertheless, there may be a potential in using the approach also for other kinds of very noisy or irregular data, like medical imagery.

11.5. Other potential applications

Let us finally mention a few other problem areas, where there may also be potential applications of the approach.

Texture analysis. A basic problem in many shape-from-texture algorithms concerns how to detect texture elements (Julesz and Bergen 1983; Julesz 1986; Blostein and Ahuja 1987; Vorhees and Poggiò 1987). Since the scale-space primal sketch does not require any prior scale information, and scale levels can be automatically adapted to size variations in image data, it constitutes a potentially useful tool for such analysis.

Figures 11.32–11.33 show experimental results for one synthetic and two realistic images. Note that in both cases a set of blobs is extracted with a size gradient that can be used as a cue[8] to the three-dimensional structure.

---

[8] A shape-from-texture method using a (simplified) blob detection method of this type is presented in chapter 14.

Figure 11.32. Multi-scale blob detection on a synthetic texture image generated from perspective projection of a planar surface with a sinusoidal grey-level pattern; (left) grey-level image with added white Gaussian noise with standard deviation 10 % of grey-level range, (middle) the 75 most significant dark blobs, (right) the 75 most significant bright blobs.

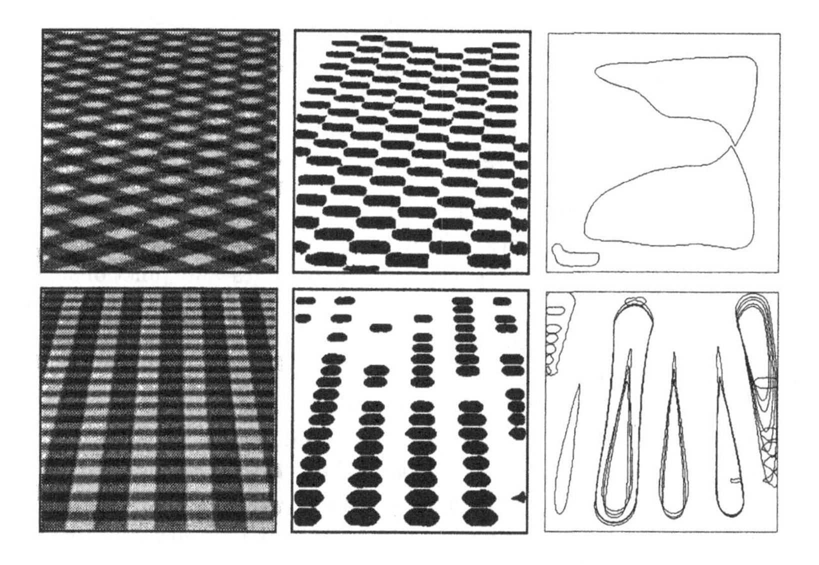

Figure 11.33. Multi-scale blob detection applied to two different views of a real-world surface texture; (left) grey-level image, (middle and right) the 100 most significant dark blobs (marked either as blob regions or boundaries).

11.5.1. Perceptual grouping

In the presented experiments, we have seen that the blobs extracted from the scale-space primal sketch often induce intuitively reasonable groupings of various patterns. For example, in figure 11.33(a) only the individual squares were ranked as important, while in figures 11.33(d)–(f) vertical stripes were also found. See also the dot pattern example in figure 11.34. Note that the grouping process is not given by any set of pre-specified logical rules, but by a differential equation combined with a set of geometric constructions.

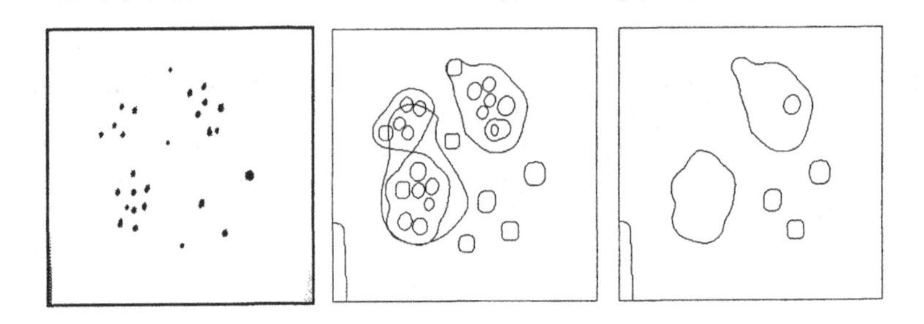

Figure 11.34. Multi-scale blob detection on a dot pattern image; (left) original grey-level image, (middle) low threshold on the significance measure, (right) high threshold on the significance measure. Note that all dots are detected and that a number of intuitive groupings are performed.

11.5.2. Object detection and matching

The blobs delivered from the scale-space primal sketch can serve as coarse landmarks for different types of matching purposes. The relations given by, say, matches between a blob and a set of edges and junctions, provides a sparse set of features, which could be used for restricting the search space and hence the combinatorial explosion in model matching. Another possible application is to use the blobs for initiating object models, like deformable models (Kass *et al.* 1987; Terzopoulos *et al.* 1988; Pentland 1990), or geon primitives (Biedermann 1985; Dickinson *et al.* 1990). Experimental work indicates that the approach may be useful for establishing coarse correspondences in sequence data; see also (Koller *et al.* 1992). An attractive property of blobs is that they are conceptually easy to match over time based on spatial overlap.

12

Summary and discussion

The proposed representation is similar to the primal sketch idea suggested by Marr (1976, 1982), in the sense that it is a two-dimensional representation of the significant grey-level structures in the image. It is also computed under extremely weak assumptions. In addition, it is a region-based, not an edge-based representation, and it is more *qualitative*, without strong assumptions about the shapes of the primitives. It consists of blobs (extremum regions) at multiple scales in scale-space, and allows for

- automatic detection of salient (stable) scales, if they exist,
- ranking of blob-like structures in order of significance, and
- generation of hypotheses for grouping and segmentation.

This implies that candidate regions are generated for further processing, as well as information about the scale. We have seen that the representation gives clues to subsequent analysis, and that it can guide focus-of-attention mechanisms. At the same time it is obtained with no *a priori* assumptions, and in principle without tuning parameters. The only free parameter is the number of blobs to be selected for further analysis (or display).

Extraction of structure—transformational invariance. The underlying principle used for extracting blob structures is that structure should be invariant under transformations in parameter space: The suggested method consists of three steps;

1. vary the parameters systematically,
2. detect locally stable states (intervals),
3. choose a representative descriptor as an abstraction of each stable interval, and pass only this information on to the higher level modules.

In this specific case, the parameter that is varied is the scale parameter in the scale-space representation, and the significance measure is defined in terms of a four-dimensional volume in scale-space. The methodology can, however, be applied also in other types of situations. One example, concerning junction classification, is described in section 11.3.

12.1. Scale-space experiences

Let us point out a few aspects of scale-space representation that have been given little or insufficient attention in the literature, but have to be dealt with when building a representation of the type proposed.

12.1.1. Suppression of local extrema due to noise

First, it is noteworthy that the amount of noise in real images usually leads to a large number of local extrema. These extrema may disappear rather early if they are subsumed by some more prominent extremum. However, if located regions with smoothly varying grey-level, they will exist over a large range of scale. This effect is alleviated but not remedied by annihilation between nearby noise extrema. Even though the amplitude can be expected to decrease rapidly, it is not clear that a globally valid threshold can be set on objective grounds. This problem is related to the issue of estimating the noise level in an image, which hardly can be addressed without some constraining assumptions, such as in (Voorhees and Poggio 1987).

12.1.2. Stable scale is a local property

Another property, indicated in section 7.6, is that images of scenes of even moderate complexity rarely have a global scale, at which all structure above the noise level is present. Stable scales are local properties associated with objects, not with entire images. This aspect is explicitly dealt with in the representation.

Figure 12.1 shows an unusual situation, where one could possibly say that there is a globally stable scale for the entire image. In a graph showing the logarithm of the number of local extrema as a function of effective scale, this property manifests itself as a plateau. For realistic images of moderate complexity it will, however, usually not be possible to find such globally stable states. Even if there were a number of prominent plateaus corresponding to regionally stable structures at different scales, by adding a large number of such profiles one would anyway obtain a relatively smoothly decreasing curve. Hence, in general, such plateaus can only be expected regionally or if all structures in the image have the same size.

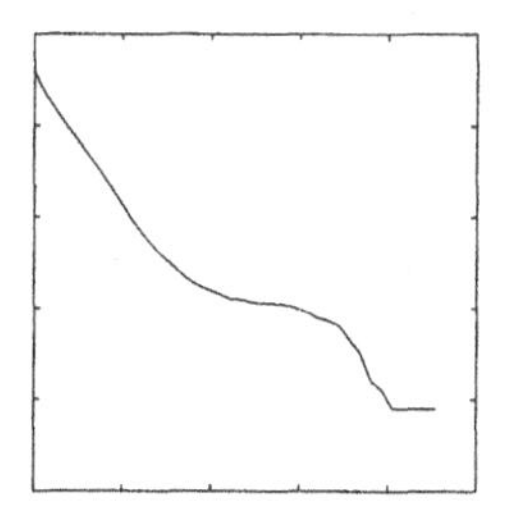
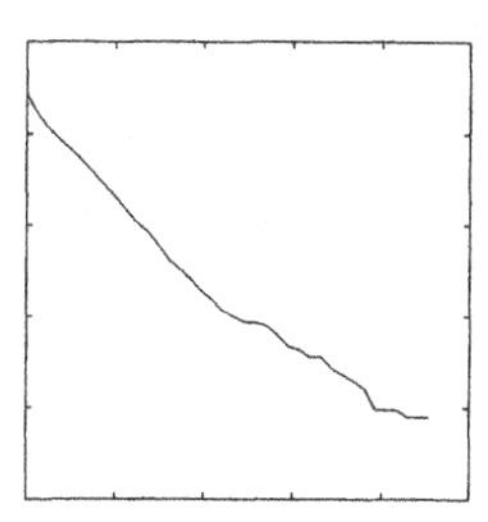

Figure 12.1. (a) An unusual situation, where one could possibly talk about a globally stable scale for an entire image. (b) This property manifests itself as a plateau in a graph showing the logarithm of the number of local extrema as function of (effective) scale. For realistic images of moderate complexity it will, however, usually not be possible to find such globally stable states. (c) Corresponding results for the Godthem Inn image shown in figure 10.18.

12.1.3. Stable scale is a multi-valued function

Moreover, given some region in space, there may be several stable scales associated with that region, corresponding to structures at different scales. Therefore, the task of finding a best scale for treating a certain *point* is an impossible problem (except in very simple images in which only one such stable scale is associated with each point in the image).

In this context it should be remarked that the scale value given by assumption 10.2 does not necessarily reflect the size of the corresponding blob region in the image. Although, in general, large values of the scale parameter can be expected to correspond to large scale structures, there is no *direct* relationship. Under certain conditions (typically when there are no superimposed finer scale structures) a large scale structure may, in fact, be assigned a small scale value. Therefore, the scale value obtained from assumption 10.2 should rather be interpreted as an abstract scale parameter, *reflecting the smallest amount of smoothing for which the blob manifests itself as a single blob entity.*

12.1.4. Decreasing amplitude of feature points

The behaviour of local extrema in scale-space has been studied also by Lifshitz and Pizer (1990). They link points across scales based on *iso-intensity,* using integral paths of the vector field

$$(L_{x_1}L_t, L_{x_2}L_t, -(L_{x_1}^2 + L_{x_2}^2)), \tag{12.1}$$

and construct a "stack" representation, in which the grey-level at which an extremum disappears is used for defining a region in the original image by local thresholding at that grey-level. The representation is demonstrated to be applicable for certain segmentation problems in medical image analysis.

However, Lifshitz and Pizer observe the serious problem of *non-containment.* It essentially means that a point, which at one scale has been classified as belonging to a certain region (associated with a local maximum), can escape from the region when the scale parameter increases.[1]. Moreover, such paths can be nested in quite a complicated way.

The main cause to problem in the iso-intensity linking is that grey-levels, corresponding to features tracked over scales, will *change*[2] under scale-space smoothing. For example, concerning a local extremum, it is a necessary consequence of the diffusion equation that the grey-level at a maximum point must decrease with scale. This problem is avoided in the scale-space primal sketch, in which the linking is explicitly based on qualitative feature points (here, local extrema). The natural generalization to other aspects of image structures is in terms of the differential singularities treated in section 6.1.4 and section 8.6.

12.2. Relations to previous work

Of course, there are also earlier attempts to derive similar representations of the grey-level landscape. Rosenfeld and his co-workers (Gross 1986; Sher and Rosenfeld 1987) have studied blob detection in pyramids, using relaxation methods. Blostein and Ahuja (1989) detect texture elements based on zero-crossings at multiple scales and a significance measure based on a background noise assumption. There is also a wealth of literature on pyramids, see (Levine 1980; Burt 1981; Burt and Adelson 1983; Crowley and Parker 1984; Crowley and Sanderson 1987). The texton theory (Julesz and Bergen 1983; Julesz 1986; Voorhees and Poggio 1987) essentially also treats the blob detection problem. There are finally a number of representations based on intensity changes (Marr 1982; Bergholm 1987; Watt 1988; Baker 1988), and approaches working at higher levels, such as the token based symbolic grouping by Saund (1990). Of interest is also the approach by Haralick *et al.* (1983), which allows a more detailed representation but only at a single spatial scale.

The suggested approach differs from these in three important aspects.

[1] More precisely, what can happen is the following: Assume that a point A is contained in a region associated with an extremum B at a certain scale, and follow these points by iso-intensity linking to corresponding point A' and B' at a coarser scale. Then, we are not guaranteed that A' is contained in the same region as B'. Even worse, these paths can be intertwined in a rather complicated way, which means that the relations between extremum regions across scales can hardly be regarded as hierarchical.

[2] A similar problem arises in the motion constraint equation in optical flow, where it is usually assumed that the intensity value of a physical point is preserved under motion. However, as Pentland (1991) has demonstrated, the photometric distortions can under certain conditions be much larger than the geometric effects due to motion.

First, it can be seen as preceding the edge-based schemes in that it selects the appropriate scales and regions, intrinsically defined by the image itself, in a complementary data-driven manner. Second, it is a hierarchical representation of image structures at *all scales* with explicit information about their significance and relations, and a competition between parts at different locations and scales. Finally, it is derived in a formal way using the well-defined notion of scale-space, which allows a precise study of events at different scales.

12.3. Grey-level blobs vs. Laplacian sign blobs

One can ask more generally: What is the relation between the suggested representation and the zero-crossings of the Laplacian? Given a function $f\colon \mathbb{R}^2 \to \mathbb{R}$, define a bright (dark) *Laplacian sign blob* as a connected region satisfying $\nabla L < 0$ (> 0). Since at any local maximum (minimum) it holds that $\nabla L < 0$ (> 0), it follows that to every grey-level blob there is a unique Laplacian sign blob of the same polarity. However, the reverse relation is not valid; given a bright (dark) Laplacian sign blob, there may be one maximum (minimum), several maxima (minima), or even no maximum (minimum) in that region.

In fact, a representation similar to the grey-level blob tree at a single scale has been studied independently by Blom (1992), who considers the nesting structure of level curves through critical points. The major differences are that Blom considers degenerate (non-Morse) critical point, and that he points out that a hexagonal discrete grid has certain theoretical advantages. In this work, grey-level volumes are associated with the different arcs of the nesting tree, and the representation is embedded in scale-space.

12.4. Invariance properties

Since the scale-space primal sketch is defined solely in terms of topological properties as local extrema, level curves through saddle points, and bifurcations between critical points, it obeys a number of natural invariance properties. Invariance with respect to *translations* and *rotations* of the spatial domain is trivial. Further, given a certain scale level, the topological relations of the grey-level blob tree are preserved under arbitrary monotone intensity transformations. Under evolution in scale-space, the invariance of the hierarchical relations is restricted to *affine intensity transformations*. Such transformations also leave the relative ranking of blobs on significance unaffected. Trivially, under *uniform rescalings* of the spatial coordinates, $x \mapsto sx$ ($s \in \mathbb{R}^+$), a singularity at a point $(x_0;\ t_0)$ in the scale-space representation of the original signal is transferred to a new point $(sx_0;\ s^2t_0)$. This means that the hierarchical relations are preserved, and the appearance and disappearance scales of the scale-space blobs are multiplied by constant factors. Concerning the ranking on significance, it is clear that the logarithmic measure τ_{eff} is invariant to uniform rescalings, and hence the scale-space lifetime. The intention with the transformation function V_{trans} is that also the integrand should be well-behaved[3] under this operation.

12.5. Alternative approaches and further work

Let us finally mention a few issues that are subject to future work:

12.5.1. Normalization

In the current implementation the normalization of the scale parameter and the grey-level blob volume has been on white noise data. The reason is that it constitutes a conservative choice, and makes theoretical analysis simple. If statistics is accumulated of how blobs in such data can be expected to behave over scales, then the result is an estimate of to how large extent accidental groupings take place in scale-space. By experiments, this normalization based on white noise images has been demonstrated to give reasonable results. Moreover, concerning scale-space lifetime, it has been theoretically shown, that for continuous signals such reference data gives rise to the same transformation function ($\tau \sim \log t$) as other self-similar distributions (see section 7.7.1). This selection should, however, not be interpreted as excluding that other approaches, which are equivalent in the continuous case, may lead to different results for discrete signals. For example, some interesting alternatives to consider would be (i) to let the vision system accumulate statistics for a large (representative) selection of different types of realistic imagery, or, (ii) if possible, consider some discrete analogue of coloured noise with a (scale invariant) Fourier spectrum of the form $|\omega|^{-N}$, where N denotes the dimension.

A possible problem with the subtraction of the mean grey-level blob volume (7.21) is that it makes the normalized grey-level blob volume sensitive to the actual scaling of the data. Therefore, to reduce this sensitivity, the tabulated values are rescaled linearly from a least squares fit between the real and the tabulated values. A possible way to avoid this problem, and also to avoid the heuristically chosen transformation function in (7.22), is by redefining the normalized grey-level blob volume as

$$V_{eff} = \frac{G_{vol}}{V_m(t)}, \tag{12.2}$$

[3]If a perfectly scale invariant reference signal could be determined, a scale invariant normalization would be trivially obtained. This is, however, very hard to accomplish on a discrete grid, which has a certain preferred scale given by the distance between grid points.

and then taking as normalized significance values

$$S_{eff} = \frac{S_{vol} - S_m(t)}{S_\sigma(t)}, \tag{12.3}$$

where S_m and S_σ denote mean values and standard deviations of scale-space blob volumes computed from reference data. This method has not yet been implemented, mainly because the simulation work for building the normalization tables is much larger.

12.5.2. Multiple blob instances

The scale-space primal sketch leads to separate systems of bright and dark scale-space blobs. Moreover, a spatial region may give rise to multiple blob responses; typically as the result of a large blob merging with a smaller blob and forming a new large scale-space blob. An obvious problem concerns how to integrate blobs of different polarity and at different scales. In general, it is argued that this problem can hardly be addressed in isolation, but has to be related to a visual task. Below, some basic properties are listed, which could be used by a reasoning system operating on this type of data.

Given a fixed level of scale, the problem of integrating bright and dark blobs can be approached by considering the grey-level blob tree, which constitutes the natural link between grey-level blobs of reverse polarity; see also Blom (1992). Concerning the behaviour over scales, it it clear that a tree describing bifurcations between scale-space blobs will be strongly coupled to the grey-level blob tree. For example, for the simple noise-free pattern in figure 7.1, a tree describing the bifurcations between scale-space blobs can be expected to be identical to the grey-level blob tree of the original signal. In the presence of noise, however, the hierarchical relations will be different. More generally, blob splits and blob creations are blob events without correspondences in the grey-level blob tree.

Other natural descriptors to define between different blobs are (i) whether two blobs overlap, and (ii) whether one blob is completely contained in another one. In this way, the problem with multiple responses from a single region may be approached. This is, however, a subject for further analysis and experimentation.

12.6. Conclusions

A multi-scale representation of grey-level image structure has been presented similar to the primal sketch idea. It can be used for extraction of important blob-like regions from an image in a solely bottom-up data-driven way, without any *a priori* assumptions about the shape of the primitives. The representation, which is essentially free from tuning parameters and ad hoc error criteria, gives a qualitative description of the grey-level landscape with information about

- approximate location,
- spatial extent, and
- an appropriate scale

for relevant regions in the image. In other words, it generates coarse but safe segmentation cues, and can serve as a hypothesis generator for higher-level processes. It has been demonstrated how such information can serve as a guide to an edge detection scheme working at a locally adapted level of scale and that it is applicable for automatic cluster detection, modality analysis of histograms, as well as junction detection and junction classification. More generally, the approach provides a mechanism for

- guiding the focus-of-attention, and
- tuning other low-level processes.

The representation is based on a well-defined notion of blob, which gives a natural geometric measure of significance. It is also based on scale-space theory, which means a well-founded treatment of structures at different scales.

The principle that is used for extracting significant image structure from scale-space is based on transformational invariance, and consists of the following steps:

- Varying the parameters systematically and trying to detect locally stable states (intervals).
- Choosing a representative descriptor as an abstraction of each stable interval, and passing only this information on to the higher level modules.

In this specific case the parameter we vary is the scale parameter in the scale-space representation. The methodology, however, can be applied also in other types of situations. A number of natural generalizations of this approach will be presented in next chapters.

Part IV

Scale selection and shape computation in a visual front-end

13

Scale selection for differential operators

The scale problem is common to most tasks involving analysis of image data—in general situations it is hardly ever possible to know in advance at what scales interesting structures can be expected to appear. Size variations of image structures can occur for several reasons, e.g. because the world contains objects of different physical size, because surface textures contain structures at different scales, and because of perspective effects and noise in the image formation process. A proper representation of scale is essential to most visual tasks requiring stable descriptors of image structure. In certain problems, such as shape-from-texture, scale variations in an image also constitute a primary cue in its own right.

The scale-space representation provides a methodology for handling such size or scale variations in images. It is based on the idea of generating a one-parameter family of gradually smoothed images, in which the fine scale details are successively suppressed. As we have seen in the first part of this book, there are several mathematical results indicating that within the class of linear transformations, the scale-space representation given by convolution with Gaussian kernels of increasing width provides a canonical way to formulate a multi-scale representation for handling image structures at different scales. One main result is that if one assumes that the first stages of visual processing, the *visual front end,* are to perform linear operations and be invariant to translations, rotations, and rescalings in space, then it can be shown that the Gaussian kernels and their associated smoothed derivatives

$$L_{x^\alpha}(\cdot;\ t) = \partial_{x^\alpha}(g * f) = (\partial_{x^\alpha} g) * f = g * (\partial_{x^\alpha} f), \qquad (13.1)$$

at various scales arise by necessity. By combining the output from such *Gaussian derivative* operators at any specific scale, smoothed differential descriptors can be defined at that scale. Defining such descriptors at *all* scales gives a multi-scale differential geometric representation of the signal; a type of representation that is useful for a large number of early visual tasks.

Although this (traditional) scale-space theory ensures that all spatial points and all scales are treated equally and consistently, it does not directly address the problem of *selecting* appropriate scales and structures from the scale-space representation for further analysis. Early work in this direction was performed by Bischof and Caelli (1988) concerning the zero-crossings of the Laplacian operator. Another approach was developed in the second and third parts of this book, where blob-like image structures were considered at different scales in scale-space, and a multi-scale tree-like representation, called the scale-space primal sketch, was constructed. A measure of significance was postulated as the volume that certain primitives of the representation, called scale-space blobs, occupy in scale-space. The scale levels in turn were determined from the scales where the scale-space blobs assumed their maximum (normalized) blob response over scales. Experimentally, the approach was demonstrated to be useful for extracting regions of interest with associated scale levels, which in turn could serve as to guide various early visual processes.

In this chapter a related parameter variation principle for finding interesting scales will be described—the evolution properties over scales of different non-linear combinations of *normalized scale-invariant derivatives* of low order computed from the scale-space representation. A heuristic principle will be presented stating that *local extrema over scales* of different combinations of *normalized scale invariant derivatives* are likely candidates to correspond to interesting structures. By theoretical considerations and numerical experiments it will be demonstrated that the proposed methodology gives useful and intuitively reasonable results in different types of situations. An attractive property of the resulting methodology is that it lends itself naturally to *two-stage algorithms* with feature detection at coarse scales followed by feature localization at finer scales.

13.1. Basic idea for scale selection

A well-known property of the scale-space representation is that the amplitude of spatial derivatives

$$L_{x^\alpha}(\cdot;\ t) = \partial_{x^\alpha} L(\cdot;\ t) = \partial_{x_1^{\alpha_1}} \ldots \partial_{x_N^{\alpha_N}} L(\cdot;\ t) \tag{13.2}$$

in general *decrease with scale,* i.e., if a signal is subjected to scale-space smoothing, then the numerical values of spatial derivatives computed from the smoothed data can be expected to decrease.[1] As an example of this, consider a sinusoidal input signal of some given frequency ω_0; for simplicity in one dimension,

$$f(x) = \sin \omega_0 x. \tag{13.3}$$

[1]This is a direct consequence of the maximum principle, which states that the value at a local maximum will not increase, and the value at a local minimum will not decrease, which means that the amplitude of the variations in a signal always decreases with scale.

It is straightforward to show that in this case the solution of the diffusion equation is given by

$$L(x;\ t) = e^{-\omega_0^2 t/2} \sin \omega_0 x. \tag{13.4}$$

Hence, the amplitude of the scale-space representation, L_{max}, as well as the amplitude of the mth order smoothed derivative, $L_{x^m,max}$, decrease exponentially with scale

$$L_{max}(t) = e^{-\omega_0^2 t/2}, \quad L_{x^m,max}(t) = \omega_0^m \, e^{-\omega_0^2 t/2}. \tag{13.5}$$

An alternative formulation of the scale-space concept is in terms of *normalized* (dimensionless) *coordinates* (section 2.5.4),

$$\xi = \frac{x}{\sigma} = \frac{x}{\sqrt{t}}. \tag{13.6}$$

One motivation for introducing such coordinates is *scale invariance* (Florack *et al.* 1992). In these coordinates the *normalized* (dimensionless) *derivative operator* is

$$\partial_\xi = \sqrt{t}\, \partial_x. \tag{13.7}$$

For the sinusoidal signal the amplitude of a normalized derivative as function of scale is given by

$$L_{\xi^m,max}(t) = t^{m/2} \, \omega_0^m \, e^{-\omega_0^2 t/2}, \tag{13.8}$$

i.e., it first increases and then decreases. It assumes a unique maximum at

$$t_{max,L_{\xi^m}} = \frac{m}{\omega_0^2}. \tag{13.9}$$

Introducing $\lambda_0 = 2\pi/\omega_0$ shows that the σ-value ($\sigma = \sqrt{t}$) for which $L_{\xi^m,max}(t)$ assumes its maximum over scales is *proportional* to the wavelength, λ_0, of the signal:

$$\sigma_{max,L_{\xi^m}} = \frac{\sqrt{m}}{2\pi}\lambda_0. \tag{13.10}$$

Note that the maximum value

$$L_{\xi^m,max}(t_{max,L_{\xi^m}}) = m^{m/2} \, e^{-m/2} \tag{13.11}$$

is *independent* of the frequency of the signal (see figure 13.1). Note also the symmetry in the situation, i.e., given any scale t_0, the maximally amplified frequency is given by $\omega_{max} = \sqrt{m/t_0}$, and for any ω_0 the scale with maximum amplification is $t_{max} = m/\omega_0^2$.

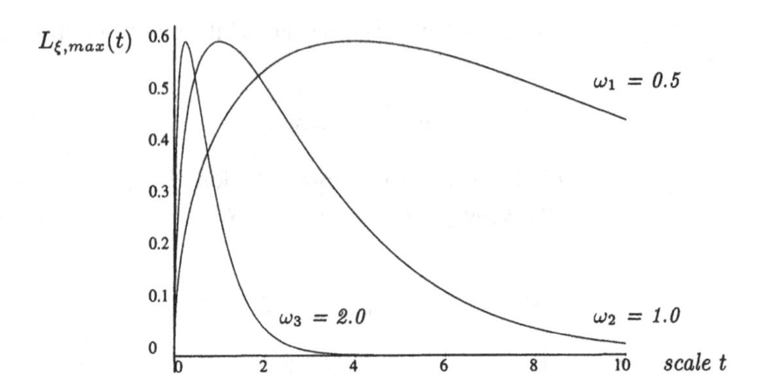

Figure 13.1. The amplitude of first order normalized derivatives as function of scale for sinusoidal input signals of different frequency ($\omega_1 = 0.5$, $\omega_2 = 1.0$ and $\omega_3 = 2.0$).

In other words, for these normalized derivatives it holds that sinusoidal signals are treated in a similar (scale invariant) way independent of their frequency (see figure 13.1).

Note that although there is an intuitive similarity between this scale response and a local Fourier transform, there are two fundamental differences; (i) the normalization factor, and (ii) this method allows for local estimates of the frequency content without any explicit setting of window size (for computing the Fourier transform).

13.2. Proposed method for scale selection

As shown above, the scale at which a normalized derivative assumes its maximum over scales is in the case of sinusoidal signals proportional to the wavelength of the signal. In this respect maxima over scales reflect the scales at which spatial variations take place in the signal. Now, I propose to generalize this observation to more complex signals, leading to a heuristic principle for scale selection, which is to be applied in situations when no other information is available. In its most general form it can be stated as follows:

> *In absence of other evidence, assume that a scale level, at which some (possibly non-linear) combination of normalized derivatives assumes a local maximum over scales, can be treated as reflecting a characteristic length of a corresponding structure in the data.*

This principle is similar although not equivalent to the method for scale selection in chapter 10, where interesting scale levels were determined from maxima over scales of a normalized blob measure. It can be theoretically justified under a number of different assumptions, and for a number of specific brightness models (see below). Its general usefulness, however, must be verified empirically, and with respect to the type of problem the principle is applied to.

13.2.1. Scaling property of maxima over scales

A basic justification for this assumption can be obtained from the fact that for a large class of (possibly non-linear) differential expressions it holds that maxima over scales have a nice behaviour under spatial rescalings of the intensity pattern. If the input image is rescaled by a constant scaling factor s, then the scale at which the maximum is assumed (measured in units of $\sigma = \sqrt{t}$) will be multiplied by the same factor.

To give a formal expression of the scaling property, let us consider the class of homogeneous polynomial differential invariants that was used for defining differential singularities in chapter 6. Given any signal $f: \mathbb{R}^2 \to \mathbb{R}$ with scale-space representation $L: \mathbb{R}^2 \times \mathbb{R}_+ \to \mathbb{R}$, consider polynomials of scale-space derivatives of the form

$$\mathcal{D}L = \sum_{i=1}^{I} c_i \prod_{j=1}^{J} L_{\bar{u}^{\alpha_{ij}}}, \tag{13.12}$$

where the multi-index notation $L_{\bar{u}^{\alpha_{ij}}} = L_{\bar{u}^{m_{ij}}\bar{v}^{n_{ij}}}$ denotes a mixed directional derivative of order $|\alpha_{ij}| = m_{ij} + n_{ij} > 0$ taken in the gradient direction $\bar{v}$ and the perpendicular direction $\bar{u}$ (see section 6.1.1). The differential expression is required to be homogeneous, i.e. the sum of the orders of differentiation must be the same for all terms $\prod_{j=1}^{J} L_{\bar{u}^{\alpha_{ij}}}$ such that for all $i \in [1..I]$ it holds that

$$\sum_{j=1}^{J} |\alpha_{ij}| = N. \tag{13.13}$$

Now, assume that a local maximum over scales in the corresponding normalized differential entity

$$\mathcal{D}_{norm}L = t^{N/2}\,\mathcal{D}L \tag{13.14}$$

is assumed at scale t_0 (and at the point x_0). Then, under a rescaling of the original signal

$$f'(sx) = f(x), \tag{13.15}$$

a maximum over scales in the corresponding normalized entity $\mathcal{D}_{norm}L'$ is assumed at scale

$$t_0' = s^2 t_0 \tag{13.16}$$

(and at the point $x_0' = sx_0$). This result can be easily verified by relating the scale-space representations of f and f' by a change of variables. Given the relation $f'(sx) = f(x)$, introduce

$$x' = sx, \quad t' = s^2 t, \tag{13.17}$$

and let the scale-space representations of f and f' be defined by

$$L(\cdot;\ t) = g(\cdot;\ t') * f, \quad L'(\cdot;\ t') = g(\cdot;\ t') * f'. \qquad (13.18)$$

Then, after some calculations, similar to those in section 2.4.8, it follows

$$\partial_t(\mathcal{D}_{norm}L)(x;\ t) = 0 \quad \Leftrightarrow \quad \partial_{t'}(\mathcal{D}_{norm}L')(x';\ t') = 0$$

where $(\mathcal{D}_{norm}L')(x';\ t') = t'^{N/2}(\mathcal{D}L')(x';\ t')$.

Although this result because of simplicity of presentation has been expressed for two-dimensional signals, corresponding results hold in one as well as higher dimensions.

In other words, if a differential expression can be defined such that it has the property of assuming a maximum over scales for a desired class of signals, then this maximum will have a nice behaviour under rescalings of the input pattern. In particular, if such maxima over scales are used for scale selection, the selected scale levels will automatically adapt to size variations in the local image structure.

13.2.2. Selecting detection scales from maxima over scales

It is proposed that such maxima over scales can be used in the stage of *detecting features,* i.e., when establishing the existence of different types of structures in the image. Basically, the scale at which a maximum over scales is attained will be assumed to give information about how large a feature is, in analogy with the common approach of taking the spatial position at which the maximum operator response is assumed as an estimate of the location of the feature.

Of course, the localization can be poor if features are detected at coarse scales. Therefore, the method needs to be complemented by a second processing stage, in which more refined processing is invoked to improve the localization (see below). In this respect, the method naturally gives rise to two-stage algorithms with feature detection at coarse scales followed by feature localization at finer scales.

An obvious problem concerns what differential expressions to use in the first step. Is any differential invariant feasible? Here, no attempt will be made to state a final answer to this question. Instead, we contend that the differential expression should at least be selected so as to capture the type of structures under consideration. Now, experimental and theoretical results will be presented demonstrating how this approach applies to a number of different feature detectors formulated in terms of polynomials of normalized Gaussian derivatives.

13.2.3. Experiments: Scale-space signatures from real data

Figure 13.2 illustrates the variation over scales of two simple differential expressions formulated in terms of normalized spatial derivatives. It displays the variation over scales of the trace and the determinant of the normalized Hessian matrix

$$\operatorname{trace} \mathcal{H}_{norm} L = t \nabla^2 L = t\,(L_{xx} + L_{yy}), \tag{13.19}$$

$$\det \mathcal{H}_{norm} L = t^2\,(L_{xx} L_{yy} - L_{xy}^2), \tag{13.20}$$

computed at two different points from a sunflower image. To reduce the sensitivity with respect to the sign of these entities, and hence the polarity of the signal, these entities have been squared before presentation. These graphs are called the *scale-space signatures* of $(\operatorname{trace} \mathcal{H}_{norm} L)^2$ and $(\det \mathcal{H}_{norm} L)^2$, respectively.

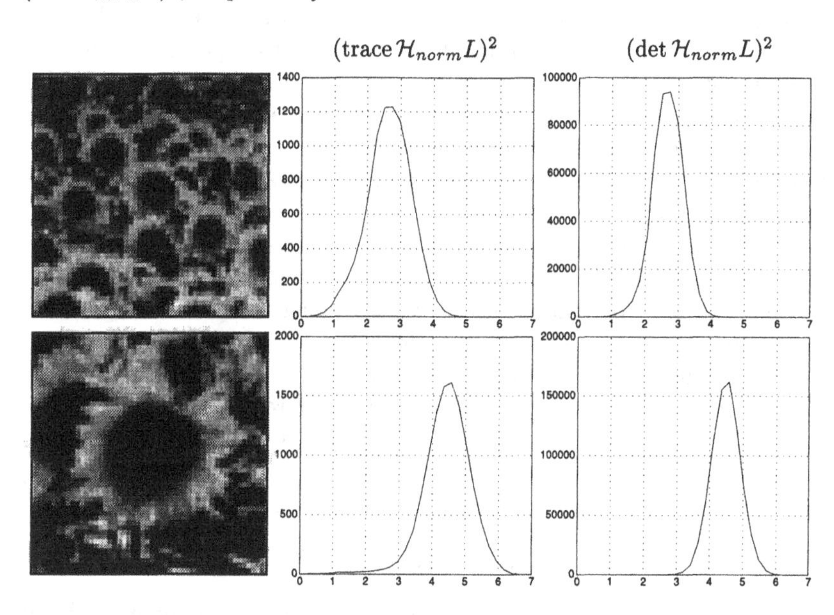

Figure 13.2. Scale-space signatures of the trace and the determinant of the normalized Hessian matrix for two details of a sunflower image; (left) grey-level image, (middle) signature of $(\operatorname{trace} \mathcal{H}_{norm} L)^2$, (right) signature of $(\det \mathcal{H}_{norm} L)^2$. (All entities are computed at the central point. The scaling of the horizontal axis is basically logarithmic, while the scaling of the vertical axis is linear.)

Clearly, the maximum over scales in the top row of figure 13.2 is obtained at a finer scale than in the bottom row. An examination of the ratio between the scale levels where the graphs attain their maxima shows that this value is roughly equal to the ratio of the diameters of the sunflowers in the centers of the two images respectively. This demonstrates that experimental results in agreement with the proposed heuristic principle can be obtained also for real-world data having a richer frequency content.

The reason why these particular differential expressions have been selected here is because they constitute differential entities useful for blob detection; see e.g. (Marr 1982; Voorhees and Poggio 1987; Blostein and Ahuja 1989). In section 13.3 it will be demonstrated how maxima over scales of these entities can be used for formulating a multi-scale blob detector with automatic scale selection. First, however, we shall develop the basic idea further.

13.2.4. Simultaneous detection of interesting points and scales

In figure 13.2, the signatures of the normalized differential entities were computed using the same spatial points at all scales. These points were in turn selected manually at the centers of the sunflowers, where the response of the selected differential operators can be expected to be close to maximal with respect to small variations of the spatial coordinates. Such points cannot, however, be expected to be known *a priori*, and it can also be expected that the point, at which the maximum is assumed, moves in space when the scale parameter in scale-space varies. This is a special case of the well-known fact that scale-space smoothing leads to shape distortions.

Therefore, a more general approach to scale selection from local extrema in the scale-space signature is by accumulating the signature of any normalized differential entity $\mathcal{D}_{norm}L$ along the path $r: \mathbb{R}_+ \to \mathbb{R}^N$ that a local extremum in $\mathcal{D}_{norm}L$ describes across scales. The mathematical framework for describing such paths has been developed in section 8.6. Formally, an *extremum path* of a differential entity $\mathcal{D}_{norm}L$ is defined (from the implicit function theorem) as a connected set of points $(r(t);\ t) \in \mathbb{R}^N \times \mathbb{R}_+$ in scale-space such that for any $t \in \mathbb{R}_+$ the point $r(t)$ is a local extremum of the mapping $x \mapsto (\mathcal{D}_{norm}L)(x;\ t)$,

$$\{(x;\ t) \in \mathbb{R}^N \times \mathbb{R}_+\} = \{(r(t);\ t) \in \mathbb{R}^N \times \mathbb{R}_+ : (\nabla(\mathcal{D}_{norm}L))(r(t);\ t) = 0\}.$$

At the point at which an extremum in the signature is assumed, it holds that the derivative along the scale direction is zero as well. Hence, it is natural to define a *normalized scale-space extremum* of a differential entity $\mathcal{D}_{norm}L$ as a point $(x_0;\ t_0) \in \mathbb{R}^N \times \mathbb{R}_+$ that is *simultaneously* a local extremum with respect to both the spatial coordinates and the scale parameter. In terms of derivatives, such points satisfy

$$\left\{ \begin{array}{l} (\nabla(\mathcal{D}_{norm}L))(x_0;\ t_0) = 0, \\ (\partial_t(\mathcal{D}_{norm}L))(x_0;\ t_0) = 0. \end{array} \right. \qquad (13.21)$$

These normalized scale-space extrema constitute natural generalizations of extrema with respect to the spatial coordinates, and can serve as natural interest points for feature detectors formulated in terms of spatial maxima of differential operators, such as blob detectors, junction detectors, symmetry detectors, etc. Different specific examples of that will be worked out in more detail in following sections.

First, however, let us note that such normalized scale-space maxima trivially inherit the invariance properties of differential singularities (section 6.1.4) and the scaling property of extrema in the scale-space signature (section 13.2.1). This means that if a normalized scale-space maximum is assumed at $(x_0;\ t_0)$ in the scale-space representation of a signal f, then in a rescaled signal f' defined by $f'(sx) = f(x)$, a corresponding maximum in the scale-space representation of f' is assumed at $(sx_0;\ s^2 t_0)$.

13.3. Blob detection

Figure 13.3 shows the result of detecting normalized scale-space extrema of the normalized Laplacian in an image of a sunflower field. Every scale-space maximum is graphically illustrated by a circle centered at the point at which the spatial maximum is assumed, and with the size determined such that the area (measured in pixel units) is equal to the scale at which the maximum is assumed. To reduce the number of blobs, a threshold on the maximum normalized response has been selected such that 250 blobs remain.

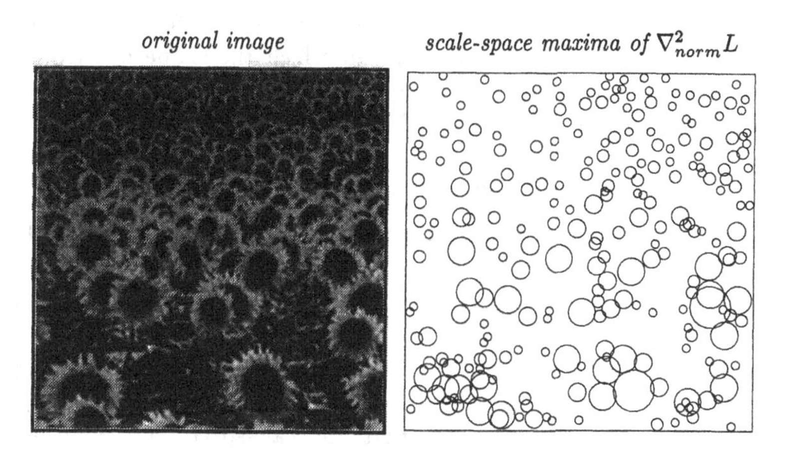

Figure 13.3. Normalized scale-space maxima computed from an image of a sunflower field. (left) Original image. (right) The 250 normalized scale-space maxima having the strongest normalized response.

Figure 13.4 shows the result of superimposing these circles onto a bright copy of the original image, as well as corresponding results for the normalized scale-space extrema of the square of the determinant of the Hessian matrix. Observe how these conceptually very simple differential geometric descriptors give a very reasonable description of the blob-like structures in the image (in particular concerning the blob size) considering how little information is used in the processing.

Figure 13.4. Normalized scale-space maxima superimposed onto a bright copy of the original image. (a) The normalized trace of the Hessian matrix. (b) The normalized determinant of the Hessian matrix.

In chapter 14 it will be shown how this idea can be extended to a more refined blob detection method, in which the shape of each blob descriptor is improved by computing local statistics of gradient directions in a neighbourhood of the blob and representing it by an ellipse.

13.3.1. Analysis of scale-space maxima for a Gaussian blob

To study the behaviour of the scale selection method analytically, consider

$$f(x_1, x_2) = g(x_1, x_2;\ t_0) = \frac{1}{2\pi t_1} e^{-(x_1^2+x_2^2)/2t_0} \tag{13.22}$$

as a simple model of a two-dimensional blob with characteristic length $\sqrt{t_0}$. From the semi-group property of the Gaussian kernel $g(\cdot;\ t_1) * g(\cdot;\ t_2) = g(\cdot;\ t_1 + t_2)$ it follows that the scale-space representation L of f is

$$L(x_1, x_2;\ t) = g(x_1, x_2;\ t_0 + t). \tag{13.23}$$

Clearly, the spatial maximum of $|\nabla^2 L|$ is assumed at $(x_1, x_2)^T = (0, 0)^T$. The corresponding normalized entity is

$$|(\nabla^2_{norm} L)(0, 0;\ t)| = \frac{t}{\pi(t_0 + t)^2} = 0. \tag{13.24}$$

Differentiation with respect to t shows that the maximum is given by

$$\partial_t(\nabla^2_{norm} L)(0, 0;\ t) = 0 \quad \Longleftrightarrow \quad t = t_0, \tag{13.25}$$

which verifies that the maximum over scales is assumed at a scale proportional to a characteristic length of the blob. Trivially, scale-space extrema of $(\nabla^2_{norm} L)^2$ are assumed at the same scale, as are the scale-space extrema of $(\det \mathcal{H}_{norm} L)^2$.

Qualitatively similar results can be obtained if the rotationally symmetric blob model (13.22) is replaced by a more general non-uniform Gaussian blob

$$f(x_1, x_2) = g(x_1;\ t_1)\, g(x_2;\ t_0) = \frac{1}{\sqrt{2\pi t_1}} e^{-x_1^2/2t_1} \frac{1}{\sqrt{2\pi t_2}} e^{-x_2^2/2t_2}$$

with characteristic lengths $\sqrt{t_1}$ and $\sqrt{t_2}$ along the two coordinate directions. From the scale-space representation of f

$$L(x_1, x_2;\ t) = g(x_1;\ t_1 + t)\, g(x_2;\ t_2 + t), \qquad (13.26)$$

it can (after a number of algebraic manipulations) be shown that for any $t_1, t_2 > 0$ there is a unique maximum over scales in

$$|(\nabla^2_{norm} L)(0, 0;\ t)| = \frac{t(t_1 + t_2 + 2t)}{2\pi(t_1 + t)^{3/2}(t_2 + t)^{3/2}}. \qquad (13.27)$$

The closed-form solution for the maximum over scales in

$$|(\det L)\mathcal{H}_{norm}(0, 0;\ t)| = \frac{t^2}{4\pi^2(t_1 + t)^2(t_2 + t)^2} \qquad (13.28)$$

is simple. It is assumed at

$$t_{\det \mathcal{H} L} = \sqrt{t_1 t_2}, \qquad (13.29)$$

verifying that the scale at which the scale-space maximum is assumed reflects a characteristic size of the blob.

13.3.2. Analysis of scale-space maxima for a periodic signal

Another interesting special case to consider is a periodic signal defined as the sum of two perpendicular sine waves,

$$f(x, y) = \sin \omega_1 x + \sin \omega_2 y \qquad (\omega_1 \leq \omega_2). \qquad (13.30)$$

Its scale-space representation is

$$L(x, y;\ t) = e^{-\omega_1^2 t/2} \sin \omega_1 x + e^{-\omega_2^2 t/2} \sin \omega_2 y, \qquad (13.31)$$

and $\nabla^2_{norm} L$ and $\det \mathcal{H}_{norm} L$ assume their spatial maxima at $(\pi/2, \pi/2)$. Setting the derivative

$$\partial_t |(\nabla^2_{norm} L)(\tfrac{\pi}{2}, \tfrac{\pi}{2};\ t)| = \partial_t (t\, (\omega_1^2 e^{-\omega_1^2 t/2} + \omega_2^2 e^{-\omega_2^2 t/2})) \qquad (13.32)$$

to zero gives

$$\omega_1^2(2-\omega_1^2 t)\,e^{-\omega_1^2 t/2}+\omega_2^2(2-\omega_2^2 t)\,e^{-\omega_2^2 t/2}=0. \tag{13.33}$$

This equation has a unique solution when the ratio ω_2/ω_1 is close to one, and three solutions when it is sufficiently large. Hence, there is a unique maximum over scales when ω_2/ω_1 is close to one, and two maxima when the ratio is sufficiently large. (The bifurcation occurs when $\omega_2/\omega_1 \approx 2.4$.) In the special case when $\omega_1 = \omega_2 = \omega_0$, the maximum over scales is assumed at

$$t_{\text{trace}\,\mathcal{H}L}=\frac{2}{\omega_0^2}. \tag{13.34}$$

Similarly, setting

$$\partial_t|(\det\mathcal{H}_{norm}L)(\pi/2,\pi/2;\ t)|=\partial_t(t^2\,\omega_1^2 e^{-\omega_1^2 t/2}\,\omega_2^2 e^{-\omega_2^2 t/2}) \tag{13.35}$$

to zero gives that the maximum over scales is assumed at scale

$$t_{\det\mathcal{H}L}=\frac{4}{\omega_1^2+\omega_2^2}. \tag{13.36}$$

Hence, for both the Gaussian blob model and the periodic sine waves, the analytical results agree with the suggested heuristic principle. When the scale parameter is measured in units of $\sigma = \sqrt{t}$, the scale level, at which the maximum over scales are assumed, is proportional to a characteristic length of corresponding structures in the signal.

13.4. Junction detection

A similar approach can be used for detecting grey-level junctions. In this section, we shall first consider experimental results with scale-space signatures accumulated at junction points. Then, it will be shown how a multi-scale junction detection method can be formulated in terms of normalized scale-space maxima of a differential invariant.

13.4.1. Selection of detection scale

A commonly used entity for junction detection is the curvature of level curves in intensity data multiplied by the gradient magnitude (section 6.2.2). A special choice is to multiply the level curve curvature with the gradient magnitude raised to the power of three. This is the smallest value of the exponent that leads to a polynomial expression

$$\tilde{\kappa}=L_{x_2}^2L_{x_1x_1}-2L_{x_1}L_{x_2}L_{x_1x_2}+L_{x_1}^2L_{x_2x_2}. \tag{13.37}$$

The corresponding normalized differential expression is obtained by replacing each derivative operator ∂_{x_i} by $\sqrt{t}\,\partial_{x_i}$,

$$\tilde{\kappa}_{norm} = t^2 \tilde{\kappa}. \tag{13.38}$$

Figure 13.5 shows the result of accumulating the signature of this normalized differential entity at two different details of a block image. (The signatures have been computed at the central point.) It can be seen that the maximum over scales in the top row is obtained at a finer scale than is the maximum in the bottom row, reflecting the fact that the corner is sharper and neighbouring junctions are closer in the first case than in the second case.

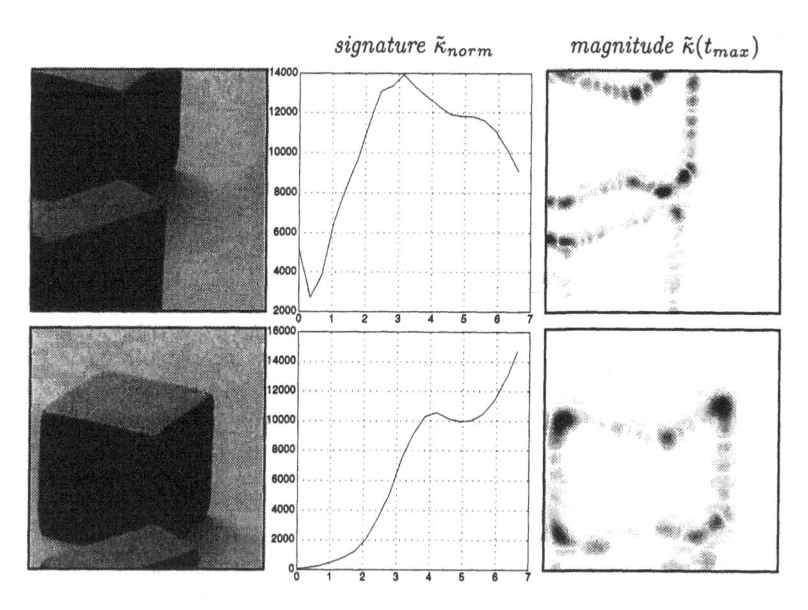

Figure 13.5. Scale-space signature of (the absolute value of) the rescaled level curve curvature $\tilde{\kappa}_{norm} = t^2\,\tilde{\kappa}$ computed at two different details of a block image; (left) grey-level image, (middle) signature of $\tilde{\kappa}_{norm}$ at the central point, (right) $\tilde{\kappa}_{norm}$ computed at the scale at which the maximum is assumed.

13.4.2. Analysis of scale-space maxima for a diffuse junction model

To model this behaviour analytically, consider

$$f(x_1, x_2) = \Phi(x_1;\ t_0)\,\Phi(x_2;\ t_0) \tag{13.39}$$

as a simple model of a *diffuse L-junction*, where $\Phi(\cdot;\ t_0)$ describes a *diffuse step edge*

$$\Phi(x_i;\ t_0) = \int_{x'=-\infty}^{x_i} g(x';\ t_0)\,dx' \tag{13.40}$$

with *diffuseness* t_0. From the semi-group property of the Gaussian kernel it follows that the scale-space representation L of f is

$$L(x_1, x_2;\ t) = \Phi(x_1;\ t_0 + t)\, \Phi(x_2;\ t_0 + t). \tag{13.41}$$

Differentiation with respect to x_1 and x_2 gives

$$\begin{aligned} L_{x_1} &= g(x_1;\ t_0 + t)\, \Phi(x_2;\ t_0 + t), \\ L_{x_2} &= \Phi(x_1;\ t_0 + t)\, g(x_2;\ t_0 + t), \\ L_{x_1x_1} &= (-x_1/(t_0 + t))\, g(x_1;\ t_0 + t)\, \Phi(x_2;\ t_0 + t), \\ L_{x_1x_2} &= g(x_1;\ t_0 + t)\, g(x_2;\ t_0 + t), \\ L_{x_2x_2} &= (-x_2/(t_0 + t))\, \Phi(x_1;\ t_0 + t)\, g(x_2;\ t_0 + t), \end{aligned} \tag{13.42}$$

and insertion of these expressions into (6.13) shows that the evolution over scales of $\tilde{\kappa}_{norm}$ computed at the origin is

$$\tilde{\kappa}_{norm}(0, 0;\ t) = \frac{t^2}{16\pi^2 (t_0 + t)^4}. \tag{13.43}$$

This expression is of the same form as (13.24), which gives

$$\partial_t(\tilde{\kappa}_{norm}(0, 0;\ t)) = 0 \quad \Longleftrightarrow \quad t = t_0. \tag{13.44}$$

Hence, the maximum over scales is assumed at a scale level proportional to the diffuseness of the junction.

A limitation of this analysis though, is that the signature is computed at a fixed point, while the maximum in $\tilde{\kappa}$ can be expected to drift due to the scale-space smoothing. Unfortunately, the equation that determines the position of the maximum in $\tilde{\kappa}$ is analytically complicated (it contains a non-linear combination of the Gaussian function, the primitive function of the Gaussian, and polynomials).

Carrying out an analytical study along non-vertical extremum paths is, however, straightforward for the previously treated non-uniform Gaussian blob model. From (13.26) we have that its scale-space representation is

$$L(x_1, x_2;\ t) = g(x_1;\ t_1 + t)\, g(x_2;\ t_2 + t). \tag{13.45}$$

Differentiation and insertion into (13.37) shows that the absolute value of the rescaled curvature as function of scale is determined by

$$\tilde{\kappa}(x_1, x_2;\ t) = \frac{x_1^2 (t_2 + t_1) + x_2^2 (t_1 + t)}{(t_1 + t)^2 (t_2 + t)^2}\, g^3(x_1;\ t_1 + t)\, g^3(x_2;\ t_2 + t),$$

and that this entity assumes its maximum value on the ellipse[2]

$$\frac{3x_1^2}{2(t_1 + t)} + \frac{3x_2^2}{2(t_2 + t)} = 1. \tag{13.46}$$

[2]The property that $\tilde{\kappa}$ is maximal along a curve and not a single point is a direct consequence of the skew invariance property of $\tilde{\kappa}$ (section 6.2.2), and the fact that the non-uniform Gaussian kernel corresponds to linear stretching of a rotationally symmetric function.

On this ellipse, the absolute value of the normalized rescaled curvature is

$$|\tilde{\kappa}_{norm}|_{ellipse} = \frac{t^2}{12e\pi^3(t_1+t)^{5/2}(t_2+t)^{5/2}}. \tag{13.47}$$

Setting the derivative of this expression with respect to the scale parameter to zero, gives that the positive root is assumed at

$$t_{\tilde{\kappa}} = \frac{t_1+t_2}{12}\left(\sqrt{1+\frac{96\,t_1t_2}{(t_1+t_2)^2}}-1\right). \tag{13.48}$$

In the rotationally symmetric case, when $t_1 = t_2 = t_0$, the maximum is assumed at scale

$$t_{\tilde{\kappa}} = \tfrac{2}{3}t_0. \tag{13.49}$$

On the other hand, if $t_2 << t_1$, then a Taylor expansion gives

$$t_{\tilde{\kappa}} = 4t_2(1+\mathcal{O}(t_2/t_1)). \tag{13.50}$$

13.4.3. Multi-scale junction detection with automatic scale selection

The scale-space signatures in figure 13.5 were accumulated at fixed points, which were given beforehand. Since such points cannot be expected to be known *a priori*, a natural extension of the approach to multi-scale junction detection is by detecting normalized scale-space maxima in the differential entity used for junction detection (here $\tilde{\kappa}$). As described earlier, this

original image *normalized scale-space maxima*

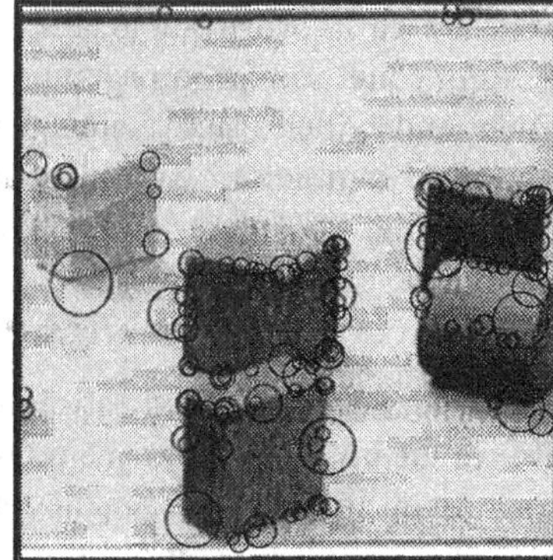

Figure 13.6. Multi-scale junction detection using normalized scale-space maxima of the rescaled level curve curvature $\tilde{\kappa}$; (left) original image, (right) the 120 scale-space maxima having the strongest maximum normalized response (superimposed onto a bright copy of the original image).

operation corresponds to tracking the spatial maxima in $\tilde{\kappa}$ across scales, and then selecting points on the extremum path where the normalized response assumes local maxima over scales.

Figure 13.6 shows the result of applying this operation to a block image. Each normalized scale-space maximum is graphically illustrated by a circle centered at the point at which the maximum is assumed, and with the size determined such that the area (measured in pixel units) is equal to the scale at which the maximum was assumed. To reduce the number of junction candidates, a threshold has been determined such that 120 blobs remain.

Of course, thresholding on the magnitude of the operator response constitutes a very simple selective mechanism in feature detection. We can nevertheless notice that this operation generates a reasonable set of junction candidates corresponding to grey-level junctions with reasonable interpretation.

13.4.4. Improving the localization using local consistency

A natural choice of spatial representative of a scale-space maximum at $(x_0;\ t_0)$ is, of course, the spatial point x_0. With respect to the problem of junction detection, this point x_0 may be treated as a first estimate of the location of the candidate junction, and the scale t_0 at which the maximum is assumed as a first estimate of the localization error.

While this junction detection method is conceptually clean, it can certainly lead to poor *localization*. In realistic images, the scale-space maxima may be assumed at rather coarse scales, where the drift due to scale-space smoothing may be substantial, and also adjacent features may have started to interfere with each other. Therefore, some postprocessing is necessary to improve the localization.

A simple method for improving the localization is as follows: Following Förstner and Gülch (1987), consider at every point $x' \in \mathbb{R}^2$ in a neighbourhood of a junction candidate x_0, the line $l_{x'}$ perpendicular to the gradient vector $(\nabla L)(x') = (L_{x_1}, L_{x_2})^T(x')$ at that point. The equation of this line is

$$D_{x'}(x) = ((\nabla L)(x'))^T(x - x') = 0. \tag{13.51}$$

A straightforward improved estimate of the location of the junction candidate can be obtained from the point that minimizes the (perpendicular) distance to all lines $l_{x'}$ in a neighbourhood of x_0. A convenient way to express such a condition computationally is to determine the point $x \in \mathbb{R}^2$ that minimizes

$$\min_{x\in\mathbb{R}^2} \int_{x'\in\mathbb{R}^2} (D_{x'}(x))^2\, w_{x_0}(x')\, dx' \tag{13.52}$$

for some window function $w_{x_0} : \mathbb{R}^2 \to \mathbb{R}$ centered at the candidate junction point x_0. Minimizing this expression corresponds to finding the point that minimizes the weighted integral of the squares of the distances from x to all $l_{x'}$ in the neighbourhood, see figure 13.7. This can be seen by noting that $D_{x'}(x)$ describes the distance from x to the line $l_{x'}$ multiplied by the gradient magnitude. Hence, every distance is given a weight proportional to the square gradient magnitude multiplied by the window function, implying that the strongest weights in the minimization are given to edge points in the neighbourhood of x_0.

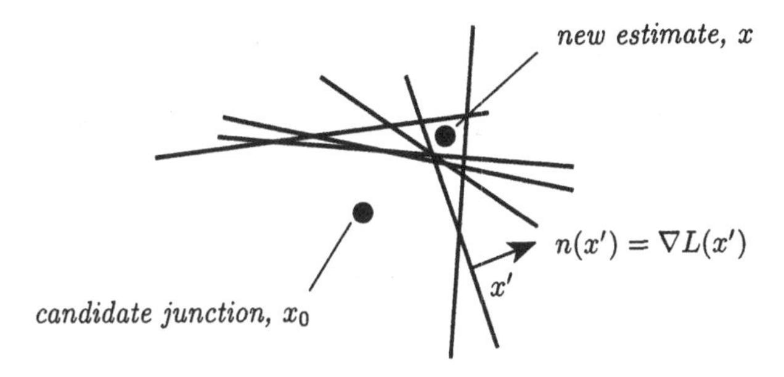

Figure 13.7. Minimizing the expression (13.52) basically corresponds to finding the point x that minimizes the distance to all edge tangents in a neighbourhood of the given candidate junction point x_0.

Explicit solution of the minimization problem in terms of local image statistics. A useful property of the least-squares minimization formulation (13.52) is that it can be solved analytically. It is straightforward to verify that after expansion, (13.52) can be written

$$\min_{x \in \mathbb{R}^2} \int_{x' \in \mathbb{R}^2} (x - x')^T \, ((\nabla L)(x')) \, ((\nabla L)(x'))^T \, (x - x') \, w_{x_0}(x') \, dx', \tag{13.53}$$

and that the minimization problem can be expressed as

$$\min_{x \in \mathbb{R}^2} x^T A \, x - 2 \, x^T b + c \quad \Longleftrightarrow \quad A \, x = b, \tag{13.54}$$

where $x = (x_1, x_2)^T$, and A, b, and c are entities determined by the local statistics of the gradient directions in a neighbourhood of x_0,

$$A = \int_{x' \in \mathbb{R}^2} (\nabla L)(x') \, (\nabla L)^T(x') \, w_{x_0}(x') \, dx', \tag{13.55}$$

$$b = \int_{x' \in \mathbb{R}^2} (\nabla L)(x') \, (\nabla L)^T(x') \, x' \, w_{x_0}(x') \, dx', \tag{13.56}$$

$$c = \int_{x' \in \mathbb{R}^2} x'^T \, (\nabla L)(x') \, (\nabla L)^T(x') \, x' w_{x_0}(x') \, dx'. \tag{13.57}$$

Provided that the 2×2 matrix A is non-degenerate, the minimum value is given by

$$d_{min} = \min_{x \in \mathbb{R}^2} x^T A x - 2 x^T b + c = c - b^T A^{-1} b, \tag{13.58}$$

and the point x that minimizes (13.52) is

$$x = A^{-1} b. \tag{13.59}$$

This means that an improved localization estimate can be computed directly from entities that can be measured in the image. What remains to be determined in this formulation is how to choose the window function, and at what scale(s) to compute the gradient vectors:

- The problem of choosing the weighting function is a special case of a common scale problem in least squares estimation: Over what spatial region should the fitting be performed? Clearly, it should be large enough such that statistics of gradient directions is accumulated over a sufficiently large neighbourhood around the candidate junction. Nevertheless, the region must not be so large that interfering structures corresponding to other junctions are included.

- The second scale problem, on the other hand, is of a slightly different nature than the previous ones—it concerns what scales should be used for *localizing image structures.* Previously, only the problem of detecting image structures has been treated.

Here, the following solutions are proposed:

Selection of window function and spatial points. When computing A, b, and c above, let the window function w_{x_0} be a Gaussian function centered at the point x_0 at which $\tilde{\kappa}_{norm}$ assumed its scale-space maximum. Moreover, let the scale value of this window function be proportional to the (detection) scale t_0 at which the scale maximum in $\tilde{\kappa}_{norm}$ was assumed.

The idea behind this approach is that the detection scale should give a representative region around the candidate junction. Experimentally, this has been demonstrated to be the case in a large number of situations.

Selection of localization scale. Clearly, the gradient estimates used for computing A, b, and c must be computed at a certain scale. To determine this *localization scale*, it is natural to select the scale that minimizes the residual d_{min} given by (13.58) as function of scale. An intuitive explanation of this criterion is that it corresponds to selecting as localization scale the scale that gives maximum consistency of the gradient directions in a neighbourhood of x_0.

The motivations behind this choice are basically as follows: At too fine scales, where a large amount of noise and interfering fine scale structures are present, the gradient directions can be expected to be randomly distributed, which in turn means that the residual will be large. On the other hand, at too coarse scales the scale-space smoothing will lead to increasing shape distortions and hence increasing residual. Selecting the minimum gives a natural trade-off between these two effects.

However, to reduce the sensitivity to the local image contrast, and to obtain an expression of dimension [length]^2, it is natural to divide d_{min} given by (13.58) by trace A. (The latter descriptor represents the weighted average of the gradient magnitude in the neighbourhood of x_0.) This gives the following modified minimization problem for determining the localization scale

$$\min_{t \in [0,t_0]} \tilde{d}_{min} = \min_{t \in [0,t_0]} \frac{d_{min}}{\text{trace}\, A} = \min_{t \in [0,t_0]} \frac{c - b^T A^{-1} b}{\text{trace}\, A}. \tag{13.60}$$

Note, in particular, that for an ideal (sharp) junction, the localization scale given by this method will always be zero[3] in the noise free case.

13.4.5. Experiments: Choice of localization scale

Figure 13.8 gives a qualitative illustration of the basic steps in this scale selection method. It shows the result of accumulating the scale-space signature of the the normalized residual $\tilde{d}_{min}$ for the two junctions considered in figure 13.5. (In each case, the scale of the window function has been set to the scale at which the maximum over scales in $\tilde{\kappa}$ was assumed.) Clearly, a distinct minimum can be seen in each graph. Moreover, the minimum over scales in the top row is assumed at a finer scale than in the bottom row, reflecting the fact that the junction in the top row is sharper than is the junction in the bottom row.

Figure 13.9 shows experimental results of applying the second stage selection of localization scale under more realistic conditions. For each one of the junction candidates detected in figure 13.6, an individual scale selection process has been invoked, and the signature of the normalized residual has been accumulated (using a window function with scale value set to the detection scale of the junction). The minimum in $\tilde{d}_{min}$ has been detected, and a new localization estimate has been computed using (13.59). Moreover, those junctions for which the new localization estimate falls outside the original region of interest have been suppressed. Finally, each remaining junction candidate has been illustrated by circle centered

[3]For a polygon-type junction (consisting of regions of uniform grey-level delimited by straight edges), the residual $\tilde{d}_{min}$ is zero if all edge tangents meet at the junction point. Hence, any amount of smoothing increases the residual, and the minimum value is obtained when the scale parameter is zero.

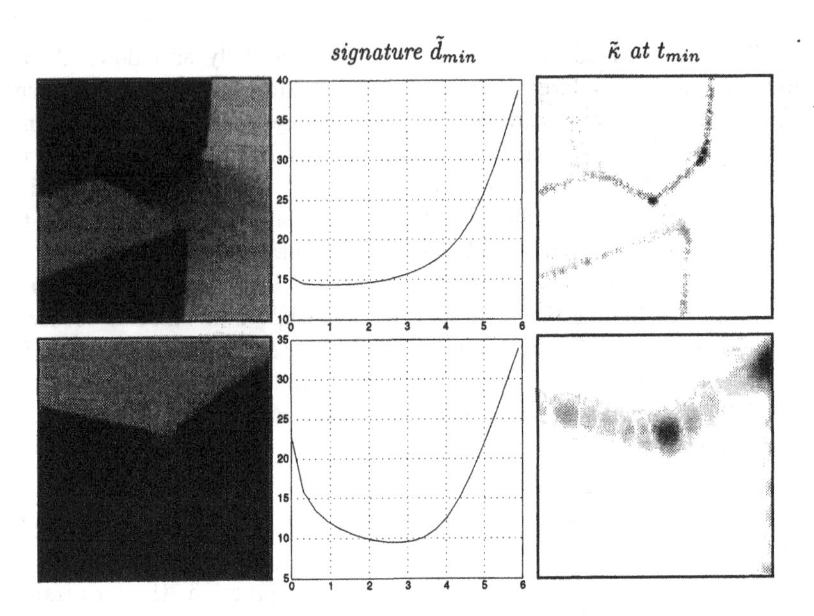

Figure 13.8. Scale-space signature of the normalized residual $\tilde{d}_{min}$ (computed using a Gaussian window function with scale level proportional to the scale at which $\tilde{\kappa}_{norm}$ assumed its maximum over scales in figure 13.5); (left) grey-level image, (middle) scale-space signature of $\tilde{d}_{min}$ accumulated at the central point, (right) $\tilde{\kappa}_{norm}$ computed at the scale where $\tilde{d}_{min}$ assumes its minimum.

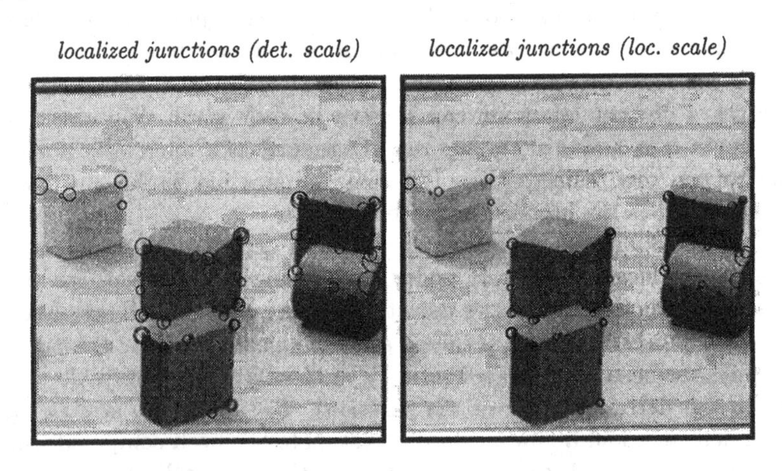

Figure 13.9. Improved localization estimates for the junction candidates in figure 13.6 (computed using the method for selection of localization scale described in section 13.4.4 and illustrated in figure 13.8). Each junction candidate is illustrated by a circle centered at the new location estimate. In the left figure, the size of the circle is determined such that the area is equal to the *detection scale,* while in the right figure the area is equal to the *localization scale.*

at the new localization estimate, and with the size determined such that the area is equal to the localization scale.

13.4.6. Composed scheme for junction detection and localization

To summarize, the proposed junction detection method is a genuine two-stage scheme with feature detection at coarse scales followed by a feature localization to finer scales. It consists of the following basic steps (illustrated in figure 13.10):

1. *Detection step.* Detect scale-space maxima of the normalized rescaled level curve curvature

   $$\tilde{\kappa}_{norm} = t^2\,\tilde{\kappa} = t^2\,(L_{x_2}^2 L_{x_1x_1} - 2L_{x_1}L_{x_2}L_{x_1x_2} + L_{x_1}^2 L_{x_2x_2})$$

   (or some other suitable normalized differential entity). This generates a set of junction candidates.

2. *Localization step.* For each junction candidate, accumulate the signature of the normalized residual

   $$\tilde{d}_{min} = \frac{c - b^T A^{-1} b}{\operatorname{trace} A}$$

   with A, b, and c computed according to (13.55)–(13.57), and using a Gaussian window function with scale value proportional to the detection scale.

   Then, at the scale at which the minimum in $\tilde{d}_{min}$ is assumed, compute an improved localization estimate using

   $$x = A^{-1}b.$$

   Optionally, repeat step 2 iteratively.

Of course, this description is only a schematic outline, and can be improved in several ways. A natural extension is to include a more refined mechanism for selecting interesting junction candidates in the first step, e.g., by linking extrema in $\tilde{\kappa}$ across scales, and then measuring significance in a similar way as done in the scale-space primal sketch (chapter 7).

Moreover, to simplify the interpretation of the output from the second step, it may in certain situations be advantageous to explicitly take care of multiple responses corresponding to the same junction. Of course, it is impossible to determine *a priori* whether two junction candidates correspond to the same physical junction or not. Nevertheless, a simple condition can be expressed in terms of mutual inclusion (in analogy with the blob-extremum matching described in section 9.2.3). If two blobs contain the centers of each other, then it may be natural to identify them,

original image

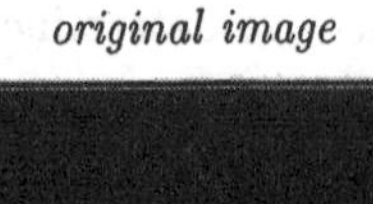

detection by scale-space maxima *suppression of overlapping blobs*

localized junctions (det. scale) *localized junctions (loc. scale)*

Figure 13.10. The major steps in the composed two-stage junction detection method consisting of detection at coarse scales followed by localization to finer scales. The "final output" is illustrated in the lower right corner, which shows localized junctions superimposed onto a bright copy of the original image.

and, for example, only consider the most significant one in a first stage of subsequent analysis. (The middle row in figure 13.10 shows an example of suppressing multiple responses in this way.)

Finally, concerning the last step, it is necessary to have a mechanism for verifying the existence of junctions and for judging whether or not the new localization estimate should be accepted. If the new estimate deviates significantly from the original estimate (for example measured relative to the detection scale), then that may be an indication that the local image structure near the candidate junction is not consistent with a simple junction model (or an edge). In the previous experiments in this section, such points were suppressed from any subsequent processing. In an active vision situation, such points may be regarded as candidate places for taking a closer look.

13.5. Edge detection

If the same type of approach is applied in edge detection, then the result is an edge detection method similar to edge focusing (Bergholm 1987; section 11.1), in the sense that edges are detected at a coarse scale and then followed to finer scales. In the original work on edge focusing, the problem of choosing the two scale levels in the algorithm was left open. Here, we obtain tools for addressing these problems.

13.5.1. Selection of detection scale

Consider the edge definition in terms of non-maximum suppression (section 6.2.1). It means that edges are defined as points for which the gradient magnitude assumes a maximum in the gradient direction. A natural measure of the strength of the edge response in this case is the normalized gradient magnitude

$$L_{\bar{v},norm} = \sqrt{t}\, L_{\bar{v}} = \sqrt{t}\, |\nabla L|, \tag{13.61}$$

where $L_{\bar{v}}$ denotes the directional derivative in the gradient direction (in analogy with the notation in chapter 6).

Analysis for a diffuse step edge. What is the effect of selecting scale levels at which this measure assumes a maximum over scales? Consider first a *diffuse step edge*

$$f(x, y) = \Phi(x;\ t_0) = \int_{x'=-\infty}^{x} g(x';\ t_0)\, dx', \tag{13.62}$$

where g is the one-dimensional Gaussian kernel, and t_0 represents the degree of diffuseness. From the scale-space representation of this signal

$$L(x, y;\ t) = \Phi(x;\ t_0 + t) = \int_{x'=-\infty}^{x} g(x';\ t_0 + t)\, dx', \tag{13.63}$$

it is clear that the normalized edge strength

$$(\partial_{\xi} L_{t_0})(0, y;\ t) = \sqrt{t}\, g(0;\ t_0 + t) \sim \frac{\sqrt{t}}{\sqrt{t_0 + t}}. \tag{13.64}$$

along the edge *increases monotonically* with scale. While having a scale selection method that would select an infinite scale for a straight edge may at first seem counter-intuitive, this choice is in a certain sense natural. For an infinite straight and diffuse edge, there are only two preferred scales, the degree of diffuseness, and the infinite extent of the edge. Obviously, it is preferable to select the latter scale, in order to obtain more reliable detection and suppress spurious structures such as noise.

Analysis for a Gaussian blob. Consider next the edges in the scale-space representation

$$L(x_1, x_2;\ t) = g(x_1, x_2;\ t_0 + t) \tag{13.65}$$

of a *Gaussian blob* $f(x_1, x_2) = g(x_1, x_2;\ t_0)$. Its edges are given by

$$\tilde{L}_{\bar{v}\bar{v}} = \frac{x^2 + y^2 - t_0 - t}{(t_0 + t)^2}\, g(x_1, x_2;\ t_0 + t) = 0, \tag{13.66}$$

and the normalized gradient magnitude computed at the edge points is *constant over scales*

$$\sqrt{t}\, L_{\bar{v}}(x_1, x_2;\ t)\Big|_{L_{\bar{v}\bar{v}}=0} = e^{-1/2}. \tag{13.67}$$

Hence, these two examples indicate a qualitative difference relative to the earlier treated blob and junction strength measures, for which the normalized magnitude of the response decreases with scale when the scale parameter gets sufficiently large.

Analysis at a bifurcation. Nevertheless, there are, of course, also situations where $L_{\bar{v},norm}$ decreases. Consider, for example, one of the edges at a *double asymmetric step edge.* When the scale parameter in scale-space becomes sufficiently large, one of the two edges must disappear. This corresponds to the annihilation of a minimum-maximum pair in the gradient magnitude. A simple model of the local behaviour of the gradient magnitude near the bifurcation can be obtained by studying the polynomial

$$L_{\bar{v}} = x_1^3 + 3x_1(t - t_b), \tag{13.68}$$

which represents the canonical type of singularity in a one-parameter family of functions, *the fold unfolding* (Poston and Stewart 1978; section 8.3), and also has the attractive property that it satisfies the diffusion equation. By differentiation one obtains

$$L_{\bar{v}\bar{v}} = 3(x_1^2 + t - t_b) = 0. \tag{13.69}$$

Setting this derivative to zero gives that the variation over scales of the normalized gradient magnitude at the edge point

$$x_{1,edge}(t) = (t_b - t)^{1/2} \quad (t \le t_b) \tag{13.70}$$

follows

$$L_{\bar{v},norm} = \sqrt{t}\, L_{\bar{v}} = 4\sqrt{t}\,(t_b - t)^{3/2}. \tag{13.71}$$

Clearly, the normalized gradient magnitude $L_{\bar{v},norm}$ decreases when the scale parameter t approaches the bifurcation scale t_b.

13.5.2. Experiments: Scale-space signatures in edge detection

Figure 13.11 shows the result of accumulating the signature of the normalized gradient magnitude at two different details of the previously treated

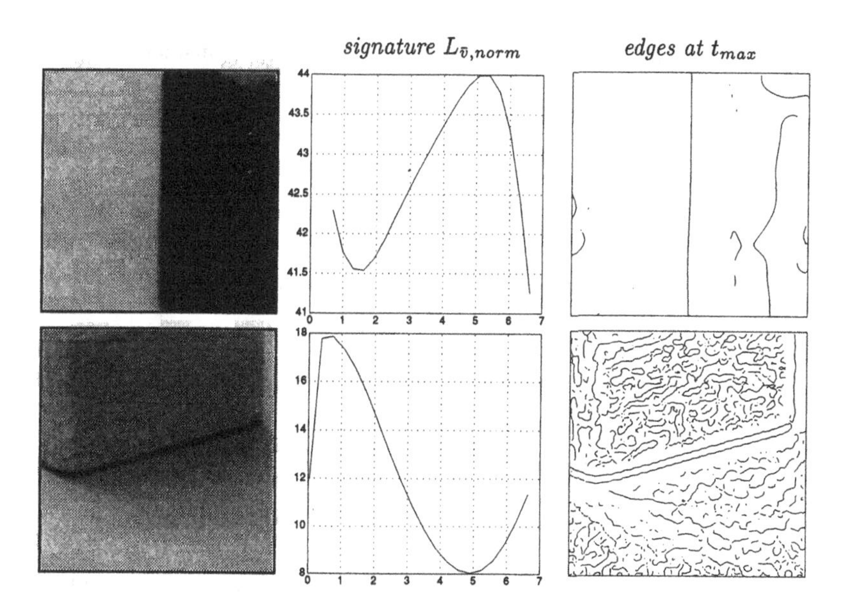

Figure 13.11. Scale-space signature of the normalized gradient magnitude $L_{\bar{v},norm} = \sqrt{t}\, L_{\bar{v}}$ at a straight edge (top row) and at a double edge (bottom row); (left) grey-level image, (middle) signature of $L_{\bar{v},norm}$ accumulated at the central point, (right) (unthresholded) edges detected at the scale at which the maximum was assumed.

block image. Note that the maximum at the step edge (top row) is assumed at a much coarser scale than is the maximum at the double edge (bottom row). Also, the relative variation over scales of the normalized gradient magnitude is much larger in the second case.

To summarize, scale selection using maxima in $L_{\bar{v},norm}$ can be expected to give rise to rather coarse scale levels, delimited from above by the bifurcation scales at which the edges disappear. Notably, the scales where these maxima are assumed do not depend directly on the diffuseness t_0 of the original edge. At double edges (and staircase edges) the behaviour is different, and relatively fine scale levels will be selected (proportional to the distance between adjacent edges).

13.5.3. Selection of localization scale

In analogy with the case of junction detection, edges computed at the detection scale may have poor localization. In particular, the localization can be expected to be poor for curved edges, where the edge drift due to scale-space smoothing may be substantial. (See, for example, the analytical expression for the edges of a Gaussian blob (13.66).) Therefore, it is natural to localize the edges detected at a coarse scale to finer scales.

A natural way of improving the localization is to proceed in analogy with the method for selection of localization scale in junction detection. Given any point x' in a neighbourhood of an edge, consider the line $l_{x'}$ given by

$$n^T(x - x') = 0, \tag{13.72}$$

where $n = (n_1, n_2)^T$ denotes the normal direction to the edge. In terms of spatial derivatives, the components of this vector (which is the normal vector to a level curve of $L_{\bar{v}\bar{v}}$) can be written

$$\begin{aligned} n_1 = \quad & L_{x_1}(L_{x_1}L_{x_1x_1x_1} + 2(L_{x_1x_1}^2 + L_{x_1x_2}^2)) \\ & + L_{x_2}(L_{x_2}L_{x_1x_2x_2} + 2(L_{x_1}L_{x_1x_1x_2} + L_{x_1x_2}(L_{x_1x_1} + L_{x_2x_2}))), \\ n_2 = \quad & L_{x_2}(L_{x_2}L_{x_2x_2x_2} + 2(L_{x_1x_2}^2 + L_{x_2x_2}^2)) \\ & + L_{x_1}(L_{x_1}L_{x_1x_1x_2} + 2(L_{x_2}L_{x_1x_2x_2} + L_{x_1x_2}(L_{x_1x_1} + L_{x_2x_2}))). \end{aligned}$$

Following previous section, consider the point x that minimizes

$$\min_x \int_{x' \in \mathbb{R}^2} (D_{x'}(x))^2 \, w_{x_0}(x') \, dx', \tag{13.73}$$

for some window function w_{x_0}. To preserve the weighting with respect to gradient magnitude, define $D_{x'}$ by

$$D_{x'}(x) = \tilde{n}^T(x - x') = 0, \tag{13.74}$$

with the gradient vector replaced by

$$\tilde{n} = |(\nabla L)(x')| \frac{n}{|n|}. \tag{13.75}$$

The minimization problem (13.74) is of the same type as (13.54) with A, b, and c defined as in (13.55)–(13.57), and with $(\nabla L)(x')$ replaced by $\tilde{n}$. From motivations analogous to section 13.4.4, let us select a localization scale that minimizes the normalized residual[4]

$$\tilde{d}_{min} = \frac{d_{min}}{\text{trace}\, A} = \frac{c - b^T A^{-1} b}{\text{trace}\, A}. \tag{13.76}$$

This corresponds to selecting a scale such that the local edge normals are maximally consistent with a local edge model, and the fine scale fluctuations in the edge normals are small.

13.5.4. Experiments: Scale-space signatures in edge localization

Figure 13.12 shows the result of applying this method for selection of localization scale to the same image data as were used in figure 13.11.

In the first row, the minimum residual was assumed at a rather fine scale indicating the sharpness of the edge. The second and the third rows show a test of the stability of the method, using the same original image data as in the first row. In the second row, where Gaussian noise of standard deviation 25% of the grey-level range has been added, the minimum residual is assumed at a coarser scale, indicating the increased need for smoothing. In the third row, where the image already has been smoothed, the minimum in the signature is obtained at a much finer scale. Finally, the fourth row shows an interesting example, where the method selects a scale level suitable for capturing a very faint shadow edge.

Concerning these experiments, it should be remarked that they are mainly intended to demonstrate the potential of applying the proposed methods for scale selection in edge detection, and that more steps[5] have to be added in order to obtain a complete algorithm. Moreover, to introduce as few commitments into the processing as possible, no selection of edges using edge focusing or thresholding on gradient magnitude has been performed. Therefore, the results should only be interpreted as locally valid in a neighbourhood of the edge through the central point.

[4]Since, A may degenerate to a rank one matrix at ideal straight edges, it is natural to replace the inverse of A by its pseudo inverse in actual computations.

[5]For example, it may be natural to consider the variation over scales of these types of descriptors on the two-dimensional surface in scale-space that is defined by the locations of edge points at all scales, i.e. by the equation $L_{\bar{v}\bar{v}}(x, y;\ t) = 0$. Moreover, a natural generalization is to let the selected scale levels vary on the edge, such that the edge can be described as a (parameterized) curve $(x(s), y(s);\ t(s))$ on that surface.

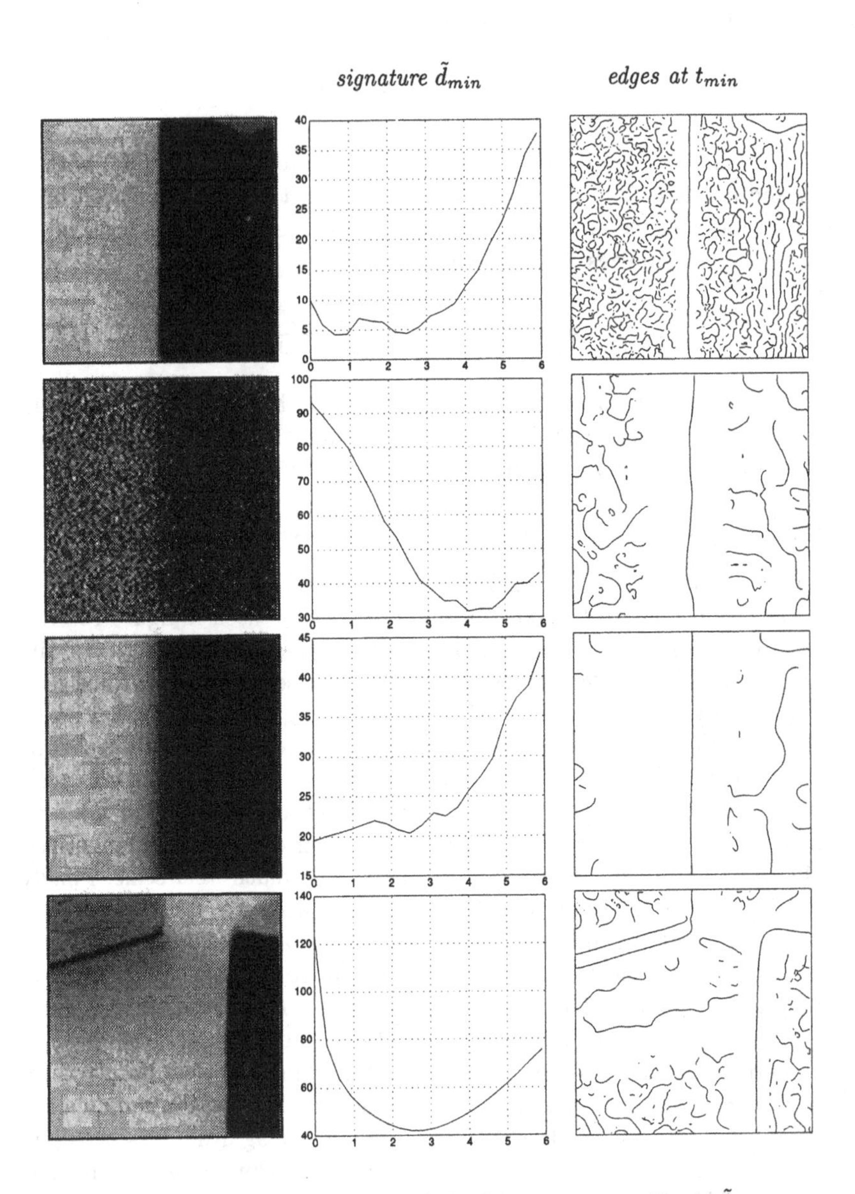

Figure 13.12. Scale-space signature of the least squares residual, $\tilde{d}_{min}$ computed at the central point using a Gaussian window function with scale level proportional to the scale at which $L_{\bar{v},norm}$ assumed its maximum over scales in figure 13.11); (left) grey-level image, (middle) scale-space signature of $\tilde{d}_{min}$ accumulated at the central point, (right) (unthresholded) edges computed at the scale where $\tilde{d}_{min}$ assumes its minimum.

13.6. Discrete implementation of normalized derivatives

Discretizing the normalized derivative operators leads to two problems; (i) how to discretize the scale-space derivatives such that scale-space properties are preserved, and (ii) how to discretize the normalization factor.

13.6.1. Discrete derivative approximations

The first problem can be solved by using the scale-space concept for discrete signals (chapters 3–4), which is given by

$$L(\cdot,\cdot;\ t) = T(\cdot,\cdot;\ t) * f(\cdot,\cdot), \tag{13.77}$$

where $T(m,n;\ t) = T_1(m;\ t)T_1(n;\ t)$ and $T_1(m;\ t) = e^{-t}I_m(t)$ is the one-dimensional discrete analogue of the Gaussian kernel defined from the modified Bessel functions I_n. The scale-space properties of L in turn transfer to any discrete derivative approximations $L_{x_1^i x_2^j}$ defined as the result of applying difference operators $\delta_{x_1^i x_2^j}$ to L (chapter 5).

13.6.2. Normalization in the discrete case

Concerning the normalization problem, it is straightforward to verify that the normalized derivative operator given by (13.7) corresponds to rescaling the Gaussian derivative kernels such that their l_1-norms remain constant over scales. For example, for the first- and the second-order derivatives it holds that

$$\begin{aligned}\int_{-\infty}^{\infty} |g_\xi(u;\ t)|\, du &= 2\sqrt{t} \int_{-\infty}^{0} g_x(u;\ t)\, du \\ &= 2\sqrt{t}\, g(0;\ t) = \sqrt{\frac{2}{\pi}},\end{aligned} \tag{13.78}$$

$$\begin{aligned}\int_{-\infty}^{\infty} |g_{\xi\xi}(u;\ t)|\, du &= 2t \int_{-\sqrt{t}}^{\sqrt{t}} g_{xx}(u;\ t)\, du \\ &= 4t\, g_x(-\sqrt{t};\ t) = \sqrt{\frac{2}{\pi e}}.\end{aligned} \tag{13.79}$$

Such normalized kernels of first order have been used, for example, in edge detection and edge classification by Korn (1988) and Zhang and Bergholm (1991, 1993), as well as in pyramids by Crowley and his co-workers (1984).

Given this analogy, it is natural to normalize the discrete derivative approximation kernels $\delta_{x_1^i x_2^j} T$ such that their l_1-norms remains constant over scales also in the discrete case. Of course, it is not necessary to compute the normalized derivative approximation kernels explicitly. In practice, concerning e.g. first order derivatives, discrete approximations to L_{x_1} and L_{x_2} can first be computed as described in section 13.6.1. Then, the result can be multiplied by the discrete normalization factor

$$\alpha_1(t) = \frac{\sqrt{2}}{\sqrt{\pi}\,(T_1(0;\ t) + T_1(1;\ t))}, \tag{13.80}$$

which approaches the continuous normalization factor $\sqrt{t}$ when the scale parameter t increases. This is the discretization method that has been used in all experiments presented here.

Observe that $\alpha_1(t)$ assumes a non-zero value when $t = 0$, in contrast to the continuous normalization factor $\sqrt{t}$, which forces any normalized derivative to be zero in unsmoothed data. This property is important to capture peaks at fine scales in the scale-space signatures of the differential entities, see figure 13.13 for an illustration.

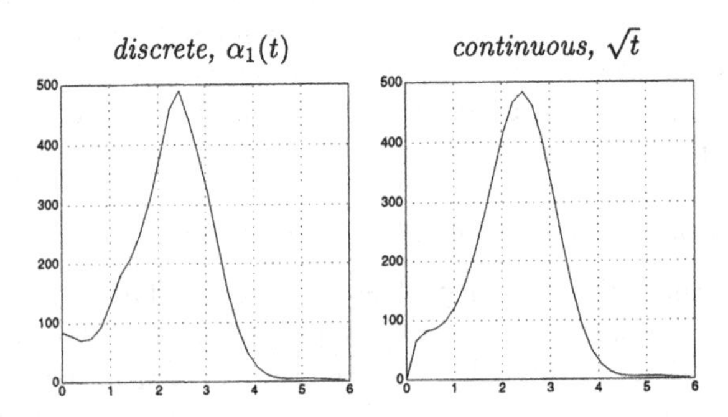

Figure 13.13. Scale-space signatures from a noisy image computed using (left) the discrete normalization factor $\alpha_1(t)$, and (right) the continuous normalization factor $\sqrt{t}$. Note that in the first case there is a peak at fine scales, while the peak is suppressed and the response tends to zero in the second case.

13.7. Interpretation in terms of self-similar Fourier spectrum

Another useful interpretation of normalized derivatives can be obtained in the context of signals having a self-similar Fourier spectrum. Consider a signal $f: \mathbb{R}^2 \to \mathbb{R}$ having a power spectrum of the form

$$S_f(\omega_1, \omega_2) = (\hat{f}\hat{f}^*)(\omega_1, \omega_2) = |\omega|^{-2\alpha} = (\omega_1^2 + \omega_2^2)^{-\alpha} \tag{13.81}$$

and study the variation over scales of the following energy measure

$$P_{L(\cdot;\ t)} = E(L_{x_1}^2(\cdot;\ t) + L_{x_2}^2(\cdot;\ t)) = \int_{x \in \mathbb{R}^2} |\nabla L(x;\ t)|^2 \, dx. \tag{13.82}$$

Using Plancherel's relation

$$\int_{\omega \in \mathbb{R}^2} \hat{h}_1(\omega)\, \hat{h}_2^*(\omega)\, d\omega = (2\pi)^2 \int_{x \in \mathbb{R}^2} h_1(x)\, h_2^*(x)\, dx, \tag{13.83}$$

(where $\hat{h}_i(\omega)$ denotes the Fourier transform of $h_i(x)$) and by letting $h_1 = h_2 = L_{x_i}$ it follows that

$$E(L_{x_i}^2(\cdot;\ t)) = \frac{1}{4\pi^2} \int_{\omega \in \mathbb{R}^2} \omega_i^2\, \hat{g}^2(\omega;\ t)\, |\omega|^{-2\alpha}\, d\omega, \tag{13.84}$$

where $\hat{g}$ denotes the Fourier transform of the Gaussian kernel. By adding (13.84) for $i = 1, 2$, by introducing polar coordinates ($\omega_1 = \rho \cos \phi, \omega_2 = \rho \sin \phi$), and by using (8.91), we get

$$P_{L(\cdot;\ t)} = \frac{1}{4\pi^2} \iint_{\rho \in [0,\, \infty[, \phi \in [0, 2\pi]} \rho^{2(1-\alpha)}\, e^{-\rho^2 t}\, \rho\, d\rho\, d\phi = \frac{1}{2\pi} \frac{\Gamma(2-\alpha)}{2\, t^{2-\alpha}}. \tag{13.85}$$

The corresponding entity based on normalized derivatives is

$$P_{norm}(\cdot;\ t) = t\, P_{L(\cdot;\ t)} \sim t^{\alpha-1}. \tag{13.86}$$

This expression is independent of scale if and only if $\alpha = 1$. In other words, in the two-dimensional case the normalized derivative model is *neutral* with respect to power spectra of the form $|\omega|^{-2}$ (i.e., Fourier spectra of the form $|\omega|^{-1}$).

It is well-known that natural images often show a qualitative behaviour similar to this (Field 1987). This is in fact a direct consequence of scale invariance; the power spectrum variation

$$S(\omega) \sim |\omega|^{-N}, \tag{13.87}$$

where N denotes the dimension of the signal, can be easily derived from the assumption that the power spectrum is to contain the same amount of energy at all frequencies.

13.8. Summary and discussion

The subject of this chapter has been to outline how the evolution properties over scales of normalized Gaussian derivatives can be used for finding interesting scales for further analysis. A heuristic principle has been presented stating that in the absence of other evidence, the characteristic size of image structures can be estimated from the scales at which normalized differential geometric descriptors assume their maxima over scales.

In particular, it has been suggested that this heuristic principle can be used in feature detection, for adaptively choosing the scales at which to detect image features. This approach has been supported by a theoretical analysis of the general scaling property of maxima in the scale-space signature, and by analytical calculations showing the effect of applying this idea to different feature detectors applied to a number of model patterns. The main support of the methodology is, however, experimental; it has been demonstrated that intuitively reasonable results can be obtained in blob detection, junction detection, and edge detection.

The methodology naturally gives rise to two-stage feature detection algorithms, where features are detected at (locally adapted) coarse scales, and then localized to finer scales. While the advantages of such a general two-stage approach are well-known, the main contribution in this respect is that mechanisms are provided for selecting both the detection scale and the localization scale.

Of course, the task of selecting "the best scale" for handling real-world image data (about which usually no or very little *a priori* information is available) is intractable if treated as a pure mathematical problem. Therefore, the proposed heuristic principle should not be interpreted as any "optimal solution", but rather as a systematic method for generating initial hypotheses in situations where no or very little information is available about what can be expected to be in the scene.

14

Direct computation of shape cues by scale-space operations

So far we have been concerned with the theory of scale-space representation and its application to feature detection in image data. A basic functionality of a computer vision system, however, is the ability to derive information about the three-dimensional shape of objects in the world.

Whereas a common approach has been to compute two-dimensional image features (such as edges) in a first processing step, and then combining these into a three-dimensional shape description (e.g., by stereo or model matching), we shall here consider the problem of deriving shape cues *directly from image data*, and by using only the types of front-end operations that can be expressed within the scale-space framework.

The basic approach we shall adopt is to observe how texture patterns on surfaces are distorted under perspective transformations. The idea is to measure this distortion in order to estimate the parameters of the transformation, and from these in turn compute the surface shape. While the transformation from the surface to the image is a non-linear mapping, we shall here simplify the problem by considering its locally linearized component (the derivative). As we shall see below, this information allows for determining local surface orientation.

To simplify the presentation, we shall restrict the treatment to the specific case with monocular data—the *shape-from-texture* problem. The general idea of studying local affine transformations as a cue to surface structure is, however, of much wider generality and can be applied in related problems, such as shape-from-stereo and shape-from-motion (see, e.g., Koenderink and van Doorn 1976, 1991).

When addressing this problem, it turns out that a number of conceptually new aspects of scale-space theory need to be introduced. For measuring local linear distortion, we shall define an image descriptor, which is not a feature computed *pointwise* (such as differential singularities), but is integrated over a *region* in scale-space. In this context the need for *two scale parameters* in the shape-from-texture problem (and related problems) is emphasized. A *local scale* parameter is needed for describing the amount of smoothing used for suppressing irrelevant fine-scale structures when computing non-linear descriptors from grey-level data. (This is the traditional scale parameter in the previously treated scale-space representation.) Then, an *integration scale* parameter is required for expressing how large the image region is that is used for integrating such pointwise descriptors into entities that can be used for estimating linear distortion and computing three-dimensional shape cues.

In the following chapter, it will also be demonstrated that once initial surface orientation estimates have been computed, it is advantageous to relax the requirement about rotationally symmetric Gaussian smoothing. By such *shape adaption* of the smoothing kernels, it is possible to improve the accuracy in the computed surface orientation estimates.

The computational model we shall arrive at is expressed completely in terms of differential invariants defined from Gaussian derivatives at multiple scales in scale-space. This makes it attractive both for theoretical analysis and for immediate implementation in a visual front end. In fact, it will be demonstrated that shape cues can be computed directly from image data using only the same types of retinotopic operations that were used for feature detection in chapter 6; (i) large support diffusion smoothing, (ii) small support derivative computations from smoothed grey-level data, and (iii) pointwise non-linear combinations of these. The only qualitative difference is that now *two stages* of that scheme will be applied.

14.1. Shape-from-texture: Review

The systematic distortion of texture patterns under perspective transformations can under certain conditions provide strong cues to surface orientation and surface shape. While the general problem of determining three-dimensional surface structure from monocular data is obviously underdetermined (because of dimensionality considerations), it may become feasible if additional *a priori* information is available. This is the subject of *shape-from-texture*: to combine measurements of image structures with different assumptions about the surface texture and the camera geometry in order to derive three-dimensional shape cues.

This idea of using perspective distortion of texture as a cue to surface shape goes back to Gibson (1950), who coined the term "texture gradient" for the systematic variation of image textures induced by perspective effects. Since then, computational studies of the shape-from-texture problem have been done by several researchers, for example, Bajcsy and Lieberman (1976), Stevens (1981), Witkin (1981), Kanatani (1984), Pentland (1986), Aloimonos (1988), Blostein and Ahuja (1989), Kanatani and Chou (1989), Blake and Marinos (1990), Brown and Shvaytser (1990), Gårding (1991, 1993), and Lindeberg and Gårding (1993).

The presentation here follows the last work. As a background, we shall first review some basic theory of shape-from-texture. Then, the specific methodology for computing surface orientation will be described.

14.1.1. Basic perspective effects

Consider a smooth surface S viewed in perspective projection, and assume that there is a pattern on the surface, which can be thought of as being painted[1] on it. Due to the perspective transformation, the surface pattern is distorted in several ways:

- the *scaling* effect—projections of distant surface structures appear smaller than projections of nearby ones,
- the *foreshortening* effect—the pattern is compressed along a certain direction, and
- the *position* effect—the orientation of projected surface structures depends upon their position relative to the image center.

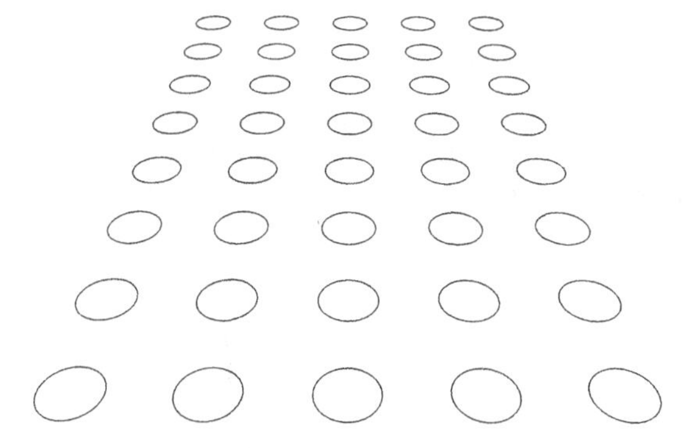

These effects are schematically illustrated in the figure above, which shows the perspective projection of a planar surface with regularly spaced circles of uniform size. Under the perspective transformation each circle transforms into an ellipse, and the above-mentioned perspective effects manifest themselves as follows:

- scaling: the length of the major (minor) axis scales inversely proportional to the distance to the surface,
- foreshortening: the ratio between the lengths of the minor and the major axes is equal to the cosine of the angle between the view direction and the surface normal, and

[1]This is the well-known painted-surface-assumption, commonly exploited in shape-from-texture algorithms. More generally, real-world textures often have fine scale depth variations, which may lead to non-trivial illumination phenomena under variations of the viewing angle. Such effects are not considered here.

- position: the direction of the major (minor) axes of the ellipse depends upon the position in the image plane.

Of course, several other effects arise as well. Some simple observations that can be made are that the *area* of the texture elements varies as does their *density*. These so-called area and density gradients are, however, not independent of the three effects listed above, which in a certain respect can be regarded as basic.

The scaling and the foreshortening effects depend directly upon the local surface geometry. The position effect, on the other hand, is mainly an artifact of mapping the image data onto a planar image, and disappears when using a spherically symmetric camera geometry. In fact, the first two effects characterize the linear component of the perspective mapping, which in turn determines the local surface orientation. In order to express these relationships precisely, let us introduce a few definitions and review some differential geometry.

14.1.2. Camera geometry

Following Gårding (1992) and using standard notation from differential geometry (see, e.g., O'Neill 1966), consider the perspective mapping of a smooth surface S onto a unit viewsphere Σ (see figure 14.1).

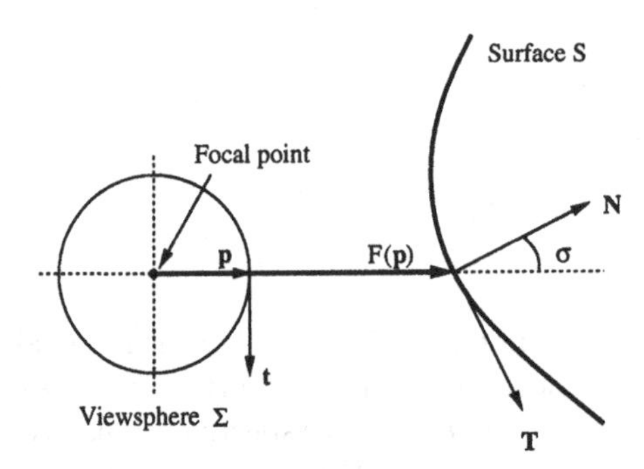

Figure 14.1. Local surface geometry and imaging model. The tangent planes to the viewsphere Σ at p and to the surface S at $F(p)$ are seen edge-on but are indicated by the tangent vectors $\bar{t}$ and $\bar{T}$. The tangent vectors $\bar{b}$ and $\bar{B}$ are not shown but are perpendicular to the plane of the drawing, into the drawing. (Adapted from Gårding 1992).

At any point p on Σ, let $(\bar{p}, \bar{t}, \bar{b})$ be a local orthonormal coordinate system defined such that $\bar{p}$ is parallel to the view direction, $\bar{t}$ is parallel to the direction of the gradient of the distance from the focal point, and $\bar{b} = \bar{p} \times \bar{t}$. Denote by $F\colon \Sigma \to S$ the perspective backprojection, and by F_{*p} the derivative of this mapping at any point p on Σ. The mapping F_{*p}, which constitutes the linear approximation of F at p, maps point in the tangent plane of Σ at p, denoted $T_p(\Sigma)$, to points in the tangent plane of S at $F(p)$, denoted $T_{F(p)}(S)$. In $T_{F(p)}(S)$, let $\bar{T}$ and $\bar{B}$ be the normalized images of $\bar{t}$ and $\bar{b}$ respectively. In the bases $(\bar{t}, \bar{b})$ and $(\bar{T}, \bar{B})$ the expression for $F_{*p}: T_p(\Sigma) \to T_{F(p)}(S)$ is

$$F_{*p} = \begin{pmatrix} r/\cos\sigma & 0 \\ 0 & r \end{pmatrix} = \begin{pmatrix} 1/m & 0 \\ 0 & 1/M \end{pmatrix}, \tag{14.1}$$

where $r = |F(p)|$ is the distance along the visual ray from the center of projection to the surface (measured in units of the focal length) and σ is the *slant* of the surface. The inverse eigenvalues of F_{*p}, m and M, have a simple geometric interpretation. They describe how a unit circle in $T_{F(p)}(S)$ is transformed when mapped to $T_p(\Sigma)$ by F_{*p}^{-1}; it becomes an ellipse with m as minor axis (parallel to the $\bar{t}$ direction) and M as major axis (parallel to the $\bar{b}$ direction).

When brightness data are available in a planar image Π rather than on the viewsphere Σ, an easy way to apply the same analysis is by decomposing the mapping $A: \Pi \to S$ from the planar image to the surface into $A = F \circ G$, where $G: \Pi \to \Sigma$ represents the so-called *gaze transformation* from any point q in the image plane to a corresponding point p on the viewsphere. Then, the mapping F_{*p} can be computed from the derivative $A_{*q} = F_{*p}\, G_{*q}$, since the derivative G_{*q} of the gaze transformation is known provided that the camera geometry is.

This differential geometric framework expresses the relations between local perspective distortion and surface shape. In particular, it shows that the derivative F_{*p} reveals the local surface orientation at any point $F(p)$ on the surface. The eigenvector $\bar{t}$ of F_{*p} corresponding to the smaller inverse eigenvalue m of F_{*p} gives the *tilt direction,* that is the direction of the gradient of the distance from Σ to the surface, and the ratio between the two inverse eigenvalues is directly related to foreshortening ϵ and surface slant σ by

$$\epsilon = \cos\sigma = \frac{m}{M}. \tag{14.2}$$

More generally, the "texture gradients" introduced by Gibson (1950) can be given precise mathematical definitions in terms of the rate of change of the components of F_{*p}. For example, the local area ratio between the image and the surface is $1/\det F_{*p} = mM$. From this entity, the *normalized area gradient* can be defined as $\nabla(mM)/(mM)$.

A detailed discussion of the shape cues that can be derived from the components of F_{*p} and its derivatives is given in (Gårding 1992).

14.2. Definition of an image texture descriptor

From the analysis in the previous section it is clear that if the linear component of the perspective mapping can be recovered, then surface orientation can be computed up to the sign of the tilt direction. If we are to make use of this property in practice, it is necessary to define an image descriptor that allows for measurements of local linear distortion.

14.2.1. The windowed second moment matrix

A useful type of descriptor for this is a *second moment matrix.* It can be thought of as a covariance matrix of a two-dimensional random variable, or as the moment of inertia of a mass distribution in the plane. It can be graphically represented by an ellipse, and as we shall see below, a linear transformation of the spatial coordinates affects the ellipse in the same way as it would affect a physical ellipse painted on a planar surface. Various types of second moment matrices have been used in texture analysis by Bigün *et al.* (1991), Rao and Schunk (1991), Kanatani (1984), Brown and Shvaytser (1990), Blake and Marinos (1990), Gårding (1991, 1993), and Super and Bovik (1992). Interestingly, the second moment matrix is also a useful image descriptor for junction detection (see section 13.4.4).

The specific type of second moment matrix to be used here is defined as follows: Let $L\colon \mathbb{R}^2 \to \mathbb{R}$ be the image brightness, and let $\nabla L = (L_x, L_y)^T$ denote its gradient. Given a symmetric and normalized window function $w\colon \mathbb{R}^2 \to \mathbb{R}$, define *the windowed second moment matrix* by

$$\mu_L(q) = \int_{x\in\mathbb{R}^2} (\nabla L(x))(\nabla L(x))^T\, w(q-x)\, dx, \tag{14.3}$$

where $q \in \mathbb{R}^2$ denotes the image point at which this descriptor is computed. For simplicity of notation, introduce an averaging operator E_q describing the windowing operation. Then, (14.3) can be written

$$\mu_L(q) = \begin{pmatrix} \mu_{11} & \mu_{12} \\ \mu_{21} & \mu_{22} \end{pmatrix} = E_q \begin{pmatrix} L_x^2 & L_xL_y \\ L_xL_y & L_y^2 \end{pmatrix} = E_q((\nabla L)(\nabla L)^T). \tag{14.4}$$

14.2.2. Interpretation in the frequency domain

A more appropriate name for μ_L would be the *windowed covariance matrix.* The reason why it is called a second moment matrix is historical, and can be understood from its interpretation in the Fourier domain.

Rename temporarily the coordinates $(x, y)^T$ to $x = (x_1, x_2)^T$, and let the window function be constant ($w(x_1, x_2) = 1$). Moreover, let $\omega = (\omega_1, \omega_2) \in \mathbb{R}^2$, and let $S_L\colon \mathbb{R}^2 \to \mathbb{R}$ denote the power spectrum of L, i.e.,

$$S_L(\omega) = \hat{L}(\omega)\, \hat{L}^*(\omega),$$

where $\hat{L}\colon \mathbb{R}^2 \to \mathbb{C}$ denotes the Fourier transform of L, and $\hat{L}^*$ is its complex conjugate. Then, by inserting $h_i = L_{x_i}$ in Plancherel's relation

(13.83) it follows that

$$\int_{x \in \mathbb{R}^2} L_{x_i} L_{x_j} \, dx = \frac{1}{(2\pi)^2} \int_{\omega \in \mathbb{R}^2} \omega_i \omega_j \, S_L(\omega) \, d\omega. \tag{14.5}$$

This relation shows that by disregarding the window function, μ_L can be interpreted as the second moment matrix of the power spectrum S_L.

14.2.3. Interpretation in terms of local directional statistics

To obtain a better intuitive understanding of this descriptor, define the entities

$$P = E_q(L_x^2 + L_y^2), \quad C = E_q(L_x^2 - L_y^2), \quad S = 2\,E_q(L_x L_y). \tag{14.6}$$

from the components of μ_L (here the argument q has been dropped to simplify the notation). Then, it can be seen that the first descriptor P is a natural measure of the strength of the operator response; it is the average of the square of the gradient magnitude in a neighbourhood of q. The two other entities C, S contain directional information, which can be summarized into the two *anisotropy* measures

$$Q = \sqrt{C^2 + S^2}, \qquad \tilde{Q} = \frac{Q}{P}. \tag{14.7}$$

Clearly, for the *normalized anisotropy* $\tilde{Q} \in [0, 1]$ it holds that

- $\tilde{Q} = 0$ if and only if $E_q(L_x^2) = E_q(L_y^2)$ and $E_q(L_x L_y) = 0$, while
- $\tilde{Q} = 1$ if and only if $(E_q(L_x L_y))^2 = E_q(L_x^2) E_q(L_y^2)$.

For example, a rotationally symmetric pattern has $\tilde{Q} = 0$, while $\tilde{Q} = 1$ for a translationally symmetric distribution. Rotational symmetry is, however, not necessary to give $\tilde{Q} = 0$. For example, any pattern with $N \geq 2$ uniformly distributed (unsigned) directions satisfies $\tilde{Q} = 0$.

It can be shown that $\mu_L(q)$ is a well-defined (coordinate independent) differential geometric entity in the image domain. It is invariant to translations, and has a nice behaviour with respect to uniform rescalings in the spatial domain and affine brightness transformations.[2] Although the matrix itself is not invariant to rotations, its eigenvectors follow any rotation of the coordinate system

$$\bar{e}_1 = \begin{pmatrix} C + Q \\ S \end{pmatrix}, \quad \bar{e}_2 = \begin{pmatrix} -S \\ C + Q \end{pmatrix}, \tag{14.8}$$

and its eigenvalues, $\lambda_1 \geq \lambda_2$, are invariant to rotations

$$\lambda_{1,2} = P \pm Q = P\,(1 \pm \tilde{Q}). \tag{14.9}$$

[2]Such transformations only affect μ_L by a uniform scaling factor.

The vector $(C, S)^T$ can be interpreted in terms of local statistics of gradient directions. The transformation (14.6) maps each gradient vector

$$\nabla L = \begin{pmatrix} L_x \\ L_y \end{pmatrix} = \rho \begin{pmatrix} \cos\alpha \\ \sin\alpha \end{pmatrix} \tag{14.10}$$

to the corresponding pointwise component of $(C, S)^T$

$$\begin{pmatrix} L_x^2 - L_y^2 \\ 2L_xL_y \end{pmatrix} = \rho^2 \begin{pmatrix} \cos^2\alpha - \sin^2\alpha \\ 2\cos\alpha\sin\alpha \end{pmatrix} = \rho^2 \begin{pmatrix} \cos 2\alpha \\ \sin 2\alpha \end{pmatrix}. \tag{14.11}$$

This means that $(C, S)^T$ can be interpreted as the weighted average of the (unsigned) gradient vectors in a local neighbourhood of q with each unit vector mapped to the double angle[3] and given a weight proportional to the square of the gradient magnitude, ρ^2, multiplied by the window function w. The direction of $\arg(C, S)^T/2$ is parallel to $\bar{e}_1$ and represents the *unsigned average direction* of the weighted gradient distribution. This can be easily seen by inserting

$$\begin{pmatrix} C \\ S \end{pmatrix} = Q \begin{pmatrix} \cos 2\beta \\ \sin 2\beta \end{pmatrix} \tag{14.12}$$

into the explicit expressions for the eigenvectors of μ_L (14.8), which gives

$$\bar{e}_1 = \begin{pmatrix} \cos\beta \\ \sin\beta \end{pmatrix}, \quad \bar{e}_2 = \begin{pmatrix} -\sin\beta \\ \cos\beta \end{pmatrix}. \tag{14.13}$$

14.2.4. Linear transformation property

The usefulness of this descriptor with respect to estimation of linear distortion can be realized from its transformation property under linear transformations. Under an invertible linear transformation of the image domain $\eta = B\xi$ (where B represents an invertible 2×2 matrix) it can be shown that if a transformed intensity pattern $R\colon \mathbb{R}^2 \to \mathbb{R}$ is defined by $L(\xi) = R(B\xi)$, then $\mu_L(q)$ transforms as

$$\mu_L(q) = B^T \mu_R(p)\, B, \tag{14.14}$$

where $\mu_R(p)$ is the second moment matrix of R at $p = Bq$ computed with respect to the "backprojected" normalized window function

$$w'(\eta - p) = (\det B)^{-1} w(\xi - q). \tag{14.15}$$

[3]This is a standard technique for defining statistical properties of periodic data; see (Mardia 1972) for an overview. Various forms of this "double angle representation" have been exploited in computer vision by, for example, Knutsson and Granlund (1983).

This result can be easily verified by inserting $\nabla L(\xi) = B^T \nabla R(B\xi)$ into (14.4), which gives

$$\mu_L(q) = \int_{\xi \in \mathbb{R}^2} w(q - \xi)\, B^T (\nabla R(B\xi))(\nabla R(B\xi))^T B\, d\xi$$
$$= B^T \left\{ \int_{\eta \in \mathbb{R}^2} w(B^{-1}(p - \eta))\, (\nabla R(\eta))(\nabla R(\eta))^T\, (\det B)^{-1}\, d\eta \right\} B.$$

The integral within brackets can be recognized as the second moment matrix of $R(p)$ if w' according to (14.15) is used as window function. Clearly, the transformed window function w' is normalized if w is, since

$$\int_{\eta \in \mathbb{R}^2} w(B^{-1}(\eta - p))(\det B)^{-1}\, d\eta = \int_{\xi \in \mathbb{R}^2} w(\xi - q)\, d\xi.$$

14.3. Deriving shape cues from the second moment matrix

With respect to shape-from-texture analysis, let now B in (14.14) represent the locally linearized perspective mapping, i.e.,

$$B = A_{*q} = F_{*p}\, G_{*q}. \tag{14.16}$$

Then, the linear transformation property can be expressed as

$$\mu_L(q) = G_{*q}^T\, F_{*p}^T\, \mu_S(F(G(q)))\, F_{*p}\, G_{*q}, \tag{14.17}$$

where $p = G(q)$ is the point on the viewsphere corresponding to an image point q, and $\mu_S(F(G(q)))$ denotes the second moment matrix defined in the tangent plane to the surface with respect to the backprojected window function $w'(\eta - F(G(q))) = (\det A_{*q})^{-1} w(\xi - q)$.

The general procedure, then, for estimating shape from texture is to combine estimates of $\mu_L(q)$ with assumptions about the structure of the surface brightness pattern $\mu_S(F(G(q)))$ to infer the structure of A_{*q}. This permits computation of F_{*p} after compensation with respect to G_{*q}.

14.3.1. Weak isotropy assumption.

A simple but often fruitful assumption is that $\mu_S(F(p))$ is proportional to the unit matrix

$$\mu_S = cI \tag{14.18}$$

for some constant $c > 0$. Such a distribution (for which $\tilde{Q} = 0$) is called *weakly isotropic*, and essentially means that there is no single dominant direction in the surface texture. Under this condition and assuming that F_{*p} is non-degenerate, μ_Σ in the tangent plane $T_p(\Sigma)$ to the viewsphere can be written

$$\mu_\Sigma(p) = c\, F_{*p}^T\, F_{*p} \tag{14.19}$$

for some $c > 0$. Hence, the eigenvectors of $\mu_\Sigma(p)$ and F_{*p} are the same, and the eigenvalues of F_{*p} are proportional to the square roots of the eigenvalues of $\mu_\Sigma(p)$. In particular, the tilt direction $\bar{t}$ is parallel to the eigenvector $\bar{e}_1$ corresponding to the maximum eigenvalue, and the slant is given by

$$\cos\sigma = \frac{m}{M} = \sqrt{\frac{\lambda_2}{\lambda_1}} = \sqrt{\frac{1-\tilde{Q}}{1+\tilde{Q}}}. \tag{14.20}$$

Hence, a direct estimate of local surface orientation is available whenever the assumption of weak isotropy can be justified.

14.3.2. Constant area assumption.

A less restrictive assumption is that the local "size" of the surface texture elements does not vary systematically. Then, $A = \det F_{*p}^{-1} = mM$ is an area measure. It can be shown (Gårding 1992) that the *normalized area gradient* defined from it relates to surface orientation by

$$\frac{\nabla A_\Sigma}{A_\Sigma} = -\tan\sigma \begin{pmatrix} 3 + r\kappa_t/\cos\sigma \\ r\tau \end{pmatrix}, \tag{14.21}$$

where κ_t is the normal curvature of the surface in the tilt direction, and τ is the geodesic torsion in the same direction. From this entity surface orientation can be recovered if the curvature is known or (assumed to be) small. In this case, there is no ambiguity in the sign of the tilt direction unlike the previous case of foreshortening.

14.3.3. Visualization by ellipses

A useful way to illustrate the second moment matrix graphically is in terms of ellipses. Since the second moment matrix is positive semidefinite, it follows that the equation

$$(\xi - q)^T \mu_L(q) (\xi - q) = 1 \quad (\xi, q \in \mathbb{R}^2) \tag{14.22}$$

defines an ellipse (possibly degenerated to a line). It is straightforward to verify that under linear transformations, this ellipse is transformed in exactly the same way as the second moment matrix (14.14). Hence, it can be justified to think of the second moment matrix as an ellipse painted on the surface.

From the analysis in section 14.1.2 it follows that the effect of the linear transformation F_{*p} on a unit circle determines F_{*p} up to the sign of tilt. Moreover, the weak isotropy assumption means that all ellipses are assumed to be circles in the tangent planes to the surface, while the constant area assumption means that the area of each ellipse is assumed to be the same.

14.4. Scale problems in texture analysis

Computation of the image second moment matrix above, or any other non-trivial texture descriptor, involves the integration of image statistics (gradient directions) over finite-sized regions in the image. This immediately leads to two fundamental scale problems; one concerning the scale(s) at which to compute the primitives for the texture descriptors (here, the first order derivatives), and one concerning the size of the regions over which to collect the statistics (here, the choice of window function).

Clearly, the image statistics must be collected from a region large[4] enough to be representative of the texture. Yet, the region must not be so large that the local linear approximation of the perspective mapping becomes invalid. For example, for an ideal texture consisting of isolated blobs, a lower limit for the extent of the integration region is determined by the size of the individual blobs, while an upper limit may be given by the curvature of the surface or interference with other nearby surface patches. This scale controlling the *window function* is referred to as *integration scale* (denoted s).

Moreover, the image statistics must be based on descriptors computed at proper scales, so that noise and "irrelevant" image structures can be suppressed. The descriptors considered here are based on first order spatial derivatives of the image brightness, and it is obvious that useful results hardly can be expected if the derivatives are computed directly from unsmoothed noisy[5] data. This scale determining the amount of *initial smoothing* in the (traditional first-stage) scale-space representation of the image is referred to as *local scale.*[6]

Note that these two scale problems are not completely unrelated, since the integration scale is an upper limit of the local scale.

14.4.1. The multi-scale windowed second moment matrix

With respect to these two scale concepts, it is natural to reinterpret the brightness function L in (14.3) as the scale-space representation of any given image f, and to associate a scale parameter with the window function w. This gives the *multi-scale windowed second moment matrix*

$$\mu_L(q;\ t, s) = \int_{x' \in \mathbb{R}^2} (\nabla L)(x';\ t)\, (\nabla L)^T(x';\ t)\, w(q - x';\ s)\, dx'. \tag{14.23}$$

[4]Note that $\mu_L(q)$ given by (14.3) contains meaningful shape information only when considered as the average over a finite neighbourhood, since the matrix $(\nabla L)(\nabla L)^T$ is always degenerate when treated pointwise.

[5]This problem disappears in ideal noise-free data if the sampling problems are handled properly.

[6]This terminology refers to local operations (derivatives).

A natural choice of window function is a Gaussian function. This choice can be motivated by the fact that the Gaussian function is rotationally symmetric and has a nice scaling behaviour, which means that the invariance properties listed in section 14.2.3 are preserved.

More importantly, however, it holds that then and only then the components of μ_L constitute scale-space representations of the corresponding components of $(\nabla L)(\nabla L)^T$. This is a direct consequence of the uniqueness of the Gaussian kernel for scale-space representation given natural front-end postulates. In the rotationally symmetric case, the following definition is therefore unique:

$$\mu_L(q;\ t, s) = \int_{x' \in \mathbb{R}^2} (\nabla L)(x';\ t)\, (\nabla L)^T(x';\ t)\, g(q - x';\ s)\, dx'. \tag{14.24}$$

Of course, separate smoothing of the individual components of a multi-dimensional entity is not guaranteed to give well-defined (coordinate independent) results.[7] From the invariance properties described in section 14.2.3, however, it follows that (14.24) is a meaningful operation.

14.4.2. Scale selection in shape-from-texture

When using the second moment matrix (14.24) in shape-from-texture computations, it is necessary to have some mechanism for combining information captured by this image descriptor at different scales, since scale levels that are suitable for estimating local linear distortion cannot be assumed to be known *a priori*. Moreover, this mechanism must be able to adapt to the local image structure, since such scale levels can be expected to vary substantially over the image, depending on the type of texture considered, the distance to the surface, and the noise in the image formation process.

Here, it is proposed that the scale selection method described in chapter 13 constitutes a mechanism for selecting locally adapted values of the integration scale parameter. The major motivation for this is that such scales can be assumed to reflect characteristic lengths of image structures (compare with the heuristic principle suggested in section 13.2 and the blob detector in section 13.3).

Experiments illustrating this property are illustrated in figure 14.2. It shows scale-space signatures of two natural measures of the strength of the response of the normalized second moment matrix, the trace and the determinant. The signatures are accumulated for the same two details of the sunflower image as were used in the experiment in figure 13.2 (page 323). Notice the qualitative similarity between scale selection using

[7] Consider, for example, straightforward smoothing of a planar curve by smoothing the coordinate functions, which is well-known to give rise to non-intuitive results.

the windowed second moment matrix and the normalized Hessian matrix respectively.

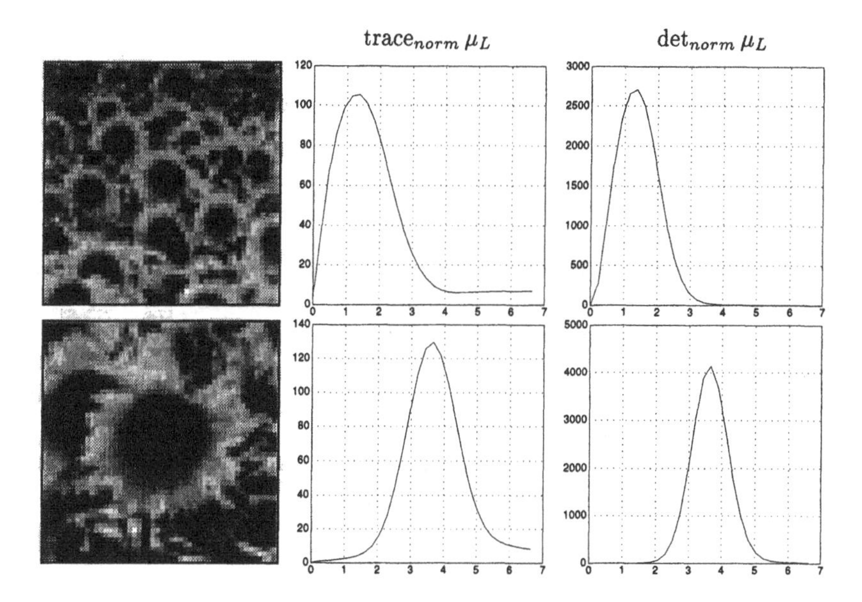

Figure 14.2. Scale-space signatures of trace and the determinant of the normalized second moment matrix computed for two details of a sunflower image; (left) grey-level image, (middle) signature of $\text{trace}_{norm}\,\mu_L$, and (right) signature of $\det_{norm}\,\mu_L$. (Compare with figure 13.2.)

14.4.3. Properties of the scale selection method

What specific differential entities should be used for scale selection when computing the second moment matrix? Here, some properties of normalized scale-space maxima will be described for signals having a slightly more complex frequency content than the single-frequency sine wave.

Sine wave patterns. Assume first for simplicity that the integration scale is much larger than the local scale in the scale selection step, and consider the sum of two *parallel* two-dimensional sine waves.

$$f_{par}(x, y) = \sin\omega_1 x + \sin\omega_2 x, \tag{14.25}$$

where $\omega_1 \leq \omega_2$. After calculations similar to those in section 13.3.2, it can be shown that there is a unique scale maximum in $\text{trace}_{norm}\,\mu_L$ when ω_2/ω_1 is close to one, while there are two scale maxima when ω_2/ω_1 is sufficiently large ($\omega_{bifurc} \approx 2.4$). A similar result holds for two *orthogonal* waves,

$$f_{orth}(x, y) = \sin\omega_1 x + \sin\omega_2 y. \tag{14.26}$$

If this function is interpreted as the locally linearized model of the projection of an isotropic periodic blob-like pattern, with foreshortening

$$\epsilon = \cos\sigma = \frac{\omega_1}{\omega_2}, \tag{14.27}$$

then the interpretation is that the response changes from one to two peaks at slant

$$\sigma_{bifurc} = \arccos(\frac{1}{\omega_{bifurc}}) \approx 65°. \tag{14.28}$$

The determinant of the normalized windowed second moment matrix, $\det_{norm} \mu_L$, behaves somewhat differently; it is identically zero for f_{par}, while there is *always* a unique peak in f_{orth}, assumed at scale

$$t_{\det \mu_L} = \frac{2}{\omega_1^2 + \omega_2^2}. \tag{14.29}$$

Of course, this does not mean that multiple scale maxima cannot occur in $\det \mu_L$. A simple example is the function

$$f_{orth\text{-}par}(x, y) = f_{par}(x, 0) + f_{par}(y, 0) \tag{14.30}$$

(with f_{par} according to (14.25)) which has two scale maxima when ω_{ratio} is sufficiently large (investigating this function leads to similar calculations as the investigation of f_{par}).

Observe the qualitative similarity with scale-space extrema of the trace and the determinant of the normalized Hessian matrix (section 13.3.2).

Isotropic patterns. More generally, for an *isotropic* pattern (with $\tilde{Q} = 0$, or equivalently, $\lambda_1 = \lambda_2$) the scale maxima of $\text{trace}_{norm}\, \mu_L$ and $\det_{norm} \mu_L$ coincide. This is easily proved from $\text{trace}\, \mu_L = \lambda_1 + \lambda_2 = 2\lambda_1$ and $\det \mu_L = \lambda_1 \lambda_2 = \lambda_1^2$, which gives

$$\partial_t \det_{norm} \mu_L = 0 \Leftrightarrow \partial_t \; \text{trace}_{norm}\, \mu_L = 0. \tag{14.31}$$

Unidirectional patterns. For a *unidirectional* pattern (with $\tilde{Q} = 1$, or equivalently, $\lambda_2 = 0$) $\det \mu_L$ is identically zero, while $\text{trace}\, \mu_L$ is non-zero. Hence, $\det \mu_L$ only responds when there are significant variations along *both* the coordinate directions, typically for blob-like signals.

Invariance under linear transformations. More importantly, the scale maxima of $\det \mu_L$ are invariant[8] with respect to affine transformations of

[8]Here, it is assumed that L represents an unsmoothed signal and that the scale maximum occurs at zero scale. This restriction is can, however, be easily relaxed in the affine scale-space framework (see section 15.2), where linear shape adaption of the smoothing kernels is permitted.

the image coordinates. Given an invertible linear transformation $L(\xi) = R(B\xi)$ it follows from (14.14) that $\det \mu_L = (\det B)^2 \det \mu_R$, which gives

$$\partial_t \det \mu_L = 0 \Leftrightarrow \partial_t \det \mu_R = 0. \tag{14.32}$$

Because of these properties, $\det \mu_L$ will be favoured as the main descriptor of these two in shape-from-texture.

14.4.4. Effect of window functions

For a perfectly periodic texture, a translationally invariant texture descriptor is obtained by integrating the local image statistics over one period. In general, however, it cannot be assumed that a texture is perfectly periodic or that its wavelength is known. To study the effect of using finite-size window functions with different integration scales, consider first a one-dimensional phase shifted sinusoidal signal

$$f(x) = \sin(\omega_0 x + \phi) \tag{14.33}$$

as a model for the essential variations in $\det \mu_L$ for a periodic isotropic pattern. The normalized integrated response corresponding to (14.3) is

$$P_{norm}(t) = \int_{x=-\infty}^{\infty} w(x)\, L_\xi^2(x;\ t)\, dx = \frac{t\,\omega_0^2\, e^{-\omega_0^2 t}}{2}(1 + e^{-2\omega_0^2 s} \cos 2\phi).$$

If the integration scale is proportional to the local scale $s = \gamma_1^2 t$, then the relative ringing is obviously proportional to

$$e^{-2\gamma_1^2} \tag{14.34}$$

at scale

$$t_0 = \frac{1}{\omega_0^2} \tag{14.35}$$

(which is the scale at which the maximum normalized response is assumed). This expression gives an indication of how large variations a given ratio γ_1 between the integration scale and the local scale can be expected to give rise to. For example, $\gamma_1 = \sqrt{2}$ gives a relative ringing of about 2%, and $\gamma_1 = 2$ a relative ringing of 0.03%. Assuming that γ_1 is reasonably large, it can be shown that the spatial sensitivity of the selected scale levels shows a similar qualitative behaviour.

14.4.5. Shape distortions due to scale-space smoothing

The scale-space smoothing also leads to shape distortions. This fact is well-known, for example, in edge detection. In texture analysis an immediate effect is that the anisotropy in the image domain is affected, which affects the estimates of slant and tilt.

Decreasing anisotropy: Systematic underestimate of slant. To model this effect, consider first a non-uniform Gaussian blob

$$f(x,y) = g(x;\ l_1^2)\, g(y;\ l_2^2) \quad (l_1 \geq l_2 > 0), \tag{14.36}$$

as a simple linearized model of the projection of a rotationally symmetric Gaussian blob with foreshortening

$$\epsilon = \cos\sigma = \frac{l_2}{l_1}. \tag{14.37}$$

From the semi-group property of the Gaussian kernel, it follows that the scale-space representation of f at scale t is

$$L(x,y;\ t) = g(x;\ l_1^2 + t)\, g(y;\ l_2^2 + t), \tag{14.38}$$

which means that under scale-space smoothing foreshortening varies as

$$\epsilon(t) = \sqrt{\frac{l_2^2 + t}{l_1^2 + t}}, \tag{14.39}$$

i.e., it increases and tends to one, which means that eventually the image will be interpreted as flat.

Increasing anisotropy: Suppression of fine scale structures. Scale-space smoothing, may however, also *increase* the anisotropy, typically by suppression of fine-scale isotropic image structures superimposed onto a coarser-scale anisotropic pattern. To model this behaviour, consider the again the sine-wave pattern

$$f(x,y) = \sin\omega_1 x + \sin\omega_2 y \quad (\omega_1 \leq \omega_2). \tag{14.40}$$

Assuming that the integration scale s is large relative to the wavelengths of this signal (i.e., that $\omega_1^2 s >> 1$, it can be shown that the normalized anisotropy is

$$\tilde{Q}_L(t) = \frac{(\omega_2/\omega_1)^2\, e^{-(\omega_2^2-\omega_1^2)t} - 1}{(\omega_2/\omega_1)^2\, e^{-(\omega_2^2-\omega_1^2)t} + 1}. \tag{14.41}$$

For small t it holds that

$$\tilde{Q}_L(t) = \frac{\omega_2^2 - \omega_1^2}{\omega_2^2 + \omega_1^2}\,(1 - \frac{2\omega_1^2\omega_2^2}{\omega_1^2 + \omega_2^2}\, t + \mathcal{O}(t^2)). \tag{14.42}$$

Also here the normalized anisotropy decreases for small t, while it tends to the maximum possible value ($\tilde{Q}_L = 1$) when t tends to infinity.

The intuitive explanation of this effect is simple. Under scale-space smoothing the fine-scale pattern is suppressed faster than the coarser-scale pattern. Hence, after a large amount of smoothing, only one of the two sine waves remains, and the pattern is translationally symmetric.

14.5. Computational methodology and experiments

Now, experimental results will be presented demonstrating how the suggested framework can be used for deriving shape cues from brightness data. We shall start by describing a methodology for selecting scale levels for the two scale parameters in the second moment matrix μ_L.

14.5.1. Basic scheme for computing the second moment matrix

In its most general form, the adaptive scheme proposed for computing μ_L can be summarized as follows. Given any point in the image;

1. vary the two scale parameters, the local scale t and the integration scale s, according to some scheme;

2. accumulate the scale-space signature for some (normalized) differential entity;

3. detect some special property of the signature, e.g., the global maximum, or all local extrema, etc;

4. set the integration scale(s) proportional to the scale(s) where the above property is assumed;

5. compute μ_L at this fixed integration scale while varying the local scale between a minimum scale (e.g., $t = 0$) and the integration scale, and then select the most appropriate local scale(s) according to some criterion.

A specific implementation of this general scheme is described below.

14.5.2. Scale variation

A completely general implementation of step 1 would involve a full two-parameter scale variation. Here, a simpler but quite useful approach is used; the integration scale is set to a constant times the local scale,

$$s = \gamma_1^2 \, t \tag{14.43}$$

(where typically $\gamma_1 = \sqrt{2}$ or 2). This choice, which is natural in terms of scale invariance, may be justified, for example, for signals that are approximately periodic (see section 14.4.4).

More generally, this method is applicable if the texture elements are approximately uniformly distributed, and if it is known in advance over how large a region in the surface (measured in terms of the number of texture elements) the directional statistics must be accumulated to give a useful texture descriptor. Under this assumption, the main effect of the automatic scale selection method will be to adjust the scale levels with respect to the perspective distance effects and noise suppression.

In view of the heuristic scale selection principle, it means that if a normalized differential entity assumes a maximum over scales at a certain scale, the size of the integration region will be proportional to a characteristic length of the image structures that gave rise to the maximum.

14.5.3. Selection of integration scales

Concerning steps 2–3, the integration scales are set from the scale(s), denoted $s_{\det \mu_L}$, where the *normalized strength of* μ_L, represented by $\det \mu_L$, assumes a local or global *maximum.*

This choice is motivated by the observation that for simple periodic and blob-like patterns, the signature of $\det \mu_L$ has a single peak reflecting the characteristic size (area) of the two-dimensional pattern, while for $\operatorname{trace} \mu_L$ the response changes from one to two peaks with increasing (linear) distortion.

Once $s_{\det \mu_L}$ has been determined, it is advantageous to compute μ_L at a slightly larger integration scale

$$s = \gamma_2^2 \, s_{\det \mu_L} = \gamma_1^2 \gamma_2^2 \, t_{\det \mu_L} \tag{14.44}$$

(where typically $\gamma_2 = \sqrt{2}$ or 2) in order to suppress local variations in surface textures and obtain a more stable descriptor.

More formally, using $\gamma_2 > 1$ can be motivated by the fact that the estimates of the directional information in μ_L tend to be more sensitive to small window sizes than the magnitude estimates.

14.5.4. Selection of local scales

The second stage selection of local scale in step 5 aims at reducing the shape distortions due to smoothing. The *local scales* will be set to the scales, denoted t_Q, where the *normalized anisotropy,* $\tilde{Q}$, assumes a local maximum.

This is motivated by the fact that in the absence of noise and interfering finer scale structures, the main effect of the first stage scale-space smoothing is to *decrease* the anisotropy. For example, in section 14.4.5 we have seen that the aspect ratio of an elliptical Gaussian blob $f(x, y) = g(x;\, l_1^2)\, g(y;\, l_2^2)$ varies as $(l_2^2 + t)/(l_1^2 + t)$ and clearly approaches one as t is increased. On the other hand, suppressing isotropic noise and interfering finer scale structures *increases* the anisotropy. Selecting the maximum point gives a natural trade-off between these two effects. Note that under the assumption of weak isotropy a maximum in $\tilde{Q}$ is equivalent to a maximum in σ.

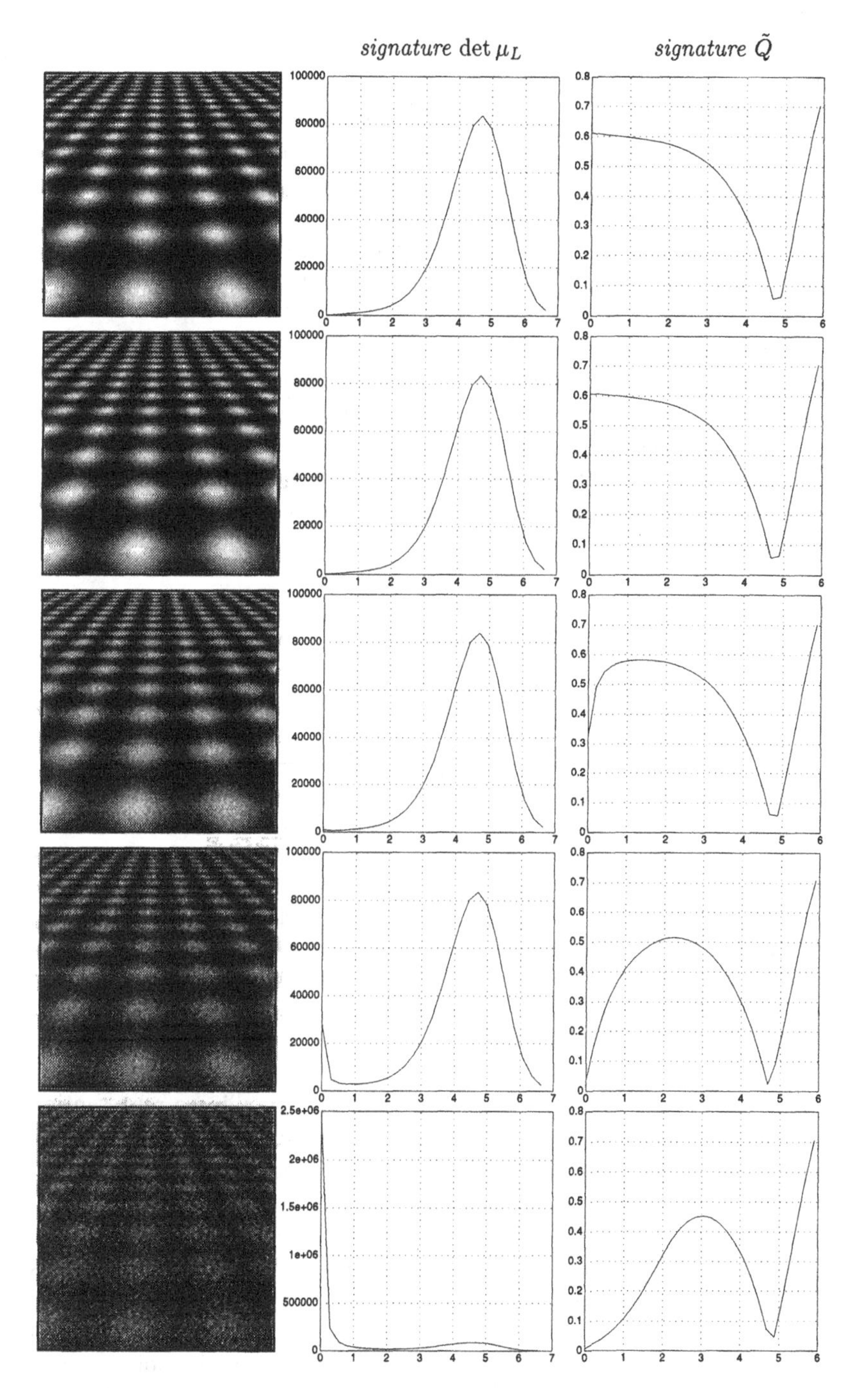

Figure 14.3. Scale-space signatures of det μ_L and $\tilde{Q}$ (accumulated at the central point) for a synthetic texture with added (white Gaussian) noise of different standard deviation. (The noise levels are $\nu = 0.0$, 1.0, 10.0, 31.6, and 100.0 from top to bottom, and the range of grey-levels is [0..255].)

14.5.5. Experiments

Figure 14.3 illustrates these effects for a synthetic image with different amounts of (added white Gaussian) noise. Note that the scale-space signature of $\det \mu_L$ has a unique maximum when the noise level ν is low, and that there are two maxima when ν is increased; one corresponding to the sinusoidal pattern and one corresponding to the noise. When ν is small, the maximum in $\tilde{Q}$ is assumed at fine scales, while with increasing ν the maximum is assumed at coarser scales.

Table 14.1 gives numerical values using the proposed method for scale selection. Observe the stability of $s_{\det \mu_L}$ with respect to variations in the noise level, and that the selected local scale t_Q increases with the noise level ν. Notice also how slowly $\tilde{Q}_{max}(t_Q)$ decreases with ν compared to $\tilde{Q}$ computed at $t = 0$. The accuracy in the orientation estimate assuming weak isotropy is presented by the three-dimensional angle $\Delta\phi_n$ between the estimated and true surface normal.

noise level ν	$s_{\det \mu_L}$	t_Q	$\hat{\sigma}$ at t_Q	$\hat{\theta}$ at t_Q	$\Delta\phi_n(t_Q)$	$\Delta\phi_n(t=0)$
1.0	34.9	0.0	60.1°	90.2°	0.2°	(0.2°)
10.0	34.4	2.0	58.9°	90.3°	1.1°	(15.3°)
31.6	34.1	4.2	55.3°	90.2°	4.7°	(45.3°)
100.0	31.4	8.5	52.3°	91.1°	7.8°	(53.7°)

Table 14.1. Numerical values of characteristic entities in the experiments in (the center of) figure 14.3 using different amounts of additive Gaussian noise. The table shows the scales $s_{\det \mu_L}$ and t_Q, where $\det \mu_L$ and $\tilde{Q}$ assume their maxima over scales, as well as estimates of the slant $\hat{\sigma}$ angle and the tilt $\hat{\theta}$ direction computed at $t = t_Q$. The error is given in terms of the (three-dimensional) angle $\Delta\phi_n(t_Q)$ between the true ($\sigma = 60°$, $\tau = 90°$) and the estimated surface normals (assuming weak isotropy). For comparison, the error in surface orientation corresponding to $t = 0$ is also shown.

Figure 14.4 illustrates these results graphically, by ellipses representing the second moment matrices. The shape of each ellipse is determined by (14.22), while the size is rescaled such that the area is proportional to the scale $s_{\det \mu_L}$ at which the maximum in the signature is assumed.

As a comparison, figure 14.5 displays a typical result of using non-adaptive (globally constant) scale selection. Here, useful shape descriptors are only obtained in a small part; the window size is too small in the lower part, while the first stage smoothing leads to severe shape distortions in the upper region. This result demonstrates that when the noise level is sufficiently high and the perspective effects are strong enough, *it is not possible to use the same values of the two scale parameters all over an image.*

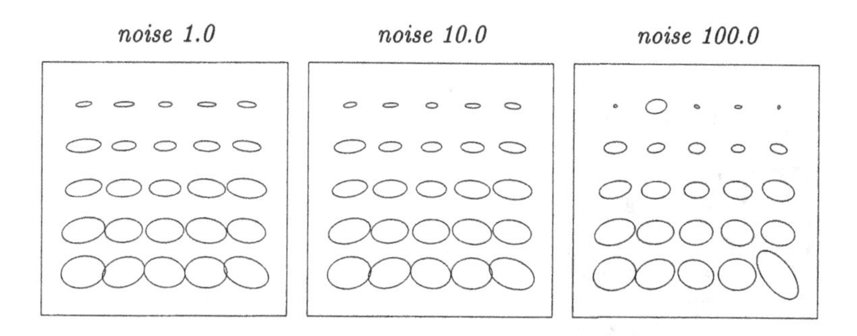

Figure 14.4. Ellipses representing μ_L computed at different spatial points using *automatic scale selection* of the local scale and the integration scale. Note the stability of these descriptors with respect to variations of the noise level.

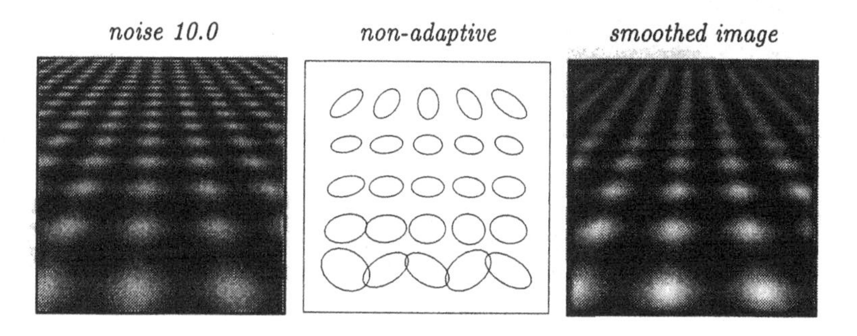

Figure 14.5. Typical example of the result of using *non-adaptive* selection of the (here constant) local and integration scales; it can be seen that geometrically useful shape descriptors are obtained only in a small part of the image.

The first row in figure 14.6 shows corresponding results for another oblique view of the planar pattern in figure 14.4, while the second row shows results computed from a curved surface (the inside of a cylinder). Note how stable these descriptors are considering that they have been computed independently and pointwise. Numerical values of slant and tilt estimates computed under the assumption of weak isotropy are given in table 14.2.

The only essential parameter in these experiments is the ratio between the integration scale and the local scale, the *relative integration scale* given by $\gamma = \gamma_1 \gamma_2$. In these experiments, it has been kept constant and rather low ($\gamma_1 = \sqrt{2}, \gamma_2 = 2$), assuming that the surface orientation estimates are to be computed from small regions, basically corresponding to individual texture elements.

The importance of this parameter is illustrated in figure 14.7, which shows ellipses computed from a real image of a wall-paper using two dif-

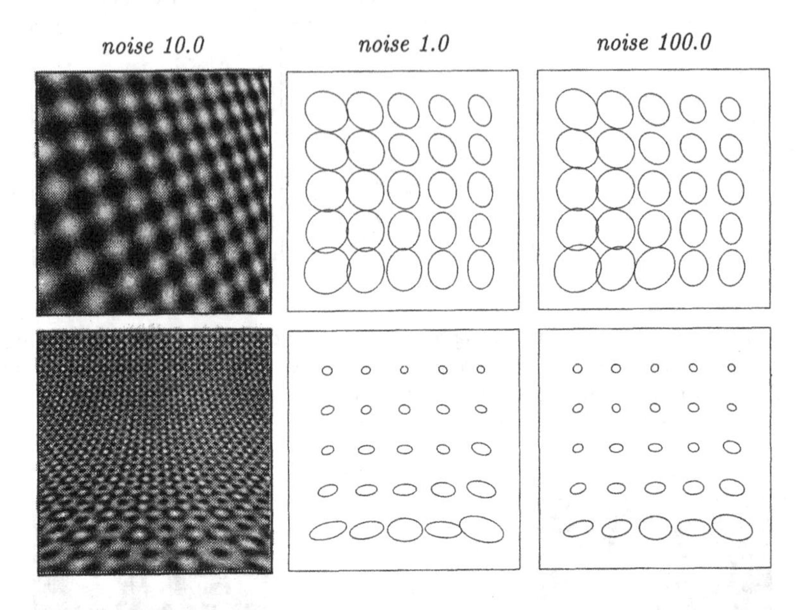

Figure 14.6. Ellipses representing μ_L computed at different spatial points using automatic scale selection of the local scale and the integration scale; (top row) an oblique view of the pattern in figure 14.4, (bottom row) a curved surface (the inside of a cylinder).

image	noise level	$\hat{\sigma}$ at t_Q	$\hat{\theta}$ at t_Q	$\Delta\phi_n$ at t_Q
"oblique"	1.0	30.2°	19.6°	0.2°
	10.0	29.3°	19.1°	0.8°
	100.0	25.0°	25.1°	5.5°
"curved"	1.0	59.5°	88.2°	1.6°
	10.0	58.1°	88.4°	2.3°
	100.0	57.4°	89.9°	2.6°

Table 14.2. Numerical values of some characteristic entities in the experiments in figure 14.6 using different amounts of additive Gaussian noise and automatic setting of the integration scale and the local scale. (The actual slant and tilt angles were ($\hat{\sigma} = 30°$, $\hat{\theta} = 20°$) and ($\hat{\sigma} = 60°$, $\hat{\theta} = 90°$) for the images named "oblique" and "curved" respectively).

ferent settings of γ_2. When using a small value of the relative integration scale, the operator responds to the orientations of the *individual flowers* painted on the surface, while when the relative integration scale is increased, the response reveals the *overall direction* of the plane (in agreement with the qualitative multi-scale behaviour that can be expected from increasing integration scales).

In the latter case, the estimated slant and tilt values at the central point were $(46.3°, 86.2°)$ (assuming weak isotropy) compared to the reference values[9] $(50.8°, 85.3°)$. This gives a difference in the (three-dimensional) normal direction of about 4.5°.

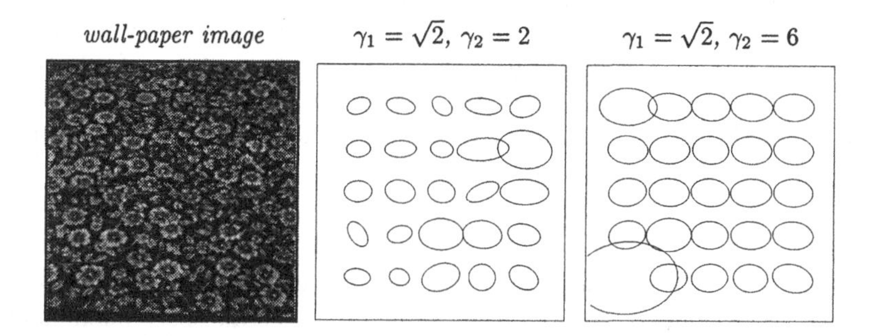

Figure 14.7. Second moment matrices computed from a real image of a wallpaper using different settings of the relative integration scale; (left) grey-level image, (middle)–(right) second moment matrices using different values of γ_2. In the first case, the ellipses reflect the directions of the individual flowers painted on the surface, while in the second case, the ellipses respond to the overall direction of the plane on which the wallpaper is mounted.

14.6. Spatial selection and blob detection

Although the above method for selecting appropriate scales for local smoothing and regional integration at a given point can be demonstrated to give stable texture descriptors on various types of real and synthetic images, the method has obvious limitations if applied blindly.

For textures that satisfy the previously stated assumptions about approximate periodicity or known relative integration scale, all points contain basically the same information. On the other hand, for textures that violate these assumptions, the image descriptors may be sensitive with respect to the points they are computed at. Obvious examples are surface patterns having uniform brightness or very sparsely distributed texture elements. Clearly, regions with uniform brightness contain no informa-

[9] Computed by stereo correspondence using the PMF algorithm developed by Pollard, Mayhew and Frisby (1985).

tion about perspective distortion, and the normalized anisotropy can be expected to be close to zero. Other effects occur at edge points, where the anisotropy can be expected to be close to one, and the effect of the perspective distortion will be small compared to the response of the edge.

A simple observation that can be made in this context is that the local distribution of gradient directions can be expected to be comparably richer at *brightness extrema*. Although the distribution cannot be assumed to be isotropic, a larger spectrum of gradient directions has to be represented. This distribution will, in turn, be biased by the perspective transformation, and reveal information about the local surface geometry.

Based on this observation, and the observation that many natural textures seem to consist of blob-like texture elements randomly scattered on the surface, we shall in this section address the problem of selecting *where* to compute the image texture descriptors. The approach that will be explored is to use blob detection as a pre-processing step to the general scheme described in previous section.

14.6.1. Blob detection: Scale-space maxima and directional statistics

In section 13.3 it was demonstrated how a conceptually simple blob detector could be formulated in terms of normalized scale-space maxima of either the trace or the determinant of the Hessian matrix. Coarse blob descriptors were obtained by representing each scale-space maximum with a circle having an area proportional to the scale at which the maximum over scales was assumed. A straightforward extension of that approach is to combine it with the general scheme for computing the second moment matrix, and then represent each blob by an ellipse instead.

Figure 14.8 shows the result of applying this operation to real and synthetic images. As a first step, normalized scale-space maxima[10] have been detected in the square of the Laplacian. Then, each scale-space maximum is illustrated by an ellipse representing μ_L computed at the point at which the scale-space maximum was assumed. The integration scale was set to $s = \gamma_1^2\gamma_2^2\, t_{\text{trace}^2\,\mathcal{H}L}$, and the local scale was determined according to step 5 in section 14.5.1.

Figure 14.9 illustrates the importance of adapting the local scale when computing μ_L. It shows ellipses computed at two different local scales; the local scale that maximizes $\det \mu_L$, and the local scale that maximizes the normalized anisotropy.

[10]Before the extrema detection, the Laplacian has been smoothed with an integration scale proportional to the local scale ($s = \gamma_1^2 t$ where $\gamma_1 = 2$). This is not an essential step of the method, but only a simple way to reduce the number of blobs. Alternative possibilities are, of course, to rank the blobs on the magnitude of the normalized response (as described in section 13.3), or to apply the more refined blob linking approach (described in chapter 7).

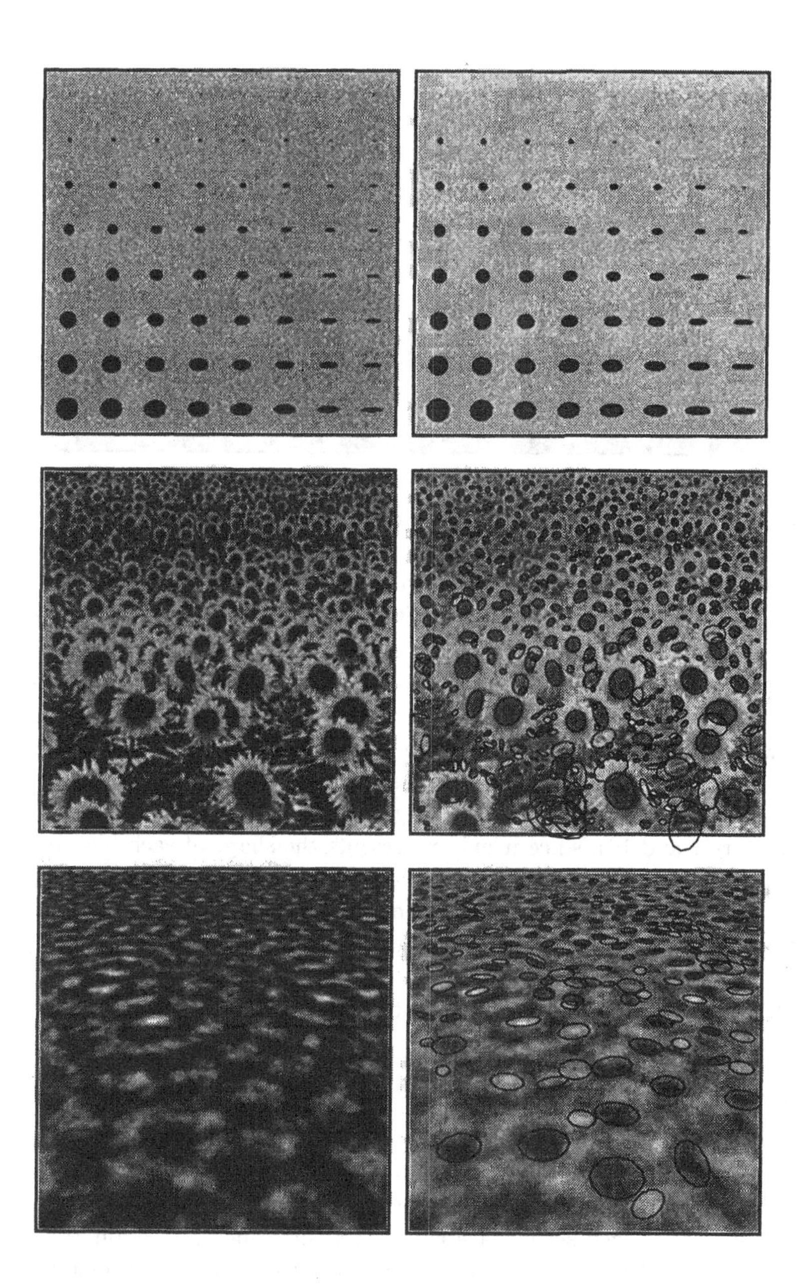

Figure 14.8. Multi-scale blob detection using scale-space extrema of the square of the Laplacian of the Gaussian combined with locally adapted computation of the second moment matrix; (left column) original images, (right column) ellipses representing second moment matrices superimposed onto bright copies of the original grey-level images.

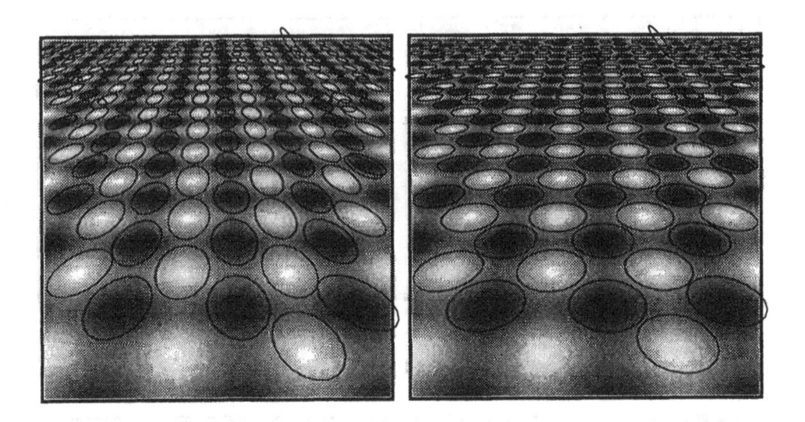

Figure 14.9. Illustration of the importance of adapting the local scale. (left) Second moment matrices computed at the local scales that maximize the determinant of the normalized second moment matrix. (right) Second moment matrices computed at the local scales that maximize the normalized anisotropy.

Note the ability of the method to zoom in to different scales, and how well the computed ellipses describe the blobs in the image, considering how little information is used in the processing.

This *multi-scale blob detector* has obvious limitations compared to other approaches (Voorhees and Poggio 1987; Blostein and Ahuja 1989; chapters 7 and 10), since it only represents the shape of each blob by an ellipse. However, it is well suited as a pre-processing step for shape-from-texture, since it produces precisely the information needed for estimating local linear distortion and size changes.

14.7. Estimating surface orientation

By combining the theories developed in the previous sections, the method for estimating local surface orientation in perspective images of textured surfaces can be summarized as follows:

1. Compute local texture descriptors μ_L as described in section 14.5. This can either be done at selected spatial positions corresponding to normalized scale-space extrema as described in section 14.6, or at a uniform grid of points generated by some default principle.

2. Determine points where surface orientation estimates are to be computed. This set of points can be the same as that used for computing the texture descriptors, or it can be smaller, e.g. a uniform grid. Associate with each point a (Gaussian) window that specifies the weighting of the texture descriptors in the neighborhood of the

point. The scale of this window function will be referred to as the *texel grouping scale* (see below).

3. Estimate surface orientation:

   a. Apply the assumption of *weak isotropy*, which leads to a direct estimate of surface orientation up to the sign of tilt.

   b. Apply the assumption of *constant area*, which permits a unique estimate of surface orientation if the curvature can be neglected in (14.21). For practical computations, estimate the blob area by the scale at which the scale-space maximum is assumed. Then, fit a locally linearized model within the window function given by step 2.

   c. Optionally, compute surface orientation estimates using other assumptions about the surface texture as well.

14.7.1. Texel grouping scale

The second moment descriptor computed by spatial and scale selection can be informally thought of as a single "texture element". To estimate surface orientation at a specified point, such local texture descriptors in the neighbourhood of the point must somehow be combined. In the case of a perfectly regular surface texture, the shape of this texture element can be relied upon to provide information about local projective distortion. Most natural textures, however, exhibit a considerable degree of randomness in their structure, and it is therefore necessary to consider more than one texture element in order to detect the systematic geometric distortions due to the perspective effects. While attempts have been made at modeling such randomness statistically (Witkin 1981; Kanatani and Chou 1989; Blake and Marinos 1990), here, a simpler approach is applied of reducing variance by integration. For this reason, the concept of *texel grouping scale* has been introduced in the scheme above. It refers to the scale used for combining texture descriptors computed at different spatial points into entities to be used for computing geometric shape descriptors.

If the texture descriptors are combined by weighted averaging, then the texel grouping scale is closely related to the relative integration scale. More precisely, from the semi-group property of Gaussian smoothing, it follows that, if the local smoothing scale t is held constant, then the second moment matrix at any coarse integration scale, s_2, can be computed from the second moment matrices at any finer integration scale, s_1, by

$$\mu_L(\cdot;\ t, s_2) = g(\cdot;\ s_2 - s_1) * \mu_L(\cdot;\ t, s_1). \qquad (14.45)$$

Hence, if the local scale is constant (e.g. zero) then in the method of estimating surface orientation from the weak isotropy assumption, the texel grouping scale is equivalent to the relative integration scale. Nevertheless, it is worth making a distinction between these two concepts, since the cascade smoothing property is not applicable when estimating surface orientation from area gradient (step 3b above).

Here, no method for automatic selection of the texel grouping scale is proposed. Instead, the surface orientation estimates will be computed at sparse regular grids with the texel grouping scale proportional to the size of the grid cell. An alternative approach is, of course, to let the texel grouping scale be proportional to the selected integration scale.

14.8. Experiments

Figure 14.10 shows results of applying this composed method to two noisy synthetic images and one real image, all with known camera geometry and surface orientation. From top to bottom, the rows show the grey-level image, the detected blobs, the true surface orientation, and the surface orientation estimates assuming weak isotropy and constant area.

Each surface orientation is graphically indicated by a disk representing the local surface orientation viewed in *parallel projection* along the visual ray through the center of the image. With this convention, the shape of each disk represents local surface orientation regardless of internal camera geometry and the position in the image.

For the synthetic planar sine wave image the estimates from the weak isotropy and constant area assumptions are very accurate ($(\hat{\sigma}, \hat{\theta}) = (60.7°, 90.1°)$ and $(61.6°, 89.0°)$ respectively, to be compared to the true orientation at the center $(\sigma, \theta) = (60.0°, 90.0°)$).

For the synthetic image of the curved cylinder (with true orientation $(55°, 90°)$, noise 25%) the estimate computed from foreshortening gives $(56.8°, 90.4°)$, while the area gradient underestimates slant $(36.2°, 86.1°)$, since the curvature is non-zero (compare with (14.21)).

For the real (planar) wall-paper image, the weak isotropy assumption gives $(47.9°, 84.6°)$ and the area gradient $(51.4°, 76.6°)$ to be compared to the reference value $(50.8°, 85.3°)$ computed by stereo correspondence.

Figure 14.10. (Opposite page): Estimating local surface orientation in; (left column) a synthetic image of a planar surface with 1.4% noise, (middle column) a synthetic image of a cylindrical surface with 25% noise, and (right column) a real image of a planar surface with known orientation. The rows show from top to bottom; (top row) original grey-level image, (middle top row) ellipses representing blobs detected by normalized scale-space extrema, (middle row) reference surface orientation, (middle bottom row) surface orientation estimate computed from the weak isotropy assumption, (bottom row) surface orientation estimate computed from the constant area assumption.

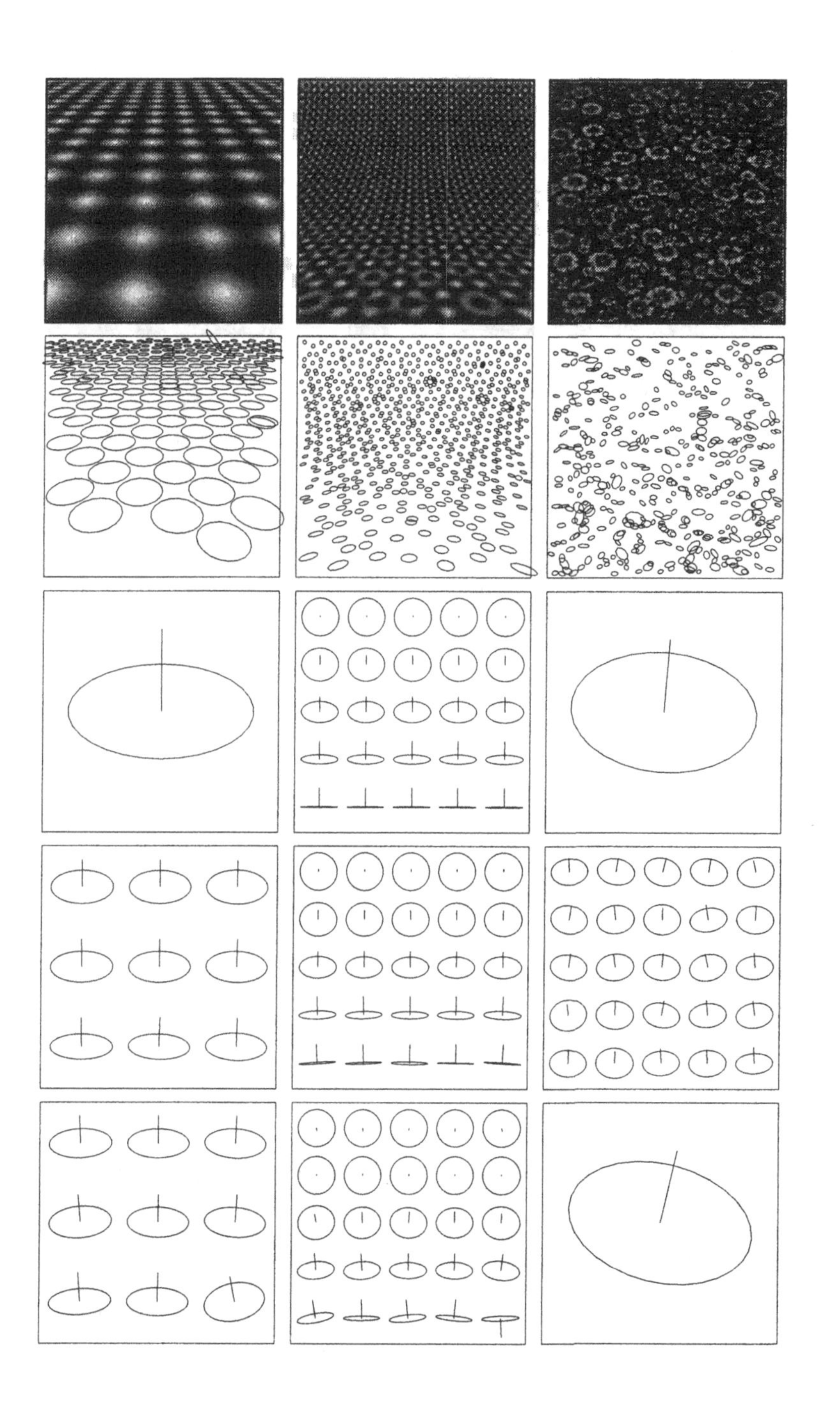

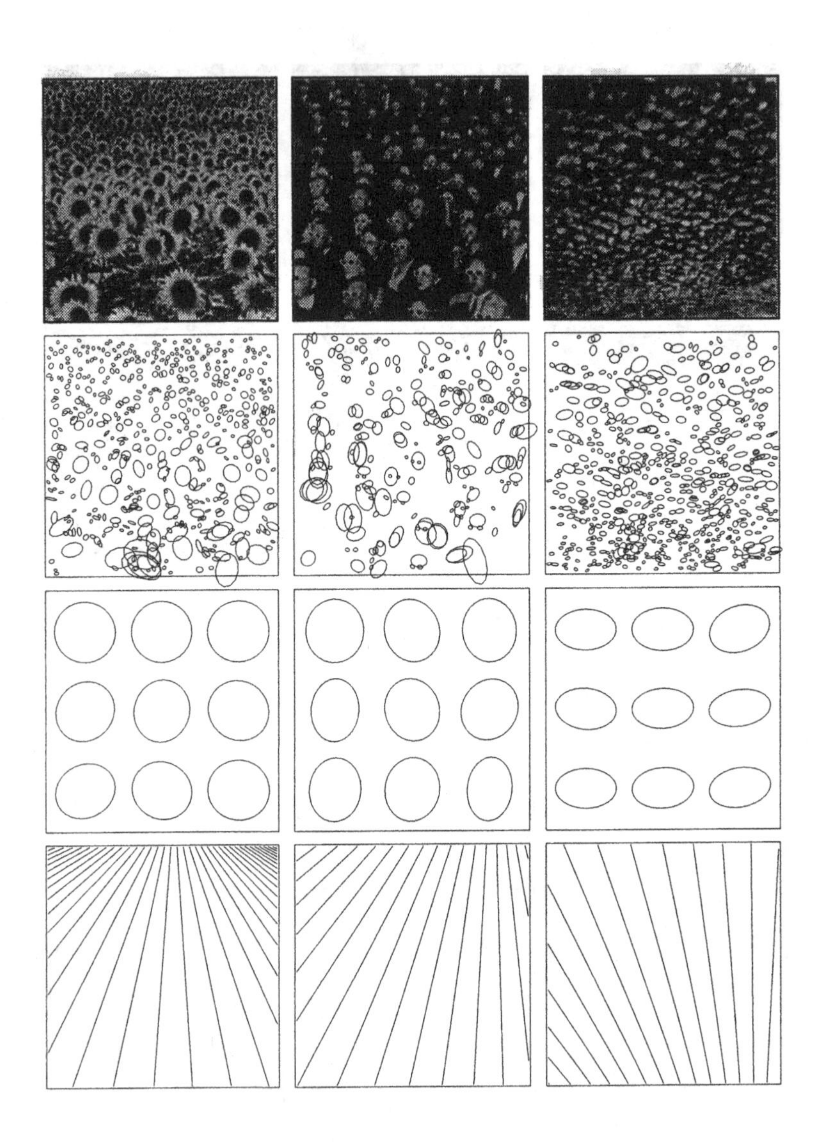

Figure 14.11. Estimation of surface orientation in real images (from Blostein and Ahuja 1989); (top row) original grey-level image, (middle top row) Blobs representing second moment matrices computed with spatial selection. (middle bottom row) Estimated foreshortening, here represented by weighted averages of the second moment descriptors associated with each blob. (bottom row) Estimated area gradient, visualized by lines aligned with the tilt direction converging to a point on the horizon.

More experiments are given in figure 14.11, which shows three images from Blostein and Ahuja (1989). In this case, the camera geometry is unknown, and it is impossible to compute absolute surface orientation estimates in the same way as in figure 14.10. Instead, the estimates from foreshortening are illustrated by ellipses, and the estimates from the area gradient are illustrated by a plane determining the position of the horizon.

It is interesting to note that in these examples, the estimates from the weak isotropy assumption respond to the orientation of the individual texture elements (e.g., the sunflowers), whereas the constant area assumption reflects the orientation of the underlying surface.

14.9. Summary and discussion

The purpose with this chapter has been to demonstrate how three-dimensional surface information can be computed using scale-space operations. In particular, the shape-from-texture problem has been addressed within a framework that, in fact, models all the steps in the computation of local slant and tilt information from raw (discrete) brightness data. The method is general in the sense that the *description stage* is decoupled from any specific *assumptions* about the type of texture considered. All such choices are postponed to the interpretation stage, where different assumptions can be selected depending upon the type of situation. At a technically more detailed level, the main contributions are:

- The need for (at least) two scale parameters in the shape-from-texture problem is emphasized, a *local scale* describing the amount of smoothing when computing derivatives of image brightness, and an *integration scale* describing the window over which statistics of non-linear descriptors is integrated.

- The notions of local scale and integration scale are formalized into a multi-scale representation of image directional statistics, called the multi-scale windowed second moment matrix.

- The framework is formulated completely in terms of *Gaussian derivatives* and *differential invariants* defined from these. This makes the computational model tractable for theoretical analysis, and allows for direct implementations in a visual front-end. Also, the computational efficiency is improved relative to many other approaches based on filter banks of differently oriented filters.

- A mechanism for automatic scale adaption in shape-from-texture is presented, analysed, and demonstrated to give useful results. As a side effect, a method for multi-scale blob detection is obtained.

- It is theoretically explained and experimentally demonstrated how the proposed framework can be used for *deriving shape cues* under two specific types of assumptions about the surface texture; weak isotropy and constant area. The extension to other assumptions is straightforward.

- It is explained in *differential geometric* terms how this representation relates to earlier methods based on directional sensitive filters.
- The *discrete* aspects of implementation are carefully treated.

Concerning limitations of the work, it should be noted that no genuine two-parameter scale variation has been implemented in the selection of integration scale. The presented implementation gives best results when either the textures are approximately periodic, the relative integration scale is known (given by γ_1 and γ_2), or the image contains relatively distinct blob-like structures.

Nevertheless, given a reasonable estimate of the integration scale, computing the slant and tilt information from the local scale that maximizes the *normalized anisotropy* $\tilde{Q}$ of the second moment matrix has been experimentally demonstrated to give useful results in a large number of situations.

Multiple scale maxima and coherent surfaces. Handling the relation between local scale and integration scale is closely related to another problem that has not been formalized here; the issue of *surface grouping.* In real-life situations it is usually necessary to apply the surface estimation technique only to some coherent subset of the texture descriptors found in an image, rather than to all of them.

In the experiments presented here, only the most dominant scale was selected at each image point. A more general approach is to detect all maxima, and then combine different estimates by spatial grouping. For example, a noisy image of a slanted pattern may give rise to maxima in the signature at fine scales due to the noise, in addition to the maxima at coarser scales corresponding to the surface texture (see figure 14.2). Separate estimation of surface orientation invoked by the fine-scale maxima would then correctly indicate a fronto-parallel surface corresponding to the noise in the image plane.

Direct computation. A remark may be necessary concerning the title of this chapter: There have been different interpretations in the computer vision community of what should be meant by "direct computation".

The presented shape-from-texture method is direct in the sense that no explicit search is necessary when computing surface orientation estimates. Of course, maxima have to be detected in the scale-space signature. Such operations can, however, be easily performed locally and be made parallel if dedicated hardware is devoted to each scale level. Moreover, the method is expressed completely in terms of a straightforward bottom-up flow of uncommitted visual front-end operations (see figure 14.12). These are the motivations for calling the method direct.

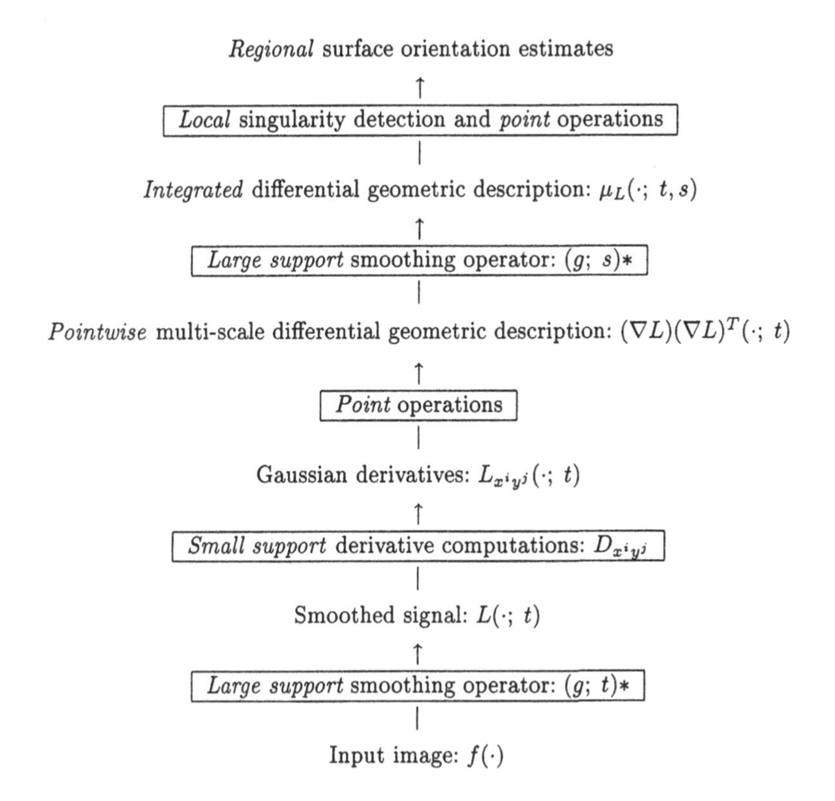

Figure 14.12. Schematic overview of the different types of computations required for computing multi-scale differential geometric texture descriptors and surface orientation estimates using the proposed framework.

Generalizations. The approach of modelling perspective distortion by local affine transformations can be applied to other "shape-from-X" problems as well, see, for example, Wildes (1991) or Jones and Malik (1992) concerning stereo. The basic idea in shape-from-stereo is to assume that two measurements are made of the same surface structure (matching and vergence), and to approximate the transformation from the left to the right image by a local affine transformation. Then, second moment matrices in the left and the right images can be related according to (14.14), and local surface orientation be computed if the camera orientations are known. The difference with respect to shape-from-texture is that specific assumptions about the surface structure are no longer needed. A shape-from-disparity-gradient method that operates in a similar way as the shape-from-texture method described in this chapter can be found in (Gårding and Lindeberg 1993).

Another straightforward generalization is treated in the next chapter, where it is described how shape adaption of the smoothing kernels can be used for improving the accuracy in the surface orientation estimates.

15

Non-uniform smoothing

Although the linear scale-space representation generated by smoothing with the rotationally symmetric Gaussian kernel provides a theoretically well-founded framework for handling image structures at different scales, the scale-space smoothing has the negative property that it leads to shape distortions. For example, smoothing across "object boundaries" can affect both the shape and the localization of edges in edge detection. Similarly, surface orientation estimates computed by shape-from-texture algorithms are affected, since the anisotropy of a surface pattern may decrease when smoothed using a rotationally symmetric Gaussian.

15.1. Non-linear diffusion: Review

In order to reduce the shape distortion problems in edge detection, Perona and Malik (1990) proposed the use of *anisotropic diffusion.* The basic idea is to modify the conductivity $c(x;\ t)$ in the non-linear diffusion equation

$$\partial_t L = \nabla^T (c(x;\ t) \nabla L) \tag{15.1}$$

such as to favour intra-region smoothing to inter-region smoothing. In principle, they solved the diffusion equation

$$\partial_t L = \nabla^T (h(|\nabla L(x;\ t)|) \nabla L) \tag{15.2}$$

for some monotonic decreasing function $h\colon \mathbb{R}_+ \to \mathbb{R}_+$. The intuitive effect of this evolution is that the conductivity will be low where the gradient magnitude is high and vice versa.

This idea has been further developed by several authors. Nordström (1990) showed that by adding a bias term to the diffusion equation, it was possible to relate this method to earlier considered regularization approaches by Terzopoulos (1983) and Mumford and Shah (1985).

By adopting an axiomatic approach, Alvarez *et al.* (1992) have shown that given certain constraints on a visual front-end, a natural choice of non-linear diffusion equation is the equation

$$\partial_t L = |\nabla L|\, \nabla^T (\nabla L / |\nabla L|) = |\nabla L|\, \kappa(L) = L_{\bar{u}\bar{u}}, \tag{15.3}$$

where $\kappa(L)$ denotes the curvature of level curves in L (used for junction detection in chapter 6 and chapter 13), and $L_{\bar{u}\bar{u}}$ represents the second order derivative in the tangent direction to a level curve. This evolution means that level curves move in the normal direction with a velocity proportional to the curvature of the level curves. If the differential equation (15.3) is slightly modified into

$$\partial_t L = (|\nabla L|^3 \, \kappa(L))^{1/3} = (\tilde{\kappa}(L))^{1/3}, \tag{15.4}$$

(where $\tilde{\kappa}(L)$ is the rescaled level curve curvature used for junction detection in section 6.2.2 and section 11.3), then it can be shown that its solutions are relative invariant under affine transformations of the spatial coordinates (Alvarez *et al.* 1992). This property has been used by Sapiro and Tannenbaum (1993) for defining an affine invariant curve evolution scheme.

An interesting approach to describing non-linear diffusion more generally is pursued by Florack *et al.* (1993), who consider general non-linear coordinate transformations of the spatial coordinates as a means of expressing such operations. Interestingly, this approach covers several of the above mentioned methods.

15.1.1. Properties of non-linear diffusion methods

Trivially, it follows from the maximum principle that if $h > 0$ then any non-linear scale-space representation of the form (15.1) satisfies the causality requirement, or equivalently, the non-enhancement property of local extrema: $\partial_t L < 0$ at local maxima, and $\partial_t L > 0$ at local minima.

The maximum principle does, however, not extend to higher-order derivatives. This can be easily seen in the one-dimensional case

$$\partial_t L = \partial_x(h(|L_x|)L_x) = h_x(|L_x|)L_x + h(|L_x|)L_{xx}. \tag{15.5}$$

Introduce $\phi(L_x) = h(|L_x|)L_x$. Then, (15.5) can be written

$$\partial_t L = \partial_x(\phi(L_x)) = \phi'(L_x)L_{xx}, \tag{15.6}$$

and the evolution of the gradient follows

$$\partial_t L_x = \phi''(L_x)L_{xx}^2 + \phi'(L_x)L_{xxx}. \tag{15.7}$$

For a local gradient maximum with $L_{xx} = 0$ and $L_{xxx} < 0$ it holds that $\partial_t L_x > 0$ if $\phi'(L_x) < 0$. For the conductance function used by Perona and Malik (1990)

$$h(|\nabla L|) = e^{-|\nabla L|^2/k^2}, \tag{15.8}$$

where k is a free parameter, we have $\phi'(L_x) < 0$ if $L_x > k/\sqrt{2}$. In other words, gradients that are sufficiently strong will be enhanced, and gradients that are sufficiently weak will be suppressed (no matter what are the spatial extents of the image structures).

A natural way to reduce this sensitivity is by smoothing the gradient function prior to computing the conductivity for the diffusion equation. A modification of (15.2) proposed by Whitaker and Pizer (1993) reads

$$\partial_t L = \nabla^T \left(h(|(g(\cdot;\ s(t)) * \nabla L(\cdot;\ t))(x)|) \nabla L \right), \tag{15.9}$$

where the *second stage scale parameter*, s, is a decreasing function of the first stage scale parameter t. A basic property of this construction is that if the initial value of the second stage scale parameter, $s(0)$, is sufficiently large, then fine scale peaks will be suppressed by the second stage smoothing operation.

15.2. Linear shape-adapted smoothing

Improvements relative to the rotationally symmetric scale-space representation can also be obtained using linear theory. As has been argued by several authors, in certain situations it can be advantageous to use filters that correspond to different scale values along different directions; for example a large scale value along the direction of an edge, and a smaller scale value in the perpendicular direction. At junctions, on the other hand, where several directions meet, the converse behaviour can be advantageous in order to better resolve the directional information.

In this section, we shall consider the shape-from-texture problem and describe how linear shape adaption of the smoothing kernels can be used for improving the accuracy in surface orientation estimates. In fact, it will make the shape estimation method invariant with respect to the locally linearized perspective mapping.

15.2.1. Uniform smoothing as an uncommitted first processing step

In the shape-from-texture method described in chapter 14, *uniform* (rotationally symmetric) smoothing in the image domain was used throughout. This was motivated by the principle that in the absence of any *a priori* information about what can be expected to be in the image, the vision system should be as *uncommitted* as possible, and, for example, have no preferred directions. Such uniform smoothing, however, may lead to shape distortions and affect the computed surface orientation estimates (see the theoretical analysis in section 14.4.5 and the experiments in table 14.1). This problem has also been observed by others; see, for example, Stone (1990), who proposes an iterative smoothing method.

15.2.2. Basic idea for shape adaption of the smoothing kernels

Given initial slant and tilt estimates $(\hat{\sigma}, \hat{\theta})$ computed by uniform smoothing, a straightforward compensation technique is to let the scale parameters in the (estimated) tilt direction, denoted $t_{\hat{t}}$ and $s_{\hat{t}}$, and the scale parameters in the perpendicular direction, denoted $t_{\hat{b}}$ and $s_{\hat{b}}$, be related by

$$t_{\hat{t}} = t_{\hat{b}} \cos^2 \hat{\sigma}, \quad s_{\hat{t}} = s_{\hat{b}} \cos^2 \hat{\sigma}. \tag{15.10}$$

If, say, under the assumption of weak isotropy, this estimate is correct, then the slant estimate (14.39) will be *unaffected* by the non-uniform smoothing operation. To algebraically illustrate this property, assume first that the simple Gaussian blob model in (14.36)

$$f(x, y) = g(x;\ l_1^2)\, g(y;\ l_2^2) \quad (l_1 \geq l_2 > 0), \tag{15.11}$$

describes the projection of a rotationally symmetric blob, and that the characteristic lengths l_1 and l_2 are hence related by

$$l_2 = l_1 \cos \sigma, \tag{15.12}$$

where σ is the "true" slant value. Then, the scale-space representation of f using non-uniform smoothing is

$$L(x, y;\ t) = g(x;\ l_1^2 + t_{\hat{b}})\, g(y;\ l_2^2 + t_{\hat{t}}). \tag{15.13}$$

Assume first, for simplicity, that the tilt direction is correctly estimated, i.e. that $\hat{t} = e_y$ and $\hat{b} = e_x$, where e_x and e_y are the unit vectors along the coordinate directions. Then, under scale-space smoothing the foreshortening estimate corresponding to (14.39) is given by

$$\hat{\epsilon} = \epsilon(\hat{\sigma};\ t_{\hat{t}}, t_{\hat{b}}) = \sqrt{\frac{l_2^2 + t_{\hat{t}}}{l_1^2 + t_{\hat{b}}}} = |\cos \sigma| \sqrt{1 + \frac{t_{\hat{b}}}{l_1^2 + t_{\hat{b}}} \left(\frac{\cos^2 \hat{\sigma}}{\cos^2 \sigma} - 1 \right)}.$$

In other words, provided that the initial slant estimate is correct, the new slant estimate is unaffected by the smoothing operation. Moreover, the backprojected window function used for defining the surface second moment matrix in (14.14) will be *rotationally symmetric* when considered in the *tangent plane to the surface.* If (15.14) is used as the basis for an iterative method,

$$\hat{\sigma}_{n+1} = \arccos \epsilon(\hat{\sigma}_n;\ t_{\hat{t}}, t_{\hat{b}}) = h(\hat{\sigma}_n), \tag{15.14}$$

then from the derivative expression,

$$|(\partial_{\hat{\sigma}} h)(\hat{\sigma})| = |(\partial_{\hat{\sigma}} \arccos \epsilon)(\hat{\sigma};\ t_{\hat{b}} \cos \hat{\sigma}, t_{\hat{b}})| = \{\text{let } \hat{\sigma} = \sigma\} = \frac{t_b}{l_1^2 + t_b} < 1,$$

it is clear that the true value of $\hat{\sigma}$ is a *convergent fixed point* of the above equation (with $\epsilon = \cos \hat{\sigma}$), which means that the method will converge to the true solution provided that the initial estimate is sufficiently close to the true value.

Experiments. Letting the local scales be related by

$$t_{\hat{b}} = t / \cos \hat{\sigma}, \qquad t_{\hat{t}} = t \cos \hat{\sigma}, \tag{15.15}$$

where t denotes the ordinary scale parameter, and correspondingly for the integration scales, gives the following results when applied to the experiment in table 14.1:

At noise levels 10.0, 31.6, and 100.0, the slant estimates change from 58.9° to 59.9°, from 55.3° to 61.1°, and from 52.3° to 62.9° respectively (the reference value is 60°). The effect at smaller noise levels is minor, although a few overshoots occur due to the maximum anisotropy approach for selecting the local scale.

It is not argued that this method describes any "optimal" way to reduce the shape distortions, (since, for example, further precautions may need to be taken with respect to convergence). The intention is rather to demonstrate that in addition to the local adaptivity to scale levels, the shape of the smoothing kernels can be easily adjusted within the same framework. Incidentally, this also provides an alternative way to define non-uniform smoothing.

Next, a formal description is given of how this straightforward method for shape adaption can be performed under arbitrary (invertible) linear transformations. It will be proved that the transformation property (14.14) of the second moment matrix as well as the invariance of the scale-space maxima of the normalized second moment matrix (14.32) can be formulated such that they hold *exactly* also for strictly positive scales in the scale-space representation of the input image.

15.3. Affine scale-space

The linear scale-space representation treated in previous chapters was based on the rotationally symmetric Gaussian kernel. A natural generalization to consider when dealing with linear transformations of the spatial domain is the *affine scale-space representation* generated by convolution with non-uniform Gaussian kernels. Given a symmetric positive semi-definite (covariance) matrix, $\Sigma_t \in \mathrm{SPSD}(2)$,[1] the non-uniform Gaussian kernel in the two-dimensional case is defined by

$$g(x;\ \Sigma_t) = \frac{1}{2\pi\sqrt{\det \Sigma_t}} e^{-x^T \Sigma_t^{-1} x/2}, \tag{15.16}$$

where $x \in \mathbb{R}^2$. In the special case when the matrix Σ_t is a scalar entity t times the unit matrix I ($\Sigma_t = tI$), this function corresponds to the ordinary (uniform) Gaussian kernel with scale value t.

[1] The notation SPSD(2) stands for the cone of symmetric positive semidefinite 2×2 matrices.

Given any function $f\colon \mathbb{R}^2 \to \mathbb{R}$, the affine scale-space representation of f can be defined as the three-parameter family of functions $L\colon \mathbb{R}^2 \times \mathrm{SPSD}(2) \to \mathbb{R}$ given by

$$L(\cdot;\ \Sigma_t) = g(\cdot;\ \Sigma_t) * f(\cdot). \tag{15.17}$$

15.3.1. *Transformation property under linear transformations*

The basic reason for introducing the affine scale-space is that it is *closed* under linear (and affine) transformations of the spatial coordinates. Let $f_L, f_R\colon \mathbb{R}^2 \to \mathbb{R}$ be two intensity patterns related by an invertible linear transformation $\eta = B\xi$, i.e.,

$$f_L(\xi) = f_R(B\xi), \tag{15.18}$$

and define the affine scale-space representations by

$$L(\cdot;\ \Sigma_L) = g(\cdot;\ \Sigma_L) * f_L(\cdot), \tag{15.19}$$
$$R(\cdot;\ \Sigma_R) = g(\cdot;\ \Sigma_R) * f_R(\cdot), \tag{15.20}$$

where $\Sigma_L, \Sigma_R \in \mathrm{SPSD}(2)$. Then, L and R are related by

$$L(\xi;\ \Sigma_L) = R(\eta;\ \Sigma_R), \tag{15.21}$$

where

$$\Sigma_R = B\Sigma_L B^T. \tag{15.22}$$

Hence, for any matrix Σ_L there exists a matrix Σ_R such that the affine scale-space representations of f_L and f_R are equal (see the commutative diagram in figure 15.1). This property does not hold for the traditional linear scale-space representation based on the rotationally symmetric Gaussian (unless the linear transformation can be decomposed into rotations and uniform rescalings of the spatial coordinates).

$$\begin{array}{ccc}
L_L(\xi;\ \Sigma_L) & \xrightarrow{\ \eta = B\xi\ } & L_R(\eta;\ B\Sigma_L B^T) \\
\uparrow & & \uparrow \\
*g(\cdot;\ \Sigma_L) & & *g(\cdot;\ B\Sigma_L B^T) \\
| & & | \\
f_L(\xi) & \xrightarrow{\ \eta = B\xi\ } & f_R(\eta)
\end{array}$$

Figure 15.1. Commutative diagram of the non-uniform scale-space representation under linear transformations of the spatial coordinates in the original image.

Proof. To verify the transformation property, insert $d(B\xi) = \det B\, d\xi$ into the definition of L,

$$L(q;\ \Sigma_L) = \int_{\xi\in\mathbb{R}^2} g(q-\xi;\ \Sigma_L)\, f_L(\xi)\, d\xi$$
$$= (\det B)^{-1} \int_{\xi\in\mathbb{R}^2} g(B^{-1}(Bq - B\xi);\ \Sigma_L)\, f_R(B\xi)\, d(B\xi). \quad (15.23)$$

Under a linear transformation, the non-uniform Gaussian kernel transforms as

$$g(B^{-1}\zeta;\ \Sigma_L) = \frac{1}{2\pi\sqrt{\det \Sigma_L}}\, e^{-(B^{-1}\zeta)^T\Sigma_L^{-1}(B^{-1}\zeta)/2}$$
$$= \sqrt{\det BB^T}\, \frac{1}{2\pi\sqrt{B\det\Sigma_L B^T}}\, e^{-\zeta^T(B\Sigma_L B^T)^{-1}\zeta/2}, \quad (15.24)$$

which gives

$$g(B^{-1}\zeta;\ \Sigma_L) = \det B\, g(\zeta;\ B\Sigma_L B^T). \quad (15.25)$$

By inserting (15.25) in (15.23) with $\zeta = Bq - B\xi$ and by letting $\eta = B\xi$ with $p = Bq$ it follows that

$$L(q;\ \Sigma_L) = \int_{\xi\in\mathbb{R}^2} g(Bq - B\xi;\ B\Sigma_L B^T)\, f_R(B\xi)\, d(B\xi)$$
$$= \int_{\eta\in\mathbb{R}^2} g(Bq - \eta;\ B\Sigma_L B^T)\, f_R(\eta)\, d\eta \quad (15.26)$$
$$= R(Bq;\ B\Sigma_L B^T) = R(p;\ \Sigma_R),$$

which proves (15.21) and (15.22). □

15.3.2. Interpretation in terms of eigenvectors and eigenvalues

The effect of smoothing by convolution with the non-uniform Gaussian kernel can be easily understood in terms of the eigenvectors $\bar{b}$ and $\bar{t}$ and the eigenvalues $t_b > t_t > 0$ of Σ_t^{-1}. Let $u = (u_b, u_t)$ denote coordinates along the two eigenvectors respectively. The two coordinate systems are related by $u = Ux$, where U is a unitary matrix ($UU^T = U^TU = I$) such that $\Sigma_t = U^T\Lambda_t U$ and $\Sigma_t^{-1} = U^T\Lambda_t^{-1}U$, where Λ_t is a diagonal matrix with t_b and t_t along the diagonal. In the transformed coordinates, the quadratic form can be written

$$x^T\Sigma_t^{-1}x = x^TU^T\,\Lambda_t^{-1}\,Ux = u^T\Lambda_t^{-1}\,u = \frac{u_b^2}{t_b} + \frac{u_t^2}{t_t}, \quad (15.27)$$

which means that the non-uniform Gaussian kernel in this coordinate system assumes the form

$$g(u_b, u_t\ \Lambda_t) = \frac{1}{2\pi\sqrt{t_bt_t}}e^{-(u_b^2/2t_b + u_t^2/2t_t)} = \frac{1}{\sqrt{2\pi t_b}}e^{-u_b^2/2t_b}\,\frac{1}{\sqrt{2\pi t_t}}e^{-u_t^2/2t_t}. \quad (15.28)$$

In other words, convolution with (15.16) corresponds to (separable) smoothing with a one-dimensional Gaussian kernel with scale value t_b along the $\bar{b}$-direction, and a one-dimensional Gaussian kernel with scale value t_t along the $\bar{t}$-direction.

15.3.3. *Diffusion equation interpretation of affine scale-space*

Convolution with the non-uniform Gaussian kernel can also be interpreted in terms of the diffusion equation. Assume that t_t and t_b are related to a one-dimensional scale parameter t by

$$t_b = t\alpha, \qquad t_t = t/\alpha, \tag{15.29}$$

for some constant[2] $\alpha > 1$. Then, in terms of the transformed coordinates, (u_b, u_t), the non-uniform scale-space representation given by (15.17) satisfies the non-uniform diffusion equation

$$\partial_t L = \frac{1}{2}(\alpha\,\partial_{u_b u_b} + \frac{1}{\alpha}\,\partial_{u_t u_t})\,L \tag{15.30}$$

with initial condition $L(\cdot;\ 0) = f$. With $\Lambda_t = t\Lambda_0$, where Λ_0 is a diagonal matrix with entries α and $1/\alpha$, and with $\nabla_u = (\partial_{u_b}, \partial_{u_t})^T$, the matrix form of this equation is

$$\partial_t L = \frac{1}{2}\,\nabla_u^T \Lambda_0 \nabla_u L. \tag{15.31}$$

In terms of the original coordinates, (x, y), and with $\nabla = (\partial_x, \partial_y)^T$, the non-uniform diffusion equation assumes the form

$$\partial_t L = \frac{1}{2}\,\nabla^T \Sigma_0 \nabla L, \tag{15.32}$$

where $\Sigma_0 = U^T \Lambda_0 U$ is the positive definite matrix corresponding to Λ_0 rotated back to the (x, y) coordinate system (i.e. $\Sigma_t = t\Sigma_0$). In the calculations above it has without loss of generality been assumed that the determinants of Λ_0 and Σ_0 are equal to one.

15.3.4. *Fourier transform of the non-uniform Gaussian kernel*

From the Fourier transform of the one-dimensional Gaussian kernel

$$G(\omega;\ t) = \int_{x=-\infty}^{\infty} g(x;\ t)\, e^{-i\omega x}\, dx = e^{-\omega^2 t/2} \tag{15.33}$$

it directly follows that the Fourier transform of the non-uniform Gaussian kernel in the transformed (u_b, u_t) coordinates is

$$G(w_b, w_t;\ \Lambda_t) = e^{-t_b w_b^2/2}\, e^{-t_b w_t^2/2} = e^{-(t_b w_b^2 + t_t w_t^2)/2} = e^{-w^T \Lambda_t w/2}.$$

[2] In the example in Section 15.2.2 above, α corresponds to $1/\cos\sigma$.

With $w = U\omega$, and $\omega = (\omega_x, \omega_y)$ the Fourier transform in the original coordinate system can then be written

$$G(\omega;\ \Sigma_t) = e^{-\omega^T \Sigma_t \omega/2}. \tag{15.34}$$

From this expression it can be immediately realized that the semi-group property transfers as follows to the non-uniform Gaussian kernel

$$G(\omega;\ \Sigma_1) * G(\omega;\ \Sigma_2) = G(\omega;\ \Sigma_1 + \Sigma_2). \tag{15.35}$$

15.4. Definition of an affine invariant image texture descriptor

We are now prepared to define a second moment matrix based on the affine scale-space representation. Given an image $f\colon \mathbb{R}^2 \to \mathbb{R}$ with affine scale-space representation $L\colon \mathbb{R}^2 \times \mathrm{SPSD}(2) \to \mathbb{R}$, the second moment matrix based on non-uniform smoothing $\mu_L\colon \mathbb{R}^2 \times \mathrm{SPSD}(2)^2 \to \mathrm{SPSD}(2)$ can be defined by

$$\mu_L(\cdot;\ \Sigma_t, \Sigma_s) = g(\cdot;\ \Sigma_s) * ((\nabla L)(\cdot;\ \Sigma_t)\,(\nabla L)(\cdot;\ \Sigma_t)^T) \tag{15.36}$$

where Σ_s represents the covariance matrix corresponding to the integration scale, and Σ_t the covariance matrix corresponding to the local scale.

15.4.1. Transformation property under linear transformations

Under a linear transformation of the image coordinates $\eta = B\xi$, this descriptor transforms as

$$\mu_L(q;\ \Sigma_t, \Sigma_s) = B^T \mu_R(Bq;\ B\Sigma_t B^T, B\Sigma_s B^T)\, B. \tag{15.37}$$

Proof. Differentiation of (15.21) gives

$$\nabla L(\xi;\ \Sigma_t) = B^T \nabla R(B\xi;\ B\Sigma_t B^T), \tag{15.38}$$

which implies that $\mu_L(q;\ \Sigma_t, \Sigma_s)$ assumes the form

$$\int_{\xi\in\mathbb{R}^2} g(q-\xi;\ \Sigma_s)\, B^T(\nabla R(B\xi;\ B\Sigma_t B^T))\,(\nabla R(B\xi;\ B\Sigma_t B^T))^T B\, d\xi.$$

After calculations similar to those in section 15.3.1 this expression can in turn be rewritten as

$$B^T \int_{\eta\in\mathbb{R}^2} g(Bq-\eta;\ B\Sigma_s B^T)\,(\nabla R(\eta;\ B\Sigma_t B^T))\,(\nabla R(\eta;\ B\Sigma_t B^T))^T\, d\eta B$$

which proves (15.37). □

15.4.2. Invariance property of normalized scale-space maxima

Assume now that the covariance matrices are used for shape adaption, and that given a non-uniform Gaussian kernel with fixed shape and orientation, only size variations are performed. Then, Σ_t and Σ_s can be written

$$\Sigma_t = t\Sigma_1, \quad \Sigma_s = s\Sigma_2, \tag{15.39}$$

for some *constant* matrices $\Sigma_1, \Sigma_2 \in \text{SPSD}(2)$. Without loss of generality assume that $\det \Sigma_1 = \det \Sigma_2 = 1$. Then, (15.37) can be written

$$\mu_L(q;\ t\Sigma_1, s\Sigma_2) = B^T \mu_R(Bq;\ tB\Sigma_1 B^T, sB\Sigma_2 B^T)\, B. \tag{15.40}$$

The corresponding expression for the (normalized) determinants becomes

$$\begin{aligned} &\partial_t(t \det \Sigma_1 \det \mu_L(q;\ t\Sigma_1, s\Sigma_2)) = 0 \quad \Leftrightarrow \\ &\partial_t(t \det(B\Sigma_1 B^T) \det \mu_R(Bq;\ tB\Sigma_1 B^T, sB\Sigma_2 B^T)) = 0 \end{aligned} \tag{15.41}$$

showing that maxima over scales in the (normalized) determinant of the second moment matrix are preserved under linear transformations provided that the matrices Σ_1 and Σ_2 are matched accordingly. The same result applies to spatial maxima.

15.4.3. Invariance property of surface orientation estimates

Assume now that μ_L represents the second moment matrix in the image, and that μ_R describes the corresponding entity in the tangent plane to the surface. For simplicity, assume also that the surface pattern is weakly isotropic, i.e. that

$$\mu_R(Bq;\ 0, s_0 I) = c' I \tag{15.42}$$

holds for some constant c' and some sufficiently large s_0. Moreover, assume that the isotropy of such a pattern does not change under scale-space smoothing, or more realistically, that the pattern remains (approximately) weakly isotropic over some range of scales. Then,

$$\mu_R(Bq;\ t'I, s'I) = c' I, \tag{15.43}$$

holds when $t' \in I_t$ and $s' \in I_s$ for some scale ranges I_t and I_s. Now, let

$$\Sigma_1 = \Sigma_2 = \frac{(B^T B)^{-1}}{\det(B^T B)^{-1}}, \tag{15.44}$$

in (15.40). Then,

$$\mu_L(q;\ t\Sigma_1, s\Sigma_2) = B^T \mu_R(Bq;\ t'I, s'I)\, B, \tag{15.45}$$

where $t' = (\det B)^2 t$ and $s' = (\det B)^2 s$. This expression reduces to

$$\mu_L(q;\ t\Sigma_1, s\Sigma_2) = c B^T B \tag{15.46}$$

if $t' \in I_t$ and $s' \in I_s$, which means that the estimate of the anisotropy is exact.

15.4.4. *Summary: Affine invariant shape-from-texture method*

With respect to the shape-from-texture problem, this means that provided that an appropriate estimate of the surface orientation is available, this linear *shape-adapted smoothing is invariant with respect to the locally linearized perspective mapping.* Moreover, up to the first order of approximation it corresponds to the application of a rotationally symmetric local smoothing operator and integration with a rotationally symmetric window function in the tangent plane to the surface; see the commutative diagram in figure 15.2. Since also the scale maxima (and the spatial maxima) of $\det \mu_L$ satisfy this invariance property, it follows that *the entire surface estimation procedure obeys the (relative) invariance with respect to affine transformations, and hence the locally linearized perspective mapping.*

$$
\begin{array}{ccccc}
\begin{array}{c}\mu_L(\xi;\ t(B^TB)^{-1}, s(B^TB)^{-1}) \\ = c' B^T B\end{array} & - & \begin{array}{c}\eta = B\xi \\ \mu_L = B^T \mu_I B\end{array} & \rightarrow & \begin{array}{c}\mu_I(\eta;\ tI, sI) \\ = c' I\end{array} \\
\uparrow & & & & \uparrow \\
*g(\cdot;\ s(B^TB)^{-1}) & & & & *g(\cdot;\ sI) \\
\uparrow & & & & \uparrow \\
\nabla L_L \nabla L_L^T & & & & \nabla L_I \nabla L_I^T \\
| & & & & | \\
L_L(\xi;\ t(B^TB)^{-1}) & - & \eta = B\xi & \rightarrow & L_I(\eta;\ tI) \\
\uparrow & & & & \uparrow \\
*g(\cdot;\ t(B^TB)^{-1}) & & & & *g(\cdot;\ tI) \\
| & & & & | \\
f_L(\xi) & - & \eta = B\xi & \rightarrow & f_I(\eta)
\end{array}
$$

Figure 15.2. Commutative diagram for surface orientation estimation using shape adapted non-uniform smoothing, here by letting the scale matrices be proportional to $(B^TB)^{-1}$, where B describes the image f_L of a weakly isotropic pattern f_I under the linear transformation $f_L(\xi) = f_I(B\xi)$. The estimate of surface orientation (given by $\mu_L = cB^TB$) will be unaffected by the smoothing operation provided that the estimate of the linear transformation B used for adapting the smoothing kernel shape is correct. Note that the entity $(B^TB)^{-1}$ used for adapting the kernel shape is (up to a matrix inversion) directly measurable from the second moment matrix.

Although the entity B^TB used for adapting the shape of the smoothing kernel is directly measurable from the second moment matrix, a remark is necessary that there is a certain chicken-and-the-egg aspect in this treatment. The goal is to estimate the linear distortion, while the smoothing procedure requires such information to be obey the invariance properties.

Nevertheless, the result can be used for formulating an *iterative procedure*. It will satisfy the property that if it converges to the desired result, then the result will possess the invariance properties.

15.5. Outlook

The purpose of this chapter has been to give brief indications of different ways of relaxing the requirement about rotationally symmetric smoothing. The actual presentation has by intention been kept short, since the actual design and further developments of these methods constitute subjects for continued research.

A

Technical details

A.1. Implementing scale-space smoothing

According to the definition of the scale-space for discrete signals, the representation of a one-dimensional signal f at a scale-level t is given by,

$$L(x;\ t) = \sum_{n=-\infty}^{\infty} T(n;\ t)\, f(x-n), \tag{A.1}$$

where $T(n;\ t) = e^{-t} I_n(t)$. When to implement this transformation, there are a few numerical problems to consider:

- The infinite sum must be replaced with a finite one.
- Normally, the modified Bessel functions I_n are not available as standard library routines. Therefore, an algorithm is needed for generating the filter coefficients $T(n;\ t)$ given any value of t.
- A realistic signal is finite, but a finite approximation of (A.1) might need additional values.

Here, we will consider the first two problems, and not go into the (artificial) complications that arise with finite signals (see section 4.5.2 for a discussion). Instead, it will be assumed that f is defined for all those integers, where function values are required for the algorithms

Truncation. A reasonable way to approximate (A.1) is by truncating[1] the sum for some sufficiently large integer m,

$$L(x;\ t) \approx \sum_{n=-m}^{m} T(n;\ t)\, f(x-n) \tag{A.2}$$

[1]Clearly, there are also smoother ways to round off the filter coefficients than plain truncation. Moreover, this truncation violates the scale-space properties. In order to avoid these complications, it is here assumed that the upper bound on the relative truncation error ε is sufficiently small (typically 10^{-6} or 10^{-7} in the case of 32-bit floating point arithmetics). Whereas this approach may lead to larger smoothing kernels, and hence increase the computational work, the effects of it are usually marginal, since the smoothing coefficients tend to zero very rapidly.

selected such that the absolute error in L due to this approximation does not exceed a given error bound $\tilde{\varepsilon}$. If f is bounded ($|f(x)| \leq M$ for some M), then it is sufficient to select m such that

$$2 \sum_{n=m+1}^{\infty} T(n;\ t) \leq \frac{\tilde{\varepsilon}}{M} = \varepsilon_{trunc}. \tag{A.3}$$

Filter coefficient generation. An easy way to generate the filter coefficients is to start from the recurrence relation (Abramowitz and Stegun 1964: 9.6.26)

$$I_{n-1}(t) - I_{n+1}(t) = \frac{2n}{t} I_n(t). \tag{A.4}$$

One can apply Miller's algorithm (Press *et al.* 1986) and initiate the iterations with an arbitrary seed

$$I_k = 1 \quad \text{and} \quad I_{k+1} = 0 \tag{A.5}$$

for some sufficiently large start index k. As n decreases, the iterates from (A.4) will successively approach the correct solution. If d significant digits are required in I_m, it is sufficient to start the iterations at

$$k = 2(m + d\sqrt{m}). \tag{A.6}$$

The sequence of iterates can be normalized by computing $I_0(t)$ by a separate routine (e.g., by using the series expansions in Abramowitz and Stegun 1964: 9.8.1, 9.8.2). Then, once a sufficient number of filter coefficients has been computed, it is easy to determine how many are actually needed from the condition

$$\sum_{n=-m}^{m} T(n;\ t) \geq 1 - \varepsilon_{trunc}. \tag{A.7}$$

Concerning what actual value of m should be used for determining k, a very coarse initial estimate can be obtained from continuous analogy

$$\int_{\xi=-m'}^{m'} g(\xi;\ t)\, d\xi \geq 1 - \epsilon_{trunc}, \tag{A.8}$$

which in turn can be expressed in terms of the error function. Since this estimate is not guaranteed to be sufficient, a safer approach is, of course, to compute k according to (A.6) using a larger value of m.

An alternative approach is to derive an explicit overestimate of the remainder (A.3) in the infinite sum, see (Lindeberg 1988) for details.

Convolutions in the Fourier domain. Another possibility is, of course, to use the closed-form expression for the Fourier transform of the discrete analogue of the Gaussian kernel (3.52)

$$\psi_T(\theta) = \sum_{n=-\infty}^{\infty} T(n;\ t)e^{-in\theta} = e^{\alpha t(\cos\theta - 1)}, \tag{A.9}$$

and perform the convolutions in the frequency domain instead. Then, also the problems of estimating the truncation error will be substantially reduced. At coarse scales, this method will be computationally more efficient than convolutions in the spatial domain. However, some precautions have to be taken in order to reduce the wrap-around effects, for example, by extending the original image before the FFT is carried out.

Higher dimensions. In the separable case, the higher-dimensional scale-space smoothing can be implemented by applying the one-dimensional smoothing transformation along each dimension. Then, the truncation error $\epsilon_{N\text{-}D}$ in the N-dimensional case is related to the truncation error $\epsilon_{1\text{-}D}$ in one-dimensional by

$$1 - \epsilon_{N\text{-}D} = (1 - \epsilon_{1\text{-}D})^N. \tag{A.10}$$

Derivative approximations. When computing derivative approximations by applying difference operators to smoothed data, it is clear that the absolute error[2] is unaffected by the first order central difference operator with filter coefficients $(-1/2,\ 0,\ 1/2)$, while it increases with a factor of four when the second order difference operator $(1,\ -2,\ 1)$ is applied.

These effects accumulate when higher order derivative approximations are computed. Hence, the initial bound on the absolute truncation error in the smoothing step must be selected sufficiently small such that the prescribed accuracy in the derivative approximations is guaranteed.

[2]Note, however, that the relative error can be expected to increase much more due to cancellation of digits.

A.2. Polynomials satisfying the diffusion equation

This appendix lists a set of polynomials satisfying the diffusion equation (used in section 8.5). Each polynomial $p_{m,n}(x, y)$ has been generated from the monomial $x^m y^n$ by adding suitable lower order terms containing powers of t, and if necessary x and y as well, such that $p_{m,n}(x, y)$ satisfies the two-dimensional diffusion equation.

$$
\begin{aligned}
p_{0,0}(x, y;\ t) &= 1, \\
p_{1,0}(x, y;\ t) &= x, \\
p_{0,1}(x, y;\ t) &= y, \\
p_{2,0}(x, y;\ t) &= x^2 + t, \\
p_{1,1}(x, y;\ t) &= xy, \\
p_{0,2}(x, y;\ t) &= y^2 + t, \\
p_{3,0}(x, y;\ t) &= x^3 + 3xt, \\
p_{2,1}(x, y;\ t) &= x^2 y + yt, \\
p_{1,2}(x, y;\ t) &= xy^2 + xt, \\
p_{0,3}(x, y;\ t) &= y^3 + 3yt, \\
p_{4,0}(x, y;\ t) &= x^4 + 6x^2 t + 3t^2, \\
p_{3,1}(x, y;\ t) &= x^3 y + 3xyt, \\
p_{2,2}(x, y;\ t) &= x^2 y^2 + x^2 t + y^2 t + t^2, \\
p_{1,3}(x, y;\ t) &= xy^3 + 3xyt, \\
p_{0,4}(x, y;\ t) &= y^4 + 6y^2 t + 3t^2. \\
&\vdots
\end{aligned}
$$

Bibliography

M. Abramowitz and I. A. Stegun, eds., *Handbook of Mathematical Functions.* Applied Mathematics Series, National Bureau of Standards, 55 ed., 1964.

N. Ahuja and M. Tuceryan, "Extraction of early perceptual structure in dot patterns: Integrating region, boundary and component gestalt," *Computer Vision, Graphics, and Image Processing*, vol. 27, pp. 304–356, 1989.

Y. Aloimonos, "Shape from texture," *Biological Cybernetics*, vol. 58, pp. 345–360, 1988.

Y. Aloimonos, I. Weiss, and A. Bandyopadhyay, "Active vision," *Int. J. of Computer Vision*, pp. 333–356, 1989.

Y. Aloimonos, "Purposive and qualitative active vision," in *Proc. DARPA Image Understanding Workshop*, pp. 816–828, 1990.

Y. Aloimonos and D. Schulman, *Integration of Visual Modules: An Extension of the Marr Paradigm.* San Diego, California: Academic Press, 1989.

L. Alvarez, F. Guichard, P.-L. Lions, and J.-M. Morel, "Axioms and fundamental equations of image processing," Tech. Rep., Ceremade, Universit'e Paris-Dauphine, Paris, France, 1992.

T. Ando, "Totally positive matrices," *Linear Algebra and its Applications*, vol. 90, pp. 165–219, 1987.

V. I. Arnold, *Singularity Theory, Selected papers*, vol. 53 of *London Mathematical Society Lecture Note Series.* Cambridge: Cambridge University Press, 1981.

V. I. Arnold, S. M. Gusein-Zade, and A. N. Varchenko, *Singularities of Smooth Maps, Volume I.* Boston: Birkhäuser, 1985.

V. I. Arnold, S. M. Gusein-Zade, and A. N. Varchenko, *Singularities of Smooth Maps, Volume II.* Boston: Birkhäuser, 1988.

H. Asada and M. Brady, "The curvature primal sketch," *IEEE Trans. Pattern Analysis and Machine Intell.*, vol. 8, no. 1, pp. 2–14, 1986.

J. Babaud, A. P. Witkin, M. Baudin, and R. O. Duda, "Uniqueness of the Gaussian kernel for scale-space filtering," *IEEE Trans. Pattern Analysis and Machine Intell.*, vol. 8, no. 1, pp. 26–33, 1986.

R. Bajcsy and L. Lieberman, "Texture gradients as a depth cue," *Computer Graphics and Image Processing*, vol. 5, pp. 52–67, 1976.

R. Bajcsy, "Active perception," *Proc. IEEE*, pp. 996–1005, 1988.

H. H. Baker, "Surface reconstruction from image sequences," in *Proc. 2nd Int. Conf. on Computer Vision*, (Tampa, Florida), pp. 334–343, 1988.

D. Ballard, "Animate vision," *J. of Artificial Intell.*, vol. 48, pp. 57–86, 1991.

M. F. Barnsley, R. L. Devaney, B. B. Mandelbrot, H.-O. Peitgen, D. Saupe, and R. F. Voss, *The Science of Fractals.* New York: Springer-Verlag, 1988.

H. G. Barrow and J. M. Tenenbaum, "Recovering intrinsic scene characteristics from images," in *Computer Vision Systems: Proc. Workshop on Computer Vision Systems* (A.R. Hanson and E.M Riseman, eds.), pp. 3–26, 1978.

A. Bengtsson, J.-O. Eklundh, and J. Howako, "Shape representation by multiscale contour approximation," Tech. Rep. TRITA-NA-8607, Dept. of Numerical Analysis and Computing Science, Royal Institute of Technology, Dec. 1986.

A. Bengtsson and J.-O. Eklundh, "Shape representation by multiscale contour approximation," *IEEE Trans. Pattern Analysis and Machine Intell.*, vol. 13, pp. 85–94, Jan. 1991.

J. R. Bergen and E. H. Adelson, "Early vision and texture perception," *Nature*, vol. 333, pp. 363–364, 1988.

F. Bergholm, "Edge focusing," *IEEE Trans. Pattern Analysis and Machine Intell.*, vol. 9, pp. 726–741, Nov. 1987.

F. Bergholm, *On the Content of Information in Edges and Optical Flow.* PhD thesis, Dept. of Numerical Analysis and Computing Science, Royal Institute of Technology, May 1989.

V. Berzins, "Accuracy of Laplacian edge detectors," *Computer Vision, Graphics, and Image Processing*, vol. 27, pp. 195–210, 1984.

I. Biederman, "Human image understanding: Recent research and a theory," in *Human and Machine Vision II*, pp. 13–57, Academic Press, 1985.

J. Bigün and G. H. Granlund, "Optimal orientation detection of linear symmetry," in *Proc. 1st Int. Conf. on Computer Vision*, (London), pp. 433–438, IEEE Computer Society Press, 1987.

J. Bigün, G. H. Granlund, and J. Wiklund, "Multidimensional orientation estimation with applications to texture analysis and optical flow," *IEEE Trans. Pattern Analysis and Machine Intell.*, vol. 13, pp. 775–790, Aug. 1991.

P. Bijl, *Aspects of Visual Contrast Detection.* PhD thesis, University of Utrecht, University of Utrecht, Dept. of Med. Phys., Princetonplein 5, Utrecht, the Netherlands, May 1991.

P. Bijl and J. J. Koenderink, "Visibility of elliptical Gaussian blobs," *Vision Research*, vol. 33, no. 2, pp. 243–255, 1993.

F. Billmeyer and M. Saltzman, *Principles of Colour Technology.* John Wiley and Sons, 1982.

W. F. Bischof and T. Caelli, "Parsing scale-space and spatial stability analysis," *Computer Vision, Graphics, and Image Processing*, vol. 42, pp. 192–205, 1988.

A. Blake and C. Marinos, "Shape from texture: estimation, isotropy and moments," *J. of Artificial Intell.*, vol. 45, pp. 323–380, 1990.

A. Blake and C. Marinos, "Shape from texture: the homogeneity hypothesis," in *Proc. 3rd Int. Conf. on Computer Vision*, (Osaka, Japan), pp. 350–353, IEEE Computer Society Press, Dec. 1990.

A. Blake and A. Zisserman, *Visual Reconstruction.* Boston: MIT Press, 1987.

J. Blom, *Topological and Geometrical Aspects of Image Structure.* PhD thesis, Dept. Med. Phys. Physics, Univ. Utrecht, NL-3508 Utrecht, Netherlands, 1992.

D. Blostein and N. Ahuja, "Representation and three-dimensional interpretation of image texture: An integrated approach," in *Proc. 1st Int. Conf. on Computer Vision*, (London), pp. 444–449, IEEE Computer Society Press, 1987.

D. Blostein and N. Ahuja, "A multiscale region detector," *Computer Vision, Graphics, and Image Processing*, vol. 45, pp. 22–41, 1989.

D. Blostein and N. Ahuja, "Shape from texture: integrating texture element extraction and surface estimation," *IEEE Trans. Pattern Analysis and Machine Intell.*, vol. 11, pp. 1233–1251, Dec. 1989.

C. de Boor, *A Practical Guide to Splines*, vol. 27 of *Applied Mathematical Sciences.* New York: Springer-Verlag, 1978.

R. J. Boscovich, *Theoria Philosopiæ Naturalis*, Venice 1758. Transl. *A Theory of Natural Philosophy.* M.I.T. Press, Cambridge Mass., 1966.

R. A. Brooks, "Intelligence without representation," *J. of Artificial Intell.*, vol. 47, pp. 139–159, 1991.

L. G. Brown and H. Shvaytser, "Surface orientation from projective foreshortening of isotropic texture autocorrelation," *IEEE Trans. Pattern Analysis and Machine Intell.*, vol. 12, pp. 584–588, June 1990.

J. W. Bruce and P. J. Giblin, *Curves and Singularities.* Cambridge: Cambridge University Press, 1984.

K. Brunnström, J.-O. Eklundh, and T. Lindeberg, "On scale and resolution in the analysis of local image structure," in *Proc. 1st European Conf. on Computer Vision* (O. Faugeras, ed.), vol. 427 of *Lecture Notes in Computer Science*, pp. 3–12, Springer-Verlag, Apr. 1990. (Antibes, France).

K. Brunnström, J.-O. Eklundh, and T. Lindeberg, "Scale and resolution in active analysis of local image structure," *Image and Vision Computing*, vol. 8, pp. 289–296, Nov. 1990.

K. Brunnström, T. Lindeberg, and J.-O. Eklundh, "Active detection and classification of junctions by foveation with a head-eye system guided by the scale-space primal sketch," in *Proc. 2nd European Conf. on Computer Vision* (G. Sandini, ed.), vol. 588 of *Lecture Notes in Computer Science*, pp. 701–709, Springer-Verlag, May 1992. (Santa Margherita Ligure, Italy).

K. Brunnström, *Active visual exploration of static scenes.* PhD thesis, Royal Institute of Technology, Stockholm, Sweden, 1993. In preparation.

P. J. Burt, "Fast filter transforms for image processing," *Computer Vision, Graphics, and Image Processing*, vol. 16, pp. 20–51, 1981.

P. J. Burt and E. H. Adelson, "The Laplacian pyramid as a compact image code," *IEEE Trans. Communications*, vol. 9:4, pp. 532–540, 1983.

P. J. Burt and R. J. Kolczynski, "Enhanced image capture through fusion," in *Proc. 4th Int. Conf. on Computer Vision*, (Berlin, Germany), pp. 173–182, May 1993.

T. Caelli, "Three processing characteristics of visual texture segmentation," *Spatial Vision*, vol. 1, pp. 19–30, 1985.

F. W. C. Campbell and J. Robson, "Application of Fourier analysis to the visibility of gratings," *J. Physiol.*, vol. 197, pp. 417–424, 1977.

J. Canny, "A computational approach to edge detection," *IEEE Trans. Pattern Analysis and Machine Intell.*, vol. 8, no. 6, pp. 679–698, 1986.

V. Cantoni and S. Levialdi, eds., *Pyramidal Systems for Computer Vision.* Berlin: Springer-Verlag, 1986.

M. J. Carlotto, "Histogram analysis using a scale-space approach," *IEEE Trans. Pattern Analysis and Machine Intell.*, vol. 9, pp. 121–129, 1987.

A. Chehikian and J. L. Crowley, "Fast computation of optimal semi-octave pyramids," in *Proc. 7th Scandinavian Conf. on Image Analysis*, (Aalborg, Denmark), pp. 18–27, Aug 1991.

J. J. Clark, "Singularity theory and phantom edges in scale-space," *IEEE Trans. Pattern Analysis and Machine Intell.*, vol. 10, no. 5, pp. 720–727, 1988.

H. Cramer and M. R. Leadbetter, *Stationary and Related Stochastic Processes.* New York: John Wiley and Sons, 1967.

J. L. Crowley, *A Representation for Visual Information.* PhD thesis, Carnegie-Mellon University, Robotics Institute, Pittsburgh, Pennsylvania, 1981.

J. L. Crowley and A. C. Parker, "A representation for shape based on peaks and ridges in the Difference of Low-Pass Transform," *IEEE Trans. Pattern Analysis and Machine Intell.*, vol. 6, no. 2, pp. 156–170, 1984.

J. L. Crowley and R. M. Stern, "Fast computation of the Difference of Low Pass Transform," *IEEE Trans. Pattern Analysis and Machine Intell.*, vol. 6, pp. 212–222, 1984.

J. L. Crowley and A. C. Sanderson, "Multiple resolution representation and probabilistic matching of 2-D gray-scale shape," *IEEE Trans. Pattern Analysis and Machine Intell.*, vol. 9, no. 1, pp. 113–121, 1987.

S. M. Culhane and J. K. Tsotsos, "An attentional prototype for early vision," in *Proc. 2nd European Conf. on Computer Vision*, (Santa Margherita Ligure, Italy), pp. 551–562, May 1992.

G. Dahlquist, Å. Björk, and N. Anderson, eds., *Numerical Methods.* Prentice-Hall, 1974.

P.-E. Danielsson and O. Seger, "Rotation invariance in gradient and higher order derivative detectors," *Computer Vision, Graphics, and Image Processing*, vol. 49, pp. 198–221, 1990.

I. Daubechies, "Orthonormal bases of compactly supported wavelets," *Comm. on Pure and Applied Mathematics*, vol. XLI, pp. 909–996, 1988.

R. Deriche, "Using Canny's criteria to derive a recursively implemented optimal edge detector," *Int. J. of Computer Vision*, vol. 1, pp. 167–187, 1987.

R. Deriche, "Separable recursive filtering for efficient multi-scale edge detection," in *Proc. International Workshop on Industrial Applications of Machine Vision and Machine Intell.*, (Tokyo, Japan), Feb. 2–5 1987. Seiken Symposium.

R. Deriche and G. Giraudon, "Accurate corner detection: An analytical study," in *Proc. 3rd Int. Conf. on Computer Vision*, (Osaka, Japan), pp. 66–70, 1990.

S. J. Dickinson, A. P. Pentland, and A. Rosenfeld, "Qualitative 3-D reconstruction using distributed aspect graph matching," in *Proc. 3rd Int. Conf. on Computer Vision*, (Osaka, Japan), pp. 257–262, 1990.

L. Dreschler and H.-H. Nagel, "Volumetric model and 3D-trajectory of a moving car derived from monocular TV-frame sequences of a street scene," *Computer Vision, Graphics, and Image Processing*, vol. 20, no. 3, pp. 199–228, 1982.

R. W. Ehrich and P. F. Lai, "Elements of a structural model of texture," in *Proc. PRIP*, pp. 319–326, IEEE CS Press, 1978.

Euclid, *The Optics*, c. 300 B.B. Transl. P. Ver Eecke, *Euclide, L'Optique et la Catoptrique*. Albert Blanchard, Paris, 1959.

D. J. Field, "Relations between the statistics of natural images and the response properties of cortical cells," *J. of the Optical Society of America*, vol. 4, pp. 2379–2394, 1987.

L. M. J. Florack, B. M. ter Haar Romeny, J. J. Koenderink, and M. A. Viergever, "General intensity transformations," in *Proc. 7th Scandinavian Conf. on Image Analysis* (P. Johansen and S. Olsen, eds.), (Aalborg, Denmark), pp. 338–345, Aug 1991.

L. M. J. Florack, B. M. ter Haar Romeny, J. J. Koenderink, and M. A. Viergever, "Scale and the differential structure of images," *Image and Vision Computing*, vol. 10, pp. 376–388, July/August 1992.

L. M. J. Florack, B. M. ter Haar Romeny, J. J. Koenderink, and M. A. Viergever, "Images: Regular tempered distributions," in *Proc. NATO workshop 'Shape in Picture* (Y.O. Ying, A. Toet, and H. Heijmanns, eds.), NATO ASI Series F, (Driebergen, Netherlands), Springer Verlag, New York, September 1992. (In press).

L. M. J. Florack, B. M. ter Haar Romeny, J. J. Koenderink, and M. A. Viergever, "Cartesian differential invariants," *J. of Mathematical Imaging and Vision*, 1993. (In press).

L. M. J. Florack, B. M. ter Haar Romeny, J. J. Koenderink, and M. A. Viergever, "General intensity transformations and differential invariants," *J. of Mathematical Imaging and Vision*, 1993. (In press).

L. M. J. Florack, B. M. ter Haar Romeny, J. J. Koenderink, and M. A. Viergever, "Non-linear diffusion by metrical affinity". In preparation, 1993.

M.A. Förstner and E. Gülch, "A fast operator for detection and precise location of distinct points, corners and centers of circular features," in *ISPRS Intercommission Workshop*, 1987.

D. Forsyth and A. Zissermann, "Mutual illumination," in *Proc. IEEE Comp. Soc. Conf. on Computer Vision and Pattern Recognition*, (San Diego, California), pp. 466–473, Jun. 1989.

D. Forsyth and A. Zissermann, "Shape from shading in the light of mutual illumination," in *Proc. Fifth Alvey Vision Conf.*, (Reading, September), pp. 193–198, 1989.

J. Fourier, *The Analytical Theory of Heat*. New York: Dover Publications, Inc., 1955. Replication of the English translation that first appeared in 1878 with previous corrigenda incorporated into the text, by Alexander Freeman, M.A. Original work: "Theorie Analytique de la Chaleur," Paris, 1822.

W. T. Freeman and E. H. Adelson, "Steerable filters for early vision, image analysis and wavelet decomposition," in *Proc. 3rd Int. Conf. on Computer Vision*, (Osaka, Japan), IEEE Computer Society Press, Dec. 1990.

D. Gabor, "Theory of communication," *J. of the IEE*, vol. 93, pp. 429–457, 1946.

J. Gårding, "Properties of fractal intensity surfaces," *Pattern Recognition Letters*, vol. 8, pp. 319–324, Dec. 1988.

J. Gårding, *Shape from surface markings*. PhD thesis, Dept. of Numerical Analysis and Computing Science, Royal Institute of Technology, Stockholm, May 1991.

J. Gårding, "Shape from texture for smooth curved surfaces in perspective projection," *J. of Mathematical Imaging and Vision*, vol. 2, pp. 329–352, 1992.

J. Gårding, "Shape from texture and contour by weak isotropy," *J. of Artificial Intell.*, 1993. (In press).

J. Gårding and T. Lindeberg, "Direct computation of shape cues by multi-scale retinotopic processing," Tech. Rep. TRITA-NA-P9304, Dept. of Numerical Analysis and Computing Science, Royal Institute of Technology, Feb. 1993. (Submitted).

C. G. Gibson, *Singular Points of Smooth Mappings*. Research Notes in Mathematics, London: Pitman Publishing, 1979.

J. Gibson, *The Perception of the Visual World*. Houghton Mifflin, Boston, 1950.

J. Gibson, *The Ecological Approach to Visual Perception*. Houghton Mifflin, Boston, 1979.

M. Golubitsky and D. G. Schaeffer, *Singularities and Groups in Bifurcation Theory I*, vol. 51 of *Applied Mathematical Sciences*. New York: Springer–Verlag, 1985.

J. H. Grace and A. Young, *Algebra of Invariants*. Bronx, New York: Chelsea Publishing Company, 1965.

G. H. Granlund, "In search of a general picture processing operator," *Computer Vision, Graphics, and Image Processing*, vol. 8, pp. 155–173, 1978.

R. M. Gray, "On the asymptotic eigenvalue distribution of Toeplitz matrices," *IEEE Trans. Information Theory*, vol. 18, no. 6, pp. 725–730, 1972.

U. Grenander and G. Szegö, *Toeplitz Forms and Their Applications*. Los Angeles, California: Univ. of California Press, 1958.

W. E. L. Grimson and E. C. Hildreth, "Comments on digital step edges from zero crossings of second directional derivatives," *IEEE Trans. Pattern Analysis and Machine Intell.*, vol. 7, no. 1, pp. 121–127, 1985.

W. A. van de Grind, J. J. Koenderink, and A. J. van Doorn, "The distribution of human motion detector properties in the monocular visual field," *Vision Research*, vol. 26, no. 5, pp. 797–810, 1986.

A. D. Gross, "Multiresolution object detection and delineation," Tech. Rep. TR-1613, Computer Vision Laboratory, University of Maryland, Maryland, 1986.

S. Grossberg, "Neural dynamics of brightness perception: Features, boundaries, diffusion, and resonance," *Percept. Psychophys.*, vol. 36, no. 5, pp. 428–456, 1984.

W. Hackbush, *Multi-Grid Methods and Applications.* New York: Springer-Verlag, 1985.

A. R. Hanson and E. M. Riseman, "Processing cones: A parallel computational structure for scene analysis," Tech. Rep. 74C-7, Computer and Information Science, Univ. of Massachusetts, Amherst, Massachusetts, 1974.

R. M. Haralick, L. T. Watson, and T. J. Laffey, "The topographic primal sketch," *Int. J. of Robotics Research*, vol. 2, no. 1, pp. 50–72, 1983.

R. M. Haralick, "Digital step edges from zero-crossings of second directional derivatives," *IEEE Trans. Pattern Analysis and Machine Intell.*, vol. 6, 1984.

D. Hilbert, "Über die vollen Invariantensystemen," *Mathematische Annalen*, vol. 42, pp. 313–373, 1893.

E. Hille and R. S. Phillips, *Functional Analysis and Semi-Groups*, vol. XXXI. American Mathematical Society Colloquium Publications, 1957.

I. I. Hirschmann and D. V. Widder, *The Convolution Transform.* Princeton, New Jersey: Princeton University Press, 1955.

A. Horii, "Depth from defocusing," Tech. Rep. TRITA-NA-P9216, Dept. of Numerical Analysis and Computing Science, Royal Institute of Technology, June 1992.

A. Horii, "The focusing mechanism in the KTH head-eye system," Tech. Rep. TRITA-NA-P9215, Dept. of Numerical Analysis and Computing Science, Royal Institute of Technology, June 1992.

R. A. Hummel, "Representations based on zero crossings in scale space," in *Proc. IEEE Comp. Soc. Conf. on Computer Vision and Pattern Recognition*, pp. 204–209, 1986.

R. A. Hummel, "The scale-space formulation of pyramid data structures," in *Parallel Computer Vision* (L. Uhr, ed.), (New York), pp. 187–223, Academic Press, 1987.

R. A. Hummel and R. Moniot, "Reconstructions from zero-crossings in scale-space," *IEEE Trans. Acoustics, Speech and Signal Processing*, vol. 37, no. 12, pp. 2111–2130, 1989.

P. Johansen, S. Skelboe, K. Grue, and J. D. Andersen, "Representing signals by their top points in scale-space," in *Proc. 8:th Int. Conf. on Pattern Recognition*, (Paris, France), pp. 215–217, Oct 27-31 1986.

P. Johansen, "On the classification of toppoints in scale space," *J. of Mathematical Imaging and Vision*, 1993. To appear.

D. G. Jones and J. Malik, "Determining three-dimensional shape from orientation and spatial frequency disparities," in *Proc. 2nd European Conf. on Computer Vision* (G. Sandini, ed.), vol. 588 of *Lecture Notes in Computer Science*, pp. 661–669, Springer-Verlag, May 1992.

J. Jones and L. Palmer, "The two-dimensional spatial structure of simple receptive fields in cat striate cortex," *J. of Neurophysiology*, vol. 58, pp. 1187–1211, 1987.

J. Jones and L. Palmer, "An evaluation of the two-dimensional Gabor filter model of simple receptive fields in cat striate cortex," *J. of Neurophysiology*, vol. 58, pp. 1233–1258, 1987.

B. Julesz and J. R. Bergen, "Textons, the fundamental elements in preattentive vision and perception of textures," *The Bell System Technical J.*, vol. 62, no. 6, pp. 1619–1645, 1983.

B. Julesz, "Texton gradients: The texton theory revisited," *Biological Cybernetics*, vol. 54, pp. 245–251, 1986.

K. Kanatani, "Detection of surface orientation and motion from texture by a stereological technique," *J. of Artificial Intell.*, vol. 23, pp. 213–237, 1984.

K. Kanatani and T. C. Chou, "Shape from texture: general principle," *J. of Artificial Intell.*, vol. 38, pp. 1–48, 1989.

K. Kanatani, *Group-Theoretical Methods in Image Understanding*, vol. 20 of *Series in Information Sciences.* Spatial Vision, 1990.

S. Karlin, *Total Positivity.* Stanford Univ. Press, 1968.

M. Kass, A. Witkin, and D. Terzopoulos, "Snakes: Active contour models," *Int. J. of Computer Vision*, vol. 1, pp. 321–331, 1987.

B. B. Kimia, A. Tannenbaum, and S. W. Zucker, "Toward a computational theory of shape: An overview," in *Proc. 1st European Conf. on Computer Vision*, (Antibes, France), pp. 402–407, April 1990.

L. Kitchen and A. Rosenfeld, "Gray-level corner detection," *Pattern Recognition Letters*, vol. 1, no. 2, pp. 95–102, 1982.

A. Klinger, "Pattern and search statistics," in *Optimizing Methods in Statistics* (J.S. Rustagi, ed.), (New York), Academic Press, 1971.

C. B. Knudsen and H. I. Christensen, "On methods for efficient pyramid generation," in *Proc. 7th Scandinavian Conf. on Image Analysis*, (Aalborg, Denmark), pp. 28–39, Aug 1991.

H. Knutsson and G. H. Granlund, "Texture analysis using two-dimensional quadrature filters," in *Proc. IEEE Comp. Soc. Workshop on Computer Architecture for Pattern Analysis and Image Database Management*, 1983.

J. J. Koenderink and A. J. van Doorn, "Geometry of binocular vision and a model for stereopsis," *Biological Cybernetics*, vol. 21, pp. 29–35, 1976.

J. J. Koenderink and A. J. van Doorn, "Visual detection of spatial contrast; influence of location in the visual field, target extent and illuminance level," *Biological Cybernetics*, vol. 30, pp. 157–167, 1978.

J. J. Koenderink, "The structure of images," *Biological Cybernetics*, vol. 50, pp. 363–370, 1984.

J. J. Koenderink and A. J. van Doorn, "Dynamic shape," *Biological Cybernetics*, vol. 53, pp. 383–396, 1986.

J. J. Koenderink and A. J. van Doorn, "Representation of local geometry in the visual system," *Biological Cybernetics*, vol. 55, pp. 367–375, 1987.

J. J. Koenderink and W. Richards, "Two-dimensional curvature operators," *J. Optical Society of America*, vol. 5:7, pp. 1136–1141, 1988.

J. J. Koenderink, *Solid Shape*. MIT Press, Cambridge, Mass., 1990.

J. J. Koenderink and A. J. van Doorn, "Affine structure from motion," *J. of the Optical Society of America*, pp. 377–385, 1991.

J. J. Koenderink and A. J. van Doorn, "Generic neighborhood operators," *IEEE Trans. Pattern Analysis and Machine Intell.*, vol. 14, pp. 597–605, June 1992.

K. Koffka, *Principles of Gestalt Psychology*. New York: Harcourt Brace, 1935.

D. Koller, K. Daniilides, T. Thórhallson, and H.-H Nagel, "Model-based object tracking in traffic scenes," in *Proc. 2nd European Conf. on Computer Vision*, (Santa Margherita Ligure, Italy), 1992.

A. F. Korn, "Toward a symbolic representation of intensity changes in images," *IEEE Trans. Pattern Analysis and Machine Intell.*, vol. 10, no. 5, pp. 610–625, 1988.

M. D. Levine, "Region analysis using a pyramid data structure," in *Structured Computer Vision* (S. Tanimoto and A. Klinger, eds.), (New York), pp. 57–100, Academic Press, 1980.

L. M. Lifshitz and S. M. Pizer, "A multiresolution hierarchical approach to image segmentation based on intensity extrema," Tech. Rep., Departments of Computer Science and Radiology, University of North Carolina, Chapel Hill, N.C., U.S.A, 1987.

L.M. Lifshitz and S.M. Pizer, "A multiresolution hierarchical approach to image segmentation based on intensity extrema," *IEEE Trans. Pattern Analysis and Machine Intell.*, vol. 12, no. 6, pp. 529–541, 1990.

T. Lindeberg, "On the construction of a scale-space for discrete images," Tech. Rep. TRITA-NA-P8808, Dept. of Numerical Analysis and Computing Science, Royal Institute of Technology, 1988.

T. Lindeberg, "Scale-space for discrete signals," *IEEE Trans. Pattern Analysis and Machine Intell.*, vol. 12, pp. 234–254, Mar. 1990.

T. Lindeberg and J.-O. Eklundh, "Scale detection and region extraction from a scale-space primal sketch," in *Proc. 3rd Int. Conf. on Computer Vision*, (Osaka, Japan), pp. 416–426, Dec. 1990.

T. Lindeberg, *Discrete scale space theory and the scale space primal sketch.* PhD thesis, Dept. of Numerical Analysis and Computing Science, Royal Institute of Technology, Stockholm, May 1991.

T. Lindeberg and J.-O. Eklundh, "On the computation of a scale-space primal sketch," *J. of Visual Communication and Image Representation*, vol. 2, pp. 55–78, Mar. 1991.

T. Lindeberg and J.-O. Eklundh, "Analysis of aerosol images using the scale-space primal sketch," *Machine Vision and Applications*, vol. 4, pp. 135–144, Aug. 1991.

T. Lindeberg and J.-O. Eklundh, "The scale-space primal sketch: Construction and experiments," *Image and Vision Computing*, vol. 10, pp. 3–18, Jan. 1992.

T. Lindeberg, "Scale-space behaviour of local extrema and blobs," *J. of Mathematical Imaging and Vision*, vol. 1, pp. 65–99, Mar. 1992.

T. Lindeberg, "Detecting salient blob-like image structures and their scales with a scale-space primal sketch—A method for focus-of-attention," *Int. J. of Computer Vision*, vol. 11, no. 3, pp. 283–318, 1993. (In press).

T. Lindeberg, "Discrete derivative approximations with scale-space properties: A basis for low-level feature extraction," *J. of Mathematical Imaging and Vision*, vol. 3, pp. 349–376, 1993. (In press).

T. Lindeberg, "Effective scale: A natural unit for measuring scale-space lifetime," *IEEE Trans. Pattern Analysis and Machine Intell.*, vol. 15, Oct. 1993. (In press).

T. Lindeberg, "Scale-space for N-dimensional discrete signals," in *Shape in Picture: Proc. NATO workshop on Shape in Picture* (Y. O. Ying, A. Toet, and H. Heijmanns, eds.), NATO ASI Series F, (Driebergen Netherlands), Springer Verlag, New York, Sept. 1992. (In press).

T. Lindeberg, "Scale-space behaviour and invariance properties of differential singularities," in *Shape in Picture: Proc. NATO workshop on Shape in Picture* (Y. O. Ying, A. Toet, and H. Heijmanns, eds.), NATO ASI Series F, (Driebergen Netherlands), Springer Verlag, New York, Sept. 1992. (In press).

T. Lindeberg, "On scale selection for differential operators," in *The 8th Scandinavian Conf. on Image Analysis* (K. Heia K. A. Høgdra, B. Braathen, ed.), (Tromsø, Norway), pp. 857–866, Norwegian Society for Image Processing and Pattern Recognition, May 1993.

T. Lindeberg and J. Gårding, "Shape from texture from a multi-scale perspective," in *Proc. 4th Int. Conf. on Computer Vision* (H.-H. Nagel et. al., ed.), (Berlin, Germany), pp. 683–691, May 1993.

T. Lindeberg and L. Florack, "On the decrease of resolution as a function of eccentricity for a foveal vision system," Tech. Rep. TRITA-NA-P9229, Dept. of Numerical Analysis and Computing Science, Royal Institute of Technology, Oct. 1992. (Submitted).

D. G. Lowe, *Perceptual Organization and Visual Recognition.* Boston: Kluwer Academic Publishers, 1985.

D. G. Lowe, "Organization of smooth image curves at multiple scales," in *Proc. 2nd Int. Conf. on Computer Vision*, (Tampa, Florida), pp. 558–567, Dec. 1988.

Y.-C. Lu, *Singularity Theory and an Introduction to Catastrophe Theory.* New York: Springer-Verlag, 1976.

Lucretius, *On the Nature of the Universe*, c. 100–55 B.B. Transl. R. E. Latham, Penguin Books, 1951.

J. Malik, "Interpreting line drawings of curved objects," *Int. J. of Computer Vision*, no. 1, pp. 73–104, 1987.

J. Malik and P. Perona, "A computational model of texture segmentation," in *Proc. IEEE Comp. Soc. Conf. on Computer Vision and Pattern Recognition*, (San Diego, Ca.), 1989.

S. G. Mallat, "A theory for multiresolution signal decomposition: The wavelet representation," *IEEE Trans. Pattern Analysis and Machine Intell.*, vol. 11, no. 7, pp. 674–694, 1989.

S.G. Mallat, "Multifrequency channel decompositions of images and wavelet models," *IEEE Trans. Acoustics, Speech and Signal Processing*, vol. 37, pp. 2091–2110, 1989.

S. G. Mallat and S. Zhong, "Characterization of signals from multi-scale edges," *IEEE Trans. Pattern Analysis and Machine Intell.*, vol. 14, pp. 710–723, 1992.

K. V. Mardia, *Statistics of Directional Data.* Academic Press, London, 1972.

D. Marr, "Early processing of visual information," *Phil. Trans. Royal Soc (B)*, vol. 27S, pp. 483–524, 1976.

D. C. Marr and E. C. Hildreth, "Theory of edge detection," *Proc. Royal Society London* B, vol. 207, pp. 187–217, 1980.

D. Marr, *Vision.* W.H. Freeman, New York, 1982.

J. C. Maxwell, "On hills and dales," *The London, Edinghburgh and Dublin Philosophical Magazine and J. of Science*, vol. 40, no. 269, pp. 421–425, 1870. Reprinted in Niven, W.D The Scientific Papers of James Clark Maxwell, Vol II 1956, Dover Publications New York.

P. Meer, E. S. Baugher, and A. Rosenfeld, "Frequency domain analysis and synthesis of image pyramid generating kernels," *IEEE Trans. Pattern Analysis and Machine Intell.*, vol. 9, pp. 512–522, 1987.

P. Meer and I. Weiss, "Smoothed differentiation filters for images," *J. of Visual Communication and Image Representation*, vol. 3, no. 1, pp. 58–72, 1992.

Y. Meyer, *Ondolettes et Operateurs.* Hermann, 1988.

O. M. Miller and R. J. Voskuil, *Thematic-Map Generalization*, Geographical Review, vol. 54, pp. 13–19, 1964.

F. Mokhtarian and A. Mackworth, "Scale-based description and recognition of planar curves and two-dimensional objects," *IEEE Trans. Pattern Analysis and Machine Intell.*, vol. 8, pp. 34–43, 1986.

F. Mokhtarian, "Multi-scale representation of space curves and three-dimensional objects," in *Proc. IEEE Comp. Soc. Conf. on Computer Vision and Pattern Recognition*, (Ann Arbor, MI), pp. 298–303, June 1988.

P. Morrison and P. Morrison, *Powers of Ten.* Scientific American Books, Inc., New York 1982.

D. Mumford, *Tata Lectures on Theta I.* Boston, Massachusetts: Birkhäuser, 1983.

D. Mumford and J. Shah, "Boundary detection by minimizing functionals," in *Proc. IEEE Comp. Soc. Conf. on Computer Vision and Pattern Recognition*, 1985.

S. K. Nayar, K. Ikeuchi, and T. Kanade, "Shape from interreflections," in *Proc. 3rd Int. Conf. on Computer Vision*, (Osaka, Japan), pp. 2–11, Dec. 1990.

J. A. Noble, "Finding corners," *Image and Vision Computing*, vol. 6, no. 2, pp. 121–128, 1988.

N. Nordström, "Biased anisotropic diffusion: A unified regularization and diffusion approach to edge detection," *Image and Vision Computing*, vol. 8, pp. 318–327, Nov. 1990.

E. Norman, "A discrete analogue of the Weierstrass transform," *Proc. American Mathematical Society*, vol. 11, pp. 596–604, 1960.

B. O'Neill, *Elementary Differential Geometry.* Academic Press, Orlando, Florida, 1966.

K. Pahlavan and J.-O. Eklundh, "A head-eye system—Analysis and design," *CVGIP: Image Understanding*, vol. 56, no. 1, pp. 41–56, 1992.

K. Pahlavan, T. Uhlin, and J.-O. Eklundh, "Active vision as a methodology," in *Active Vision* (Y. Aloimonos, ed.), Advances in Computer Science, Lawrence Erlbaum Associates, 1992. (To appear).

A. Papoulis, *Probability, Random Variables and Stochastic Processes.* McGraw-Hill, 1972.

A. P. Pentland, "Shading into texture," *J. of Artificial Intell.*, vol. 29, pp. 147–170, 1986.

A. P. Pentland, "Extraction of deformable part models," in *Proc. 1st European Conf. on Computer Vision*, (Antibes, France), pp. 397–401, 1990.

A. P. Pentland, "Photometric motion," *IEEE Trans. Pattern Analysis and Machine Intell.*, vol. 13, no. 9, pp. 879–890, 1991.

P. Perona and J. Malik, "Scale-space and edge detection using anisotropic diffusion," *IEEE Trans. Pattern Analysis and Machine Intell.*, vol. 12, no. 7, pp. 629–639, 1990.

P. Perona, "Steerable-scalable kernels for edge detection and junction analysis," in *Proc. 2nd European Conf. on Computer Vision*, (Santa Margherita Ligure, Italy), pp. 3–18, May 1992.

S. B. Pollard, J. E. W. Mayhew, and J. P. Frisby, "PMF: A stereo correspondence algorithm using a disparity gradient limit," *Perception*, vol. 14, pp. 449–470, 1985.

T. Poston and I. Stewart, *Catastrophe Theory and its Applications.* London: Pitman, 1978.

W. H. Press, B. P. Flannery, S. A. Teukolsky, and W. T. Vetterling, *Numerical Recipes in C.* Cambridge: Cambridge University Press, 1986.

A. R. Rao and B. G. Schunk, "Computing oriented texture fields," *CVGIP: Graphical Models and Image Processing*, vol. 53, pp. 157–185, Mar. 1991.

S. O. Rice, "Mathematical analysis of random noise," *The Bell System Technical J.*, vol. XXIV, no. 1, pp. 46–156, 1945.

K. Rohr, "Modelling and identification of characteristic intensity variations," *Image and Vision Computing*, no. 2, pp. 66–76, 1992.

A. Rosenfeld and M. Thurston, "Edge and curve detection for visual scene analysis," *IEEE Trans. Computers*, vol. 20, no. 5, pp. 562–569, 1971.

A. Rosenfeld, *Multiresolution Image Processing and Analysis*, vol. 12 of *Springer Series in Information Sciences.* Springer-Verlag, 1984.

W. Rudin, *Principles of Mathematical Analysis.* McGraw-Hill, 1976.

J. Ruskin, *Modern Painters*, London 1843. Included in: *The Works of John Ruskin*. E. T. Cook and A. Wedderburn, eds., vol. 39, London and New York, 1903–1912.

A. H. Salden, B. M. ter Haar Romeny, and L. M. J. Florack, "A complete and irreducible set of local orthogonally invariant features of 2-dimensional images". Tech. Rep., Dept. Medical and Physiological Physics, Utrecht University, Netherlands, 1992.

A. H. Salden, L. M. J. Florack, and B. M. ter Haar Romeny, "Differential geometric description of 3D scalar images". Tech. Rep., Dept. Medical and Physiological Physics, Utrecht University, Netherlands, 1992.

G. Sapiro and A. Tannenbaum, "Affine invariant scale-space," *Int. J. of Computer Vision*, vol. 11, no. 1, pp. 25–44, 1993.

E. Saund, "Symbolic construction of a 2-D scale-space image," *IEEE Trans. Pattern Analysis and Machine Intell.*, vol. 12, no. 8, pp. 817–831, 1990.

I. J. Schoenberg, "Über Variationsvermindernde Lineare Transformationen," *Mathematische Zeitschrift*, vol. 32, pp. 321–328, 1930.

I. J. Schoenberg, "Contributions to the problem of approximation of equidistant data by analytic functions," *Quarterly of Applied Mathematics*, vol. 4, pp. 45–99, 1946.

I. J. Schoenberg, "Some analytical aspects of the problem of smoothing," in *Courant Anniversary Volume, Studies and Essays*, (New York), pp. 351–370, 1948.

I. J. Schoenberg, "On Pòlya frequency functions. ii. Variation-diminishing integral operators of the convolution type," *Acta Sci. Math. (Szeged)*, vol. 12, pp. 97–106, 1950.

I. J. Schoenberg, "On smoothing operations and their generating functions," *Bull. Amer. Math. Soc.*, vol. 59, pp. 199–230, 1953.

L. Schwartz, *Théorie des Distributions*, vol. I, II of *Actualités scientifiques et industrielles; 1091,1122*. Paris: Publications de l'Institut de Mathématique de l'Université de Strasbourg, 1950–1951.

J. Serra, *Image Analysis and Mathematical Morphology*. London: Academic Press, 1982.

J. Shah, "Segmentation by non-linear diffusion," in *Proc. IEEE Comp. Soc. Conf. on Computer Vision and Pattern Recognition*, pp. 202–207, 1991.

A. C. Sher and Rosenfeld A., "Detecting and extracting compact textured objects using pyramids," Tech. Rep. TR-1789, Computer Vision Laboratory, University of Maryland, Maryland, 1987.

F. Sjöberg and F. Bergholm, "Extraction of diffuse edges by edge focusing," *Pattern Recognition Letters*, vol. 7, pp. 181–190, Mar. 1988.

M. R. Spiegel, *Mathematical Handbook of Formulas and Tables*. Schaum's Outline Series in Mathematics, McGraw-Hill, 1968.

M. Spivak, *Differential Geometry*, vol. 1–5. Berkeley, California, USA: Publish or Perish, Inc., 1975.

K. A. Stevens, "The information content of texture gradients," *Biological Cybernetics*, vol. 42, pp. 95–105, 1981.

J. V. Stone, "Shape from texture: textural invariance and the problem of scale in perspective images of surfaces," in *Proc. British Machine Vision Conf.*, (Oxford, England), Sept. 1990.

G. Strang, *Introduction to Applied Mathematics.* Massachusetts: Wellesley-Cambridge Press, 1986.

J. O. Strömberg, "A modified Franklin system and higher order splines as unconditional basis for Hardy spaces," in *Proc. Conf. in Harmonic Analysis in Honor of Antoni Zygmund* (Beckner W. et al., ed.), vol. II, Wadworth Mathematical Series, 1983.

B. J. Super and A. C. Bovik, "Shape-from-texture by wavelet-based measurement of local spectral moments," in *Proc. IEEE Comp. Soc. Conf. on Computer Vision and Pattern Recognition*, (Champaign, Illinois), pp. 296–301, June 1992.

R. Szeliski, "Fast shape from shading," in *Proc. 1st European Conf. on Computer Vision*, (Antibes, France), pp. 359–368, April 1990.

S. Tanimoto and A. Klinger, eds., *Structured Computer Vision.* New York: Academic Press, 1980.

D. Terzopoulos, "Multilevel computational processes for visual surface reconstruction," *Computer Vision, Graphics, and Image Processing*, vol. 24, pp. 52–95, 1983.

D. Terzopoulos, "Regularization of inverse visual problems involving discontinuities," *IEEE Trans. Pattern Analysis and Machine Intell.*, vol. 8, pp. 413–424, 1986.

D. Terzopoulos, A. Witkin, and M. Kass, "Symmetry-seeking models for 3-d object reconstruction," in *Proc. 1st Int. Conf. on Computer Vision*, (London, England), pp. 269–276, 1987.

D. Terzopoulos, A. Witkin, and M. Kass, "Constraints on deformable models: Recovering 3-D shape and nonrigid motion," *J. of Artificial Intell.*, vol. 36, pp. 91–123, 1988.

A. N. Tikhonov and V. Y. Arsenin, *Solution of Ill-Posed Problems.* Washington DC: Winston and Wiley, 1977.

O. Toeplitz, "Zur Theorie der quadratischen und bilinearen Formen von unendlichvielen Veränderlichen. i. Teil: Theorie der L-formen," *Mathematische Annalen*, vol. 70, pp. 351–376, 1911.

V. Torre and Th. A. Poggio, "On edge detection," *IEEE Trans. Pattern Analysis and Machine Intell.*, vol. 8, no. 2, pp. 147–163, 1986.

J. K. Tsotsos, "Analyzing vision at the complexity level," *Behavioural and Brain Sciences*, vol. 13, pp. 423–469, 1990.

L. Uhr, "Layered 'recognition cone' networks that preprocess, classify and describe," *IEEE Trans. Comput.*, pp. 759–768, 1972.

T. Valdsoo, "The jet structure from pneumatic atomization nozzles," Tech. Rep., Dept. of Aeronautics, Royal Institute of Technology, Stockholm, Sweden, 1989.

T. Valdsoo, "The spray structure from typical fuel injectors: with different liquids and varying air-assist," Tech. Rep., Dept. of Aeronautics, Royal Institute of Technology, S-100 44 Stockholm, Sweden, 1989.

T. Valdsoo, "Some characteristics of liquid jet disintegration processes in high velocity air flow," Tech. Rep., Dept. of Aeronautics, Royal Institute of Technology, S-100 44 Stockholm, Sweden, 1989.

T. Vieville and O. D. Faugeras, "Robust and fast computation of unbiased intensity derivatives in images," in *Proc. 2nd European Conf. on Computer Vision*, (Santa Margherita Ligure, Italy), pp. 203–211, May 1992.

H. Voorhees and T. Poggio, "Detecting textons and texture boundaries in natural images," in *Proc. 1st Int. Conf. on Computer Vision*, (London, England), 1987.

R. Watt, *Visual Processing: Computational, Psychophysical and Cognitive Research.* London: Lawrence Erlbaum Associates, 1988.

H. Weyl, *The Classical Groups, Their Invariants and Representations.* Princeton, NJ: Princeton University Press, 1946.

R. T. Whitaker and S. M. Pizer, "A multi-scale approach to nonuniform diffusion," *Computer Vision, Graphics, and Image Processing*, vol. 57, no. 1, 1993.

D. V. Widder, *The Heat Equation.* New York: Academic Press, 1975.

R. P. Wildes, "Direct recovery of three-dimensional scene geometry from binocular stereo disparity," *IEEE Trans. Pattern Analysis and Machine Intell.*, vol. 13, no. 8, pp. 761–774, 1981.

H. R. Wilson, "Psychophysical evidence for spatial channels," in *Physical and Biological Processing of Images* (O.J Braddick and A.C. Sleigh, eds.), (New York), Springer Verlag, 1983.

R. Wilson and A. H. Bhalerao, "Kernel design for efficient multiresolution edge detection and orientation estimation," *IEEE Trans. Pattern Analysis and Machine Intell.*, vol. 14, no. 3, pp. 384–390, 1992.

A. P. Witkin, "Recovering surface shape and orientation from texture," *J. of Artificial Intell.*, vol. 17, pp. 17–45, 1981.

A. P. Witkin, "Scale-space filtering," in *Proc. 8th Int. Joint Conf. Art. Intell.*, (Karlsruhe, West Germany), pp. 1019–1022, Aug. 1983.

A. P. Witkin and J. M. Tenenbaum, "On the role of structure in vision," in *Human and Machine Vision* (J. Beck, B. Hope, and A. Rosenfeld, eds.), (New York), Academic Press, 1983.

A. P. Witkin, D. Terzopoulos, and M. Kass, "Signal matching through scale-space," *Int. J. of Computer Vision*, vol. 1, pp. 133–144, 1987.

R. A. Young, "The Gaussian derivative theory of spatial vision: Analysis of cortical cell receptive field line-weighting profiles," Tech. Rep. GMR-4920, Computer Science Department, General Motors Research Lab., Warren, Michigan, 1985.

R. A. Young, "The Gaussian derivative model for spatial vision: I. Retinal mechanisms," *Spatial Vision*, vol. 2, pp. 273–293, 1987.

H. Yserentant, "On the multi-level splitting of finite element spaces," *Numerische Mathematik*, vol. 49, pp. 379–412, 1986.

A. L. Yuille and T. A. Poggio, "Scaling theorems for zero-crossings," *IEEE Trans. Pattern Analysis and Machine Intell.*, vol. 8, pp. 15–25, 1986.

A. L. Yuille and T. A. Poggio, "Scaling and fingerprint theorems for zero-crossings," in *Advances in Computer Vision* (C. Brown, ed.), pp. 47–78, Lawrence Erlbaum, 1988.

A. L. Yuille, "The creation of structure in dynamic shape," in *Proc. 2nd Int. Conf. on Computer Vision*, (Tampa, Florida), pp. 685–689, Dec. 1988.

W. Zhang and F. Bergholm, "An extension of Marr's "signature" based edge classification," in *Proc. 7th Scandinavian Conf. on Image Analysis*, (Aalborg, Denmark), pp. 435–443, Aug. 1991.

W. Zhang and F. Bergholm, "An extension of Marr's signature based edge classification and other methods for determination of diffuseness and height of edges, as well as line width," in *Proc. 4th Int. Conf. on Computer Vision* (H.-H. Nagel et. al., ed.), (Berlin, Germany), pp. 183–191, May 1993.

X. Zhuang and T. S. Huang, "Multi-scale edge detection with Gaussian filters," in *Int. Conf. on Acoustics, Speech and Signal Processing*, (Tokyo, Japan), pp. 2047–2050, 1986.

Index